Lecture Notes in Computer Science 16472

The series Lecture Notes in Computer Science (LNCS), including its subseries Lecture Notes in Artificial Intelligence (LNAI) and Lecture Notes in Bioinformatics (LNBI), has established itself as a medium for the publication of new developments in computer science and information technology research, teaching, and education.

LNCS enjoys close cooperation with the computer science R & D community, the series counts many renowned academics among its volume editors and paper authors, and collaborates with prestigious societies. Its mission is to serve this international community by providing an invaluable service, mainly focused on the publication of conference and workshop proceedings and postproceedings. LNCS commenced publication in 1973.

Jing Wang · Alexandre Madeira · Lei Li
Editors

Dynamic Logic

New Trends and Applications

6th International Workshop, DaLí 2025
Xi'an, China, October 20–21, 2025
Revised Selected Papers

Editors
Jing Wang
Shaanxi Normal University
Xi'an, China

Alexandre Madeira
University of Aveiro
Aveiro, Portugal

Lei Li
Shaanxi Normal University
Xi'an, China

ISSN 0302-9743 ISSN 1611-3349 (electronic)
Lecture Notes in Computer Science
ISBN 978-3-032-22625-9 ISBN 978-3-032-22626-6 (eBook)
https://doi.org/10.1007/978-3-032-22626-6

This Springer imprint is published by the registered company Springer Nature Switzerland AG
The registered company address is: Gewerbestrasse 11, 6330 Cham, Switzerland

Preface

Building on the pioneer intuitions of Floyd-Hoare Logic, Dynamic Logic was introduced in the 1970s to reason about, and verify, classic imperative programs. Since then, the original intuitions gave rise to an entire family of logics, which became increasingly popular for reasoning about a wide range of computational systems. Simultaneously, their object (i.e. the very notion of a program) evolved in unexpected ways. This leads to the emergence of a number of dynamic logics tailored to specific programming paradigms and extended to new computing domains, including probabilistic, continuous, and quantum computation. Both its theoretical relevance and applied potential have made Dynamic Logic a topic of interest for a number of scientific venues, from wide-scope software engineering conferences to events focused more specifically on modal logics. DaLí, however, is the first international workshop series entirely devoted to this area.

Previous editions of DaLí took place in 2017 in Brasília co-located with FROCOS and TABLEAUX; in 2019 in Porto co-located with Formal Methods Week; and in 2023 in Tbilisi as a stand-alone event. During the COVID-19 pandemic, DaLí was held online in 2020 and 2022. The present 6th edition of DaLí was held, on 20 and 21 of October, 2025 in Xi'an, China, and was the first edition of the event held in Asia.

Fenrong Liu (Tsinghua University, China), Luís S. Barbosa (University of Minho, Portugal), and Alexandru Baltag (University of Amsterdam, the Netherlands) gave invited talks titled Towards a PDL Framework for Reasoning about Causality, Paraconsistency and Program Logics, and Dynamic Logics for Data Exchange, respectively. Beyond the invited talks, this volume contains the revised versions of 12 submissions, out of 31, accepted for publication as full papers. All submissions were single-blindly reviewed by three referees and subject to a careful and participated discussion within the Program Committee, afterwards.

The contribution *Completeness and Decidability of Protocol-Dependent Knowledge in Gossip* presented by Wouter J. Smit was distinguished with the Best Student Paper Award, a prize sponsored by the School of Philosophy of Shaanxi Normal University.

As in previous editions, a special issue on the workshop theme, to appear in the Journal of Logical and Algebraic Methods in Programming, is currently under preparation.

The workshop was promoted and partially funded by Banksy compete2020-feder-00892000 and CIDMA with UIDP/04106/2025 and UIDB/04106/2025. We extend our gratitude to the invited speakers for their valuable contributions, and to all presenters and participants who enriched the discussions. We are also grateful to the Program Committee members and external reviewers for their diligent work, and to our sponsors for their generous support. Our special thanks go to the School of Philosophy, Shaanxi Normal University, the local host of DaLí 2025, for their support, and especially to the excellent team of student volunteers. Finally, we warmly acknowledge all authors who submitted their work to DaLí 2025—this volume stands as a testament to their effort and commitment.

January 2026

Jing Wang
Alexandre Madeira
Lei Li

Organization

Program Committee Chairs

Jing Wang	Shaanxi Normal University, China
Alexandre Madeira	University of Aveiro, Portugal

Organizing Committee Chair

Lei Li	Shaanxi Normal University, China

Steering Committee

Alexandru Baltag	University of Amsterdam, Netherlands
Luís S. Barbosa	University of Minho, Portugal
Mário Benevides	Federal University of Rio de Janeiro, Brazil
Nina Gierasimczuk	Technical University of Denmark, Denmark
Andreas Herzig	CNRS, University of Toulouse, France
Fenrong Liu	Tsinghua University, China
Alexandre Madeira	University of Aveiro, Portugal
Manuel A. Martins	University of Aveiro, Portugal
Igor Sedlár	Czech Academy of Sciences, Prague, Czechia
Sonja Smets	University of Amsterdam, Netherlands
Johan van Benthem	Stanford University, USA and University of Amsterdam, Netherlands

Program Committee

Thomas Ågotnes	University of Bergen, Norway
Carlos Areces	University of Córdoba, Argentina
Phillippe Balbiani	University of Toulouse, France
Alexandru Baltag	University of Amsterdam, Netherlands
Luis Barbosa	University of Minho, Portugal
Mário Benevides	Fluminense Federal University, Brazil
Johan van Benthem	Stanford University, USA and University of Amsterdam, Netherlands

Thomas Bolander	Technical University of Denmark, Denmark
Raul Fervari	University of Córdoba, Argentina
Sabine Frittella	LIFO, University of Orléans, France
Xiaoxuan Fu	China University of Political Science and Law, China
Sujata Ghosh	Indian Statistical Institute Chennai, India
Leandro Gomes	University of Lille, France
Reiner Hähnle	TU Darmstadt, Germany
Asta Halkjar From	Technical University of Denmark, Denmark
Hans van Ditmarsch	LORIA, France
Andreas Herzig	CNRS, University of Toulouse, France
Gabriele Kern-Isberner	TU Dortmund, Germany
Alexander Knapp	University of Augsburg, Germany
Sophia Knight	University of Minnesota, USA
Katherine Kosaian	University of Iowa, USA
Clemens Kupke	University of Strathclyde, UK
Lei Li	Shaanxi Normal University, China
Fei Liang	Shandong University, China
Fenrong Liu	Tsinghua University, China
Alexandre Madeira	University of Aveiro, Portugal
Manuel A. Martins	University of Aveiro, Portugal
Stefan Mitsch	Carnegie Mellon University, USA
Alessandra Palmigiano	Vrije Universiteit Amsterdam, Netherlands
Elaine Pimentel	University College London, UK
R Ramanujam	Institute of Mathematical Sciences, Chennai, India
Katsuhiko Sano	Hokkaido University, Japan
Igor Sedlár	Czech Academy of Sciences, Prague, Czechia
Sonja Smets	University of Amsterdam, Netherlands
Ionuţ Ţuţu	Institute of Mathematics of the Romanian Academy, Romania
Jing Wang	Shaanxi Normal University, China
Yì Nicholas Wáng	Sun Yat-sen University, China
Zhaoqing Xu	Sichuan University, China
Fan Yang	Utrecht University, Netherlands
Zhiguang Zhao	Taishan University, China

Additional Reviewers

Yiwen Ding
Timo Eckhardt
Xiaolong Liang
John Lindqvist
Mo Liu
S P Suresh
Kaibo Xie
Zuojun Xiong
Jialiang Yan

Invited Talks

Towards a Dynamic Logic Framework for Causal Reasoning

Xiaoxuan Fu[1], Fenrong Liu[2], and Zhiguang Zhao[3]

[1] China University of Political Science and Law, China
xfuuva@gmail.com
[2] Tsinghua University, China
fenrong@tsinghua.edu.cn
[3] Taishan University, China
zhaozhiguang23@gmail.com

Causality is a central concept for explaining how events bring about certain outcomes and has been extensively studied in logic and philosophy. Contemporary formal accounts of causality are largely shaped by Lewis's counterfactual analysis [5] and by the interventionist framework of Halpern and Pearl (HP) [2, 3, 6]. In particular, Pearl's do-calculus [6, 7] highlights the essentially *dynamic* character of causal reasoning: interventions are not mere observations, but operations that modify the system and determine what holds thereafter. This dynamic viewpoint naturally suggests a logical treatment in the style of Propositional Dynamic Logic (PDL) [1, 4], where actions are interpreted by the transitions they induce and formulas can be evaluated explicitly *before* and *after* an action. At the same time, it motivates a further methodological shift: instead of building semantic directly into the *syntax*—for instance, value-attribution expressions such as $X = x$ and model-indexed structural specifications—one can keep the object language purely propositional and let causal structure be carried entirely by the *semantics*. This separation makes it possible to isolate general principles of intervention and dependence in a uniform logical language, rather than in a model-specific notation.

Guided by these considerations, we develop *Causal Propositional Dynamic Logic* (CPDL), a propositional and *model-independent* framework for causal reasoning, in the sense that its syntax does not presuppose any fixed structural-equation representation. Instead, CPDL starts from a finite set of two-valued proposition letters equipped with an explicit *dependence structure* $\mathbb{D} = (\text{Prop}, <, f)$: an acyclic relation $<$ encodes structural dependence, and a partial function f assigns Boolean dependence formulas to non-basic propositions. Truth is computed *inductively*: basic (order-0) propositions are freely assigned, while higher-order propositions are determined by their dependence formulas. Indirect structural dependence is made explicit via a composite unfolding operation, so that situations are fully determined by basic truth assignments relative to $\mathbb{D}$.

Interventions are internalized as dynamic operations through three primitive events: *assertion* $(+p)$, *retraction* $(-p)$, and *dependence removal* $(\downarrow p)$. Each event fixes (or preserves) the truth value of p while simultaneously cutting all incoming dependence links into p and deleting its defining dependence equation, thereby rendering p independent.

This combined "**truth update + link cutting**" mechanism treats interventions as genuine state-transforming actions and supports explicit pre-/post-intervention reasoning. We generate a $\mathbb{D}$-**universal model** by closing the set of situations under finite sequences of such events, providing a uniform semantic domain for multi-step interventions.

The language extends PDL with event modalities $[+p]$, $[-p]$, and $[\downarrow p]$, interpreted by the corresponding update operations on situations. This yields characteristic valid principles governing event interaction, including commutativity for batches of assertions/retractions, interaction laws for repeated updates, and a reduction principle characterizing $[\downarrow p]$ in terms of conditional assertion/retraction. At the same time, the framework captures genuinely causal *non-commutativity*: when dependence removal interacts with other interventions, event order can affect outcomes, reflecting that interventions may change not only truth values but also the underlying dependence structure. The connection with standard program constructs of PDL is further clarified by viewing dependence removal as a composite program built from atomic actions and tests.

To integrate causality with epistemic reasoning, we extend the framework to *epistemic* CPDL, evaluating formulas at pairs (E, w), where E is an agent's epistemic state (a set of possible situations). Event modalities update both the actual situation and the epistemic alternatives, while public announcements $[\varphi!]$ update the epistemic state without changing the world. This setting supports systematic interaction principles between knowledge and intervention (e.g., commutation schemes of the form $K[*p]\varphi \leftrightarrow [*p]K\varphi$) and admits reduction axioms for announcements. For any fixed finite language, we establish **decidability** of validity and provide a sound and complete **axiomatization**, placing the framework on a standard proof-theoretic footing.

Beyond the core system, CPDL admits several natural extensions and applications. In particular, inverse event operators can be added to reason about possible states prior to an intervention, supporting backward-looking causal assessment. Moreover, key HP notions—such as *sufficient cause* and *actual cause*–can be reformulated within the propositional dynamic setting, and a notion of direct causal effect can be defined and lifted to complex propositions.

Compared with the HP structural-equation framework, CPDL departs in several technically significant respects. It is **propositional** from the outset: causal reasoning proceeds over two-valued proposition letters, avoiding value-attribution syntax such as $X = x$ and the attendant machinery of multi-valued domains. Causal structure is made explicit by an acyclic dependence relation together with Boolean dependence formulas, rather than by structural equations over variable ranges. Interventions are internalized as dynamic modalities whose execution may both enforce truth values and transform the dependence graph via link cutting, instead of being realized through equation replacement. Finally, the PDL-based setting provides a uniform language in which *intervention dynamics* and *epistemic change* can be studied within the same formalism, while remaining model-independent in the sense that its syntax does not presuppose any fixed structural-equation representation. Taken together, these design choices yield a logically transparent account of causality as state transformation, and they establish an interface between HP-style causal notions and the methods of dynamic and epistemic logic.

References

1. Fischer, M.J., Ladner, R.E.: Propositional dynamic logic of regular programs. J. Comput. Syst. Sci. **18**(2), 194–211 (1979)
2. Halpern, J.: Actual Causality. MIT Press, Cambridge (2016)
3. Halpern, J., Pearl, J.: Causes and explanations: a structural-model approach. Part II: explanations. Br. J. Philos. Sci. **56**(4), 889–911 (2005)
4. Harel, D., Kozen, D., Tiuryn, J.: Dynamic Logic. MIT Press, Cambridge (2000)
5. Lewis, D.: Causation. J. Philos. **70**(17), 556–567 (1973)
6. Pearl, J.: Causality: Models, Reasoning, and Inference. Cambridge University Press, Cambridge (2000)
7. Pearl, J.: The do-calculus revisited. In: Proceedings of the Twenty-Sixth AAAI Conference on Artificial Intelligence (AAAI-12), Toronto, Canada. AAAI Press (2012)

Paraconsistency and Program Logics

Luís Soares Barbosa

INESC TEC & Dep. Informatics, University of Minho, Braga, Portugal
lsb@di.uminho.pt

Abstract

Program logics provide methods and tools for systematic reasoning about how a program works and to prove, in a compositional way, a number of guarantees about its behaviour. Modelling complex information systems, however, often entails the need for dealing with scenarios of inconsistency in which several requirements either reinforce or contradict each other.

Actually, in the practice of Software Engineering, one often faces modeling contexts in which bivalent reasoning, even if of a probabilistic nature, is not enough, entailing the need to capture both *lack* of information or its *excess*. If the first involves vagueness or uncertainty, the second may lead to potential inconsistency.

Vagueness is addressed in fuzzy logics, but potentially contradictory information arises in a number of scenarios. (e.g. knowledge representation, data integration, etc.) and cannot be swept under the carpet.

To meet such a challenge, recent joint work with Juliana Cunha and Alexandre Madeira [4, 5], proposed a variant of transition systems endowed with positive and negative accessibility relations, and a metric space over the lattice of truth values. Such structures are called *paraconsistent transition systems*, the qualifier stressing a connection to paraconsistent logic [1–3], a logic that takes inconsistent information as potentially informative.

Indeed, data is a mined field. Not only do its values and structure change, but also the logic under which this information needs to be understood changes as well. Informational states may exhibit potentially inconsistent or partially consistent data, reflecting the diversity of judgements, e.g., from different domain experts. Moreover, such information states may be linked positively, witnessing e.g. the existence of a relationship, and negatively, recording whatever prevents such a relationship. To add to the picture, the weights of such transitions are, in most cases, non-complementary, opening an inference arena encompassing both classical, vague, and even (controlled forms of) inconsistent reasoning.

This work was financed by FCT, the Portuguese Foundation for Science and Technology, within the project BANKSY, with reference COMPETE2030-FEDER-00892000 - Project 15253.

This lecture reports on this research initiative, with a specific emphasis on its potential to develop more flexible program logics. The immediate context is that of a research project aiming at developing a theory and a method for paraconsistent reasoning driven by a concrete, clinical case-study: The diagnosis of age-related macular degeneration. This is a multifactorial disease of the macule, with a complex pathophysiology, whose progression is not well understood and has to articulate several expert perspectives on the different rates/patterns detected for similar patients and even between the two eyes of the same patient.

References

1. Carnielli, W., Coniglio, M.E.: Paraconsistent Logic: Consistency, Contradiction and Negation. Logic, Epistemology, and the Unity of Science Series, vol. 40. Springer, Cham (2016). https://doi.org/10.1007/978-3-319-33205-5
2. da Costa, N.: Nota sôbre o conceito de contradição (in portuguese with english summary). Anuário da Sociedade Paranaense de Matemática **1** (1958)
3. da Costa, N.: On the theory of inconsistent formal systems. Notre Dame J. Formal Logic **15**, 497–510 (1974)
4. Cunha, J., Madeira, A., Barbosa, L.S.: Paraconsistent transition structures: compositional principles and a modal logic. Math. Struct. Comput. Sci. **35**, e14 (2025). https://doi.org/10.1017/S0960129525100170
5. Cunha, J., Madeira, A., Barbosa, L.S.: Specification of paraconsistent transition systems, revisited. Sci. Comput. Program. **240**, 103196 (2025). https://doi.org/10.1016/j.scico.2024.103196

Dynamic Logics for Data Exchange

Alexandru Baltag

University of Amsterdam, Netherlands

I focus on a relatively new family of logics, modelling communication acts in which whole ‘chunks’ of data are being exchanged: agents or groups can access all information stored at specific locations. The data that is being exchanged may be propositional, non-propositional (numerical, visual, etc.) or a mixture of both. For this, I look at extensions of Dynamic Epistemic Logic with dynamic modalities for interactions by which agents gain (or are given) access to information sources (thus being potentially able to ‘read’ all the data available at that source). Formally, these correspond to dynamic changes of a model, by which (some or all of) the epistemic relations are replaced by intersections of (some or all of) the current relations. I briefly trace the history of this line of research, then present a general setting, with motivating examples. I give complete axiomatizations and decidability results. Time-permitting, I discuss the technical challenges and innovations involved in extending these axiomatizations to operators such as common knowledge, “epistemic superiority”, or “knowledge what” (i.e., knowing a specific piece of non-propositional data, e.g. somebody’s password).

Contents

Learning About Causation: Dynamic Epistemic Modal Logic with Causal Awareness

Nina Gierasimczuk[1(✉)], Hermine J. Grosinger[2], and Katrine B. P. Thoft[1]

[1] Technical University of Denmark, Lyngby, Denmark
{nigi,kabpt}@dtu.dk
[2] Örebro University, Örebro, Sweden
hermine.grosinger@oru.se

Abstract. In this paper, we enrich the existing dynamic epistemic modal logic of causality with dynamic awareness of causal variables. In our setting the agent can become aware of (previously unavailable) causal variables. This approach allows studying important aspects of the development of agents' knowledge about causal structures. First, we propose the language and discuss some notable validities. We consider scenarios that feature the expert (teacher) who has full knowledge of the causal dependencies, and the layperson (learner) whose knowledge is limited to only the variables she is aware of. Our models and our logic are dynamic: They allow observations by both the expert and the layperson, and they accommodate gaining awareness of causal variables by the layperson. The logic also allows interventions—counterfactual manipulations of values of variables. In the epistemic causal awareness models, the different epistemic levels of the expert and the layperson correspond to implicit and explicit knowledge, respectively. Those different perspectives can also be seen with respect to the causal graphs, in a 2-layer representation (so-called hybrid causal graph) that makes use of 'shadow' causal variables. Overall, our framework connects the existing lines of research on dynamic epistemic logic of causality and dynamic awareness with learning and pro-activity in AI.

Keywords: dynamic awareness · dynamic epistemic logic · causality · hybrid causal graphs · causal learning

1 Introduction

As children, we explore the world around us by first becoming aware of new objects, 'Oh, a red button!' Next, we attempt to manipulate these objects, 'What will happen if I press the button?' The outcome can give us new information about the world, 'The button triggers a funny noise!' or 'Mommy comes quickly after I press the red button!' This awareness of our surroundings is referred to as

J. Wang et al. (Eds.): DaLí 2025, LNCS 16472, pp. 1–22, 2026.
https://doi.org/10.1007/978-3-032-22626-6_1

situation awareness in cognitive psychology, which is conceptually categorized into three levels: perception, comprehension, and projection [9]. The first level, perception, allows agents to take note of relevant elements of the environment and their attributes. For instance, the child would perceive the red button. In the second level, comprehension, agents combine the perceptions of the first level to infer dynamic relationships between the elements of the environment. For instance, the child could realize she can use her hand to press the red button. The last level, projection, allows agents to form expectations about the future behavior of elements in the environment, applying knowledge from the first two levels. For instance, the child could press the button in order to trigger a reaction from an adult.

Interestingly, a three-tier theory is also at the center of the theory of *actual causality* (as proposed in [16]). The so-called *Ladder of Causality* levels are: association, intervention, and counterfactuals. The ladder's rungs can be interpreted similarly to the situation awareness levels: the first level involves passive observations about causes and effects, the second allows intervening to obtain knowledge of causes and effects (for instance by designing controlled experiments), and the third level allows for predicting outcomes through hypothetical reasoning. The key difference between the two domains is that situation awareness is applied in real-time predictions of events, while the ladder of causation is used for structural reasoning about cause and effect, often in a counterfactual context.

Within Artificial Intelligence, proactive AI systems provide an important motivation to model causality, awareness, and epistemic reasoning. Inspired by human proactivity [11], proactive human-AI systems (developed, for example, in [12,14,18]) can autonomously initiate anticipatory actions through joint cognitive abilities. Enriching such systems with situational awareness and with causal and epistemic reasoning increases their proactive abilities. Consider the following example inspired by [12]: Bob has a domestic proactive robot, which *knows* that Bob is not *aware* of the medicine he has to take. The robot can proactively prevent Bob from entering an undesirable health crisis by making him aware of the medicine and its *causal* relation to his health by telling (for more examples, see [12]).

To formally explore the relationship between situation awareness and reasoning about causality, in this paper we propose to enrich reasoning about interventions in causal structures with awareness. We build upon two existing lines of research. On the one hand, we draw from the dynamic epistemic logic of causality proposed in [1], which allows for counterfactual reasoning about interventions. On the other hand, we engage with the modal logic-based dynamic approach to awareness, originally developed in [6] for the purpose of tackling the issue of logical omniscience. As a result of this merge, our logic supports two 'levels of knowledge': the fully-competent level of the so-called *implicit knowledge*, and the level of *explicit knowledge*, that is limited to the causal variables the agent is aware of. We propose a novel, interactive interpretation of the two levels: the implicit knowledge codes the perspective of the expert, and the explicit level—that of the layperson. By doing so, we can model not only the expert's 'objective' reasoning about causality, but also the expert's reasoning about the layperson's limited view (restricted to a subset of causal variables). Our framework supports learning about causality by involving a dynamic awareness-raising operator: $[+X]$, that denotes adding the variable X to the awareness of the layperson. Cognitively speaking, this addition can come about spontaneously, or it can be seen as a proactive action of the expert.

The plan of this paper is as follows. In Sect. 2, we recall the dynamic epistemic logic of reasoning about interventions in causal models proposed in [1]. In Sect. 3, we outline our original contribution of extending the logic with the awareness and the awareness-raising operators. We discuss several ways of combining knowledge and interventions and discuss notable validities. We show how they relate to some of the important themes in the field. In Sect. 4, we present a way to represent the expert's and the layperson's perspectives as a *hybrid causal graph*. Finally, in Sect. 5, we outline some algorithmic aspects of the dynamic operators of our logic. We conclude and discuss future directions in Sect. 6.

2 Preliminaries on Dynamic Epistemic Logic of Causality

Let us start by recalling the logic $\mathcal{L}_{\mathtt{PAKC}}$, originally introduced in [1]. We will do it step by step, to gradually build intuitions about the necessary building-blocks of $\mathcal{L}_{\mathtt{PAKC}}$ and, eventually, of our logic.

Throughout the paper we will assume a finite signature $S = \langle U, V, \mathcal{R}\rangle$, where $U = \{U_1, \ldots, U_m\}$ is a set of exogenous variables and $V = \{V_1, \ldots, V_n\}$ is the set of endogenous variables. Given a variable $X \in U \cup V$, $\mathcal{R}(X)$ denotes the range of X, i.e., the set of values that the variable can hold. We will use lower-case letters to denote values of variables.

Definition 1. *A* causal model *is a triple $M = \langle S, \mathcal{F}, \mathtt{Val}\rangle$, where S is a signature; $\mathcal{F} = \{f_{V_j} \mid V_j \in V\}$ is a set of structural functions: for each variable V_j, there is an $f_{V_j} : \mathcal{R}(U_1, \ldots, U_m, V_1, \ldots, V_{j-1}, V_{j+1}, \ldots, V_n) \to \mathcal{R}(V_j)$, i.e., the value of each endogenous variable is a function of all the other variables from the signature; $\mathtt{Val}$ is a function assigning to every $X \in U \cup V$ a value $\mathtt{Val}(X) \in \mathcal{R}(X)$, in a way that* complies with *$\mathcal{F}$, i.e., for all endogenous variables $V_i \in V$, we have that:*[1]

$$\mathtt{Val}(V_i) = f_{V_i}(\mathtt{Val}(U_1, \ldots, U_m, V_1, \ldots, V_{i-1}, V_{i+1}, \ldots, V_n)).$$

Definition 2. *Let us take $V_j \in V$ and rename the variables in $(U \cup V) \setminus \{V_j\}$ as $X_1, \ldots, X_{n+m-1}$. We say that V_j is* directly causally affected *by $X_i \in (U \cup V) \setminus \{V_j\}$, written $X_i \hookrightarrow_{\mathcal{F}} V_j$, iff there is a tuple*

$$(x_1, \ldots, x_{i-1}, x_{i+1}, \ldots, x_{m+n-1}) \in \mathcal{R}(X_1, \ldots, X_{i-1}, X_{x+1}, \ldots, X_{n+m-1})$$

---

[1] The expression $\mathtt{Val}(X_1, \ldots, X_n)$ stands for $\mathtt{Val}(X_1), \ldots, \mathtt{Val}(X_n)$. In general, we will also write $\mathcal{R}(X_1, \ldots, X_n)$ instead of $\mathcal{R}(X_1) \times \ldots \times \mathcal{R}(X_n)$.

and there are $x_i' \neq x_i'' \in \mathcal{R}(X_i)$*, such that*

$$f_{V_j}(x_1, \ldots, x_i', \ldots, x_{m+n-1}) \neq f_{V_j}(x_1, \ldots, x_i'', \ldots, x_{n+m-1}).$$

The relation $\hookrightarrow^+_{\mathcal{F}}$ *is the transitive closure of* $\hookrightarrow_{\mathcal{F}}$*.*

We will assume, as usual in this context, that the set of structural functions is *recursive*, i.e., the transitive closure of the causal parenthood relation does not include cycles (it is a strict partial order, see Def. 3 of [1]). Note that a casual system can be represented as a directed graph where the set of nodes coincides with the set of causal variables, and the set of edges with the $\hookrightarrow$ relation. The recursivity of the set of structural functions ensures that such a graph is acyclic. Given that the most established representation of causality uses the so-called Bayesian networks, which are in fact directed acyclic graphs (DAGs), the assumption of acyclicity is common in the literature (see, e.g., [7]). We will talk of such DAGs specifically in the context of learning and awareness in Sect. 4.

Example 1. We will make use of the example from [1], first to illustrate the state-of-the-art modeling; then, in the next sections, to showcase our extensions. The scenario goes as follows. Billie wants to switch on a sprinkler (S), which turns on when she presses the red button (B). However, that will work only if the circuit breaker (C) is closed. We can represent this situation as a causal model $\langle S, \mathcal{F}, \mathtt{Val}\rangle$ of signature $S = \langle U, V, \mathcal{R}\rangle$, where $U = \{B, C\}$, $V = \{S\}$, and for all $X \in U \cup V$, $\mathcal{R}(X) = \{0, 1\}$. The set of structural functions $\mathcal{F}$ contains only f_S, which is defined as follows $f_S(B, C) = 1$ iff $B = 1$ and $C = 1$. Let's say that, as it happens, the button is not pressed, the sprinkler is off, and the circuit breaker is open, so $\mathtt{Val}(B) = 0$, $\mathtt{Val}(C) = 0$, and $\mathtt{Val}(S) = 0$.

Definition 3. *An* assignment *on a signature* S *is an expression* $\overrightarrow{X} = \overrightarrow{x}$*, where* $\overrightarrow{X}$ *is a tuple of different variables from* $U \cup V$ *and* $\overrightarrow{x} \in \mathcal{R}(\overrightarrow{X})$*.*

The notion of an assignment can be used to express an intervention in a causal model. An intervention is akin to a thought experiment. It sets variables to prescribed values, which is a simple process for exogenous variables because they do not depend on anything else. For the endogenous variables, however, an intervention requires adjusting their structural functions by setting them to act as constant functions. This is similar to carrying out a 'controlled' experiment within the causal model.

Definition 4. *Let* $M = \langle S, \mathcal{F}, \mathtt{Val}\rangle$ *be a causal model; let* $\overrightarrow{X} = \overrightarrow{x}$ *be an assignment on* S*. The result of the* intervention *of assigning to variable* $\overrightarrow{X}$ *values* $\overrightarrow{x}$ *on* M *is* $M_{\overrightarrow{X}=\overrightarrow{x}} = \langle S, \mathcal{F}_{\overrightarrow{X}=\overrightarrow{x}}, \mathtt{Val}^{\mathcal{F}}_{\overrightarrow{X}=\overrightarrow{x}}\rangle$*, such that* $\mathcal{F}_{\overrightarrow{X}=\overrightarrow{x}}$ *is the same as* $\mathcal{F}$*, except that for every endogenous variable* X_i *in* $\overrightarrow{X}$*,* f_{X_i} *is replaced with a constant function returning* x_i*; as for* $\mathtt{Val}^{\mathcal{F}}_{\overrightarrow{X}=\overrightarrow{x}}$*, for each exogenous variable not in* $\overrightarrow{X}$*,* $\mathtt{Val}^{\mathcal{F}}_{\overrightarrow{X}=\overrightarrow{x}}$ *is as* $\mathtt{Val}$*; for each exogenous variable* X_i *in* $\overrightarrow{X}$*,* $\mathtt{Val}^{\mathcal{F}}_{\overrightarrow{X}=\overrightarrow{x}}(X_i) = x_i$*, and the values of all endogenous variables not in* $\overrightarrow{X}$ *comply with* $\mathcal{F}_{\overrightarrow{X}=\overrightarrow{x}}$*.*

Given this setup, the authors in [1] introduce the language $\mathcal{L}_{\mathsf{C}}$, which (apart from stating propositional facts about variables taking certain values) can express one-step interventions of setting the variables in $\overrightarrow{X}$ to the values $\overrightarrow{x}$, $[\overrightarrow{X} = \overrightarrow{x}]$. The syntax of this language is as follows:

$$\begin{aligned} \gamma &::= Z = z \mid \neg\gamma \mid \gamma \wedge \gamma \qquad \text{for } Z \in U \cup V \text{ and } z \in \mathcal{R}(Z) \\ \varphi &::= Z = z \mid \neg\varphi \mid \varphi \wedge \varphi \mid [\overrightarrow{X} = \overrightarrow{x}]\gamma \qquad \text{for } \overrightarrow{X} = \overrightarrow{x} \text{ on } S \end{aligned} \tag{1}$$

The semantics of $\mathcal{L}_{\mathsf{C}}$ is standard for the boolean connectives, and as follows for the remaining two clauses:

$$\begin{aligned} \langle S, \mathcal{F}, \mathtt{Val}\rangle &\models Z = z \quad &&\text{iff } \mathtt{Val}(Z) = z \\ \langle S, \mathcal{F}, \mathtt{Val}\rangle &\models [\overrightarrow{X} = \overrightarrow{x}]\gamma \quad &&\text{iff } \langle S, \mathcal{F}_{\overrightarrow{X}=\overrightarrow{x}}, \mathtt{Val}^{\mathcal{F}}_{\overrightarrow{X}=\overrightarrow{x}}\rangle \models \gamma \end{aligned}$$

Example 2. To further describe Billie's situation, if we counterfactually assume that the button is pressed, we carry out the intervention $[B = 1]$. The structural functions would not allow this intervention to turn on the sprinkler (because the circuit-breaker is open), hence, $[B = 1]S = 0$. However, had the button been pressed and the circuit-breaker been closed, the sprinkler would be on, i.e., the formula $[BC = 11]S = 1$ is true.[2]

In [1], the notion of a causal model is further extended to account for uncertainty about the values of variables. As usual in such cases, the valuation of the causal model is now replaced with a set of such valuations (a set of possible worlds), representing the uncertainty of the agent.

Definition 5. *An* epistemic causal model *is a triple* $E = \langle S, \mathcal{F}, \mathcal{T}\rangle$, *s.t.* S, $\mathcal{F}$ *are just as in the causal model in Definition 1, and* $\mathcal{T} = \{\mathtt{Val} \mid \mathtt{Val} \text{ complies with } \mathcal{F}\}$.

For this new type of model, the notion of intervention (given in Definition 4) must be adjusted, as interventions transform all worlds the agent considers possible.

Definition 6. *For an epistemic causal model* $E = \langle S, \mathcal{F}, \mathcal{T}\rangle$ *and an assignment* $\overrightarrow{X} = \overrightarrow{x}$*, the* intervention *of setting the variables in* $\overrightarrow{X}$ *to* $\overrightarrow{x}$*, results in a new epistemic causal model* $E_{\overrightarrow{X}=\overrightarrow{x}} = \langle S, \mathcal{F}_{\overrightarrow{X}=\overrightarrow{x}}, \mathcal{T}^{\mathcal{F}}_{\overrightarrow{X}=\overrightarrow{x}}\rangle$*, where* $\mathcal{F}_{\overrightarrow{X}=\overrightarrow{x}}$ *is as in Definition 4, and* $\mathcal{T}^{\mathcal{F}}_{\overrightarrow{X}=\overrightarrow{x}} := \{\mathtt{Val}^{\mathcal{F}}_{\overrightarrow{X}=\overrightarrow{x}} \mid \mathtt{Val} \in \mathcal{T}\}$.

This new model allows interpreting knowledge and knowledge-update operators. The language $\mathcal{L}_{\mathsf{C}}$ is extended to $\mathcal{L}_{\mathsf{PAKC}}$ with the following syntax:

$$\begin{aligned} \gamma &::= Z = z \mid \neg\gamma \mid \gamma \wedge \gamma \mid K\gamma \mid [\gamma!]\gamma \qquad \text{for } Z \in U \cup V \text{ and } z \in \mathcal{R}(Z) \\ \varphi &::= Z = z \mid \neg\varphi \mid \varphi \wedge \varphi \mid K\varphi \mid [\varphi!]\varphi \mid [\overrightarrow{X} = \overrightarrow{x}]\gamma \qquad \text{for } \overrightarrow{X} = \overrightarrow{x} \text{ on } S \end{aligned} \tag{2}$$

[2] We use this notation here for readability and for being concise. One may also use a vector notation such as the following: $\left[\begin{pmatrix} B \\ C \end{pmatrix} = \begin{pmatrix} 1 \\ 1 \end{pmatrix}\right] S = 1$.

The language $\mathcal{L}_{\texttt{PAKC}}$ is interpreted locally, with the formulas evaluated at a pair $(E, \texttt{Val})$, i.e., an epistemic causal model and one of its valuations $\texttt{Val} \in \mathcal{T}$. The Boolean fragment is as usual, and the other clauses have the following meaning:

$$\begin{array}{ll}
(E,\texttt{Val}) \models Z = z & \text{iff } \texttt{Val}(Z) = z \\
(E,\texttt{Val}) \models K\varphi & \text{iff } (E,\texttt{Val}') \models \varphi \text{ for every } \texttt{Val}' \in \mathcal{T} \\
(E,\texttt{Val}) \models [\psi!]\varphi & \text{iff } (E,\texttt{Val}) \models \psi \text{ implies } (E^{\psi},\texttt{Val}) \models \varphi \\
(E,\texttt{Val}) \models [\overrightarrow{X} = \overrightarrow{x}]\gamma & \text{iff } (E_{\overrightarrow{X}=\overrightarrow{x}}, \texttt{Val}^{\mathcal{F}}_{\overrightarrow{X}=\overrightarrow{x}}) \models \gamma
\end{array}$$

where $E^{\psi} = \langle S, \mathcal{F}, \mathcal{T}^{\psi}\rangle$, with $\mathcal{T}^{\psi} := \{\texttt{Val}' \in \mathcal{T} \mid (E,\texttt{Val}') \models \psi\}$.

Example 3. Returning to our running example, let us assume that all variables are set to 0, but Billie can only see the state of the button and the sprinkler, and cannot observe the circuit-breaker. To model this situation we need an epistemic causal model $E = \langle S, \mathcal{F}, \mathcal{T}\rangle$, where $\mathcal{T}$ consists of two assignments, $\texttt{Val}_1$ and $\texttt{Val}_2$, which differ only in the value of the variable C, the rest is set to 0 (see Fig. 1, on the left). Billie cannot distinguish between the two worlds $\texttt{Val}_1$ and $\texttt{Val}_2$ (in Fig. 1 they are both within the uncertainty range $\mathcal{T}$). Her knowledge about the values of variables can be expressed in the following way: $\neg K(C = 0)$ and $K(B = 0 \wedge S = 0)$.

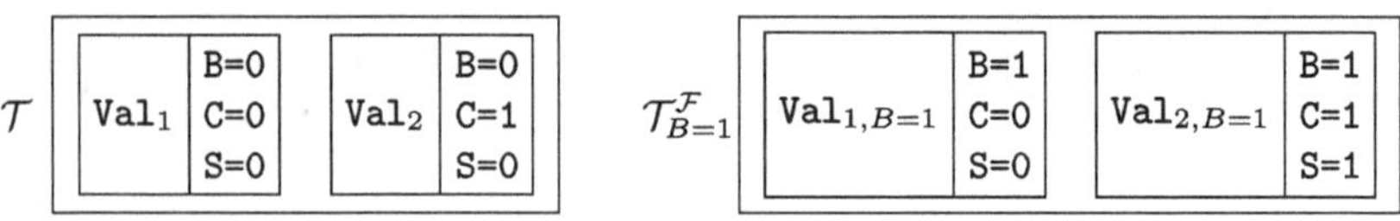

Fig. 1. Billie's valuations before (left) and after (right) the intervention $[C = 1]$.

Counterfactual reasoning about an intervention on the button, $[B = 1]$, will transform each possible world in $\mathcal{T}$ into a new one, $\texttt{Val}_{1,B=1}$ and $\texttt{Val}_{2,B=1}$ (see Fig. 1, on the right). We will then have that, $\texttt{Val}_{1,B=1}(B) = 1$ and $\texttt{Val}_{1,B=1}(C) = \texttt{Val}_{1,B=1}(S) = 0$, but $\texttt{Val}_{2,B=1}(B) = \texttt{Val}_{2,B=1}(C) = \texttt{Val}_{2,B=1}(S) = 1$. Accordingly, in $\mathcal{L}_{\texttt{PAKC}}$, we can express that Billie does not know that pressing the button would turn on the sprinkler: $\neg K[B = 1]S = 0$.

The knowledge-update operator works as a public announcement, i.e., it restricts the uncertainty range of the agent. In the case of Billie, if she received information from a fully trusted source that the circuit-breaker is open, she would be left with only $\texttt{Val}_1$, and so she would know that had she pushed the button, the sprinkler would be off: $[C = 0!]K[B = 1]S = 0$ (see Fig. 2).

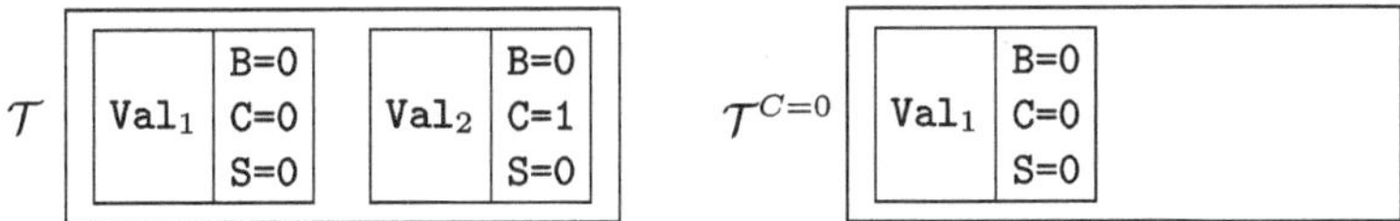

Fig. 2. Billie's valuations before (left) and after (right) the announcement $[C = 0!]$.

In [1], the authors provide a sound and complete axiomatization of $\mathcal{L}_{\texttt{PAKC}}$. They also make many interesting connections to other works in the field of causality. They note that although the logic adequately models thought experiments, it cannot account for *learning* from experiments. The problem lies in the axiom (a formula valid in their setting): $[\overrightarrow{X} = \overrightarrow{x}]K\varphi \rightarrow K[\overrightarrow{X} = \overrightarrow{x}]\varphi$, which says that whatever is known after the intervention was already known before the intervention. We will call this effect 'the principle of *no-learning*'. This inadequacy relates to the so-called 'frame problem' in Artificial Intelligence, which describes the multi-faceted difficulty with formal representation and formal reasoning about how actions affect the world [17]. Another problem of a similar kind, that concerns specifically logical modeling of knowledge in relation to awareness, is that of logical omniscience [15].

3 Causal Awareness

One might think that, intuitively, knowledge implies awareness. Note however that while we know some facts, we might be unaware of their logical consequences. This subtlety is hard to address with the standard tools of epistemic modal logic. The problem of logical omniscience, i.e., the effect of knowledge being closed on logical consequence, appears 'by default' in this logic [19] and makes the theory incompatible with cognitive reality. In [6], that puzzle was addressed by enriching the language of epistemic logic with an *awareness* operator that allows limiting the agents' 'access' (at a possible world) to only some set of formulas. This modification allows distinguishing between implicit and explicit knowledge: the former is the information that can be recovered from the epistemic situation by an agent who is perfect at both observing and reasoning, while the latter is 'filtered' through the awareness set, i.e., there might be truths that are implicitly, but not explicitly, known to the agent. The authors argue that abandoning the assumption of logical omniscience requires taking 'knowledge' to be of the explicit kind. Their account is also dynamic: facts can become explicitly known by an operation of *raising awareness* of the agent. In what follows, we will apply a similar line of reasoning to the case of causal epistemic models and the previously described $\mathcal{L}_{\texttt{PAKC}}$. Let us start with defining our new models, which extend the previously defined epistemic causal models with an awareness function $\mathcal{A}$ which specifies, for a given possible world, the causal variables of which the agent is aware.

Definition 7. *A epistemic causal awareness model is* $\mathcal{M} = \langle S, \mathcal{F}, \mathcal{T}, \mathcal{A} \rangle$*, where:* $S = \langle U, V, \mathcal{R} \rangle$ *is a signature,* $\mathcal{F}$ *is a set of structural functions over the set of endogenous variables* V*,* $\mathcal{T}$ *is a non-empty set of possible assignments complying with* $\mathcal{F}$*, and* $\mathcal{A}$ *is the awareness function* $\mathcal{A} : \mathcal{T} \to \mathcal{P}(U \cup V)$*.*[3]

As before, we should consider how counterfactual thought experiments, a.k.a. interventions, are performed on such a model. For now, we assume that the counterfactual reasoning is carried out on the level of implicit knowledge. That means that it results in the same outcome as interventions in Definition 6.

We can now extend the language $\mathcal{L}_{\texttt{PAKC}}$ to include two new awareness-related operators, where $A\varphi$ reads 'the agent is aware of φ' and $[+\overrightarrow{X}]\varphi$ reads 'after raising the awareness of the agent by the set of variables in $\overrightarrow{X}$, φ holds'. The syntax of the language of dynamic epistemic causal awareness logic $\mathcal{L}_{\texttt{PAKCA+}}$ is defined in the following way:

$$\gamma ::= Z = z \mid \neg\gamma \mid \gamma \wedge \gamma \mid K\gamma \mid [\gamma!]\gamma \mid A\gamma \mid [+\overrightarrow{X}]\gamma$$
$$\varphi ::= Z = z \mid \neg\varphi \mid \varphi \wedge \varphi \mid K\varphi \mid [\varphi!]\varphi \mid A\varphi \mid [+\overrightarrow{X}]\varphi \mid [\overrightarrow{X} = \overrightarrow{x}]\gamma \qquad (3)$$

The language $\mathcal{L}_{\texttt{PAKCA+}}$ is interpreted on pairs of epistemic causal awareness model $\mathcal{M}$ and a particular assignment $\texttt{Val}$, apart from all the clauses already defined in $\mathcal{L}_{\texttt{PAKC}}$, which are analogous, we have two new elements:

$$(\mathcal{M}, \texttt{Val}) \models A\varphi \quad \text{iff } v(\varphi) \subseteq \mathcal{A}(\texttt{Val})$$
$$(\mathcal{M}, \texttt{Val}) \models [+\overrightarrow{X}]\,\varphi \quad \text{iff } (\mathcal{M}_{+\overrightarrow{X}}, \texttt{Val}) \models \varphi$$

The first clause requires defining the function v, which on the input of a formula φ outputs all causal variables occurring in φ (this concept is analogous to the one from [21], where it was defined for propositional variables), see Definition 8 below. The second clause requires specifying the extension of the epistemic causal awareness model with the variables in $\overrightarrow{X}$—we cover this in Definition 9 (as in [6]). Let $set(\overrightarrow{X})$ be the set of all variables listed in $\overrightarrow{X}$.

Definition 8. *Let* φ *be a formula of* $\mathcal{L}_{\texttt{PAKCA+}}$*. We define inductively the set of variables of a* φ*,* $v(\varphi)$*, in the following way:*

$$v(Z = z) := \{Z\}$$
$$v(\neg\varphi) := v(\varphi)$$
$$v([\varphi_1!]\varphi_2) := v(\varphi_1) \cup v(\varphi_2)$$
$$v(\varphi_1 \wedge \varphi_2) := v(\varphi_1) \cup v(\varphi_2)$$
$$v(\bigcirc\varphi) := v(\varphi), \text{ where } \bigcirc \in \{A, K\}$$
$$v(\oplus\varphi) := set(\overrightarrow{X}) \cup v(\varphi), \text{ where } \oplus \in \{[\overrightarrow{X} = \overrightarrow{x}], [+\overrightarrow{X}]\}$$

[3] Note that here we do not make any further assumptions about the awareness function. However, in Sect. 3.1, we consider an interesting restriction.

Definition 9. *Given an epistemic causal awareness model* $\mathcal{M} = \langle S, \mathcal{F}, \mathcal{T}, \mathcal{A} \rangle$ *and a set of variables* $X \subseteq U \cup V$, $\mathcal{M}_{+\overrightarrow{X}} = \langle S, \mathcal{F}, \mathcal{T}, \mathcal{A}' \rangle$, *where* $\mathcal{A}'(\texttt{Val}) = \mathcal{A}(\texttt{Val}) \cup set(\overrightarrow{X})$, *for all* $\texttt{Val} \in \mathcal{T}$.

Example 4. Let us return to Billie and her sprinkler. Billie's gardener is an expert on watering the grounds around her mansion. He can counter-factually reason about her situation. Billie does not have to worry about the circuit breakers. She is aware of the button, and, normally, pressing it turns the sprinkler on. The gardener, on the other hand, is not limited in his awareness of the electrical wiring of the garden, and so of the circuit-breaker. Both the gardener and Billie can directly observe the button and the sprinkler, but they do not see the state of the circuit breaker. As long as the button works as intended, i.e., it turns on the sprinkler, Billie is perfectly happy with her level of awareness of the causal variables. If that fails, however, she would call on the gardener, and he could, by reasoning counterfactually realize that she should be made aware of the existence of the circuit-breaker and explain its relation to other variables (see Fig. 3).

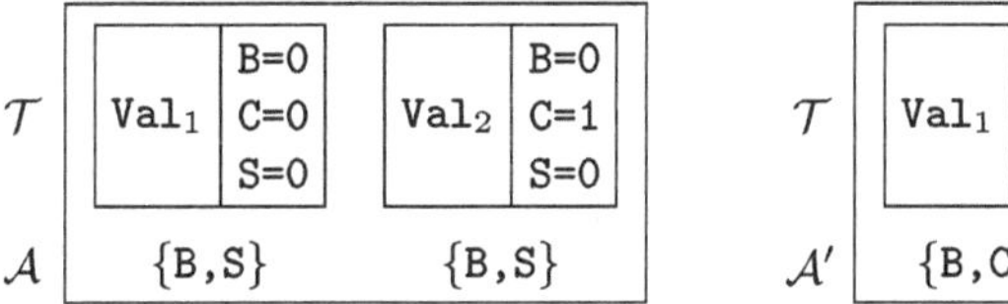

Fig. 3. Billie's valuations before (left) and after (right) raising awareness by C. $\mathcal{A}$ gives the set of causal variables Billie is aware of in a possible world.

3.1 Explicit Causal Knowledge

In our epistemic causal awareness models, the layperson's perspective is restricted by the awareness set, i.e., the variables she considers. It then makes sense to introduce a more restrictive kind of epistemic attitude, the so-called *explicit knowledge* (introduced in [10] and extensively used, e.g., in [6] and [21]). In [10], explicit knowledge of φ was defined to be equivalent to $K\varphi \wedge A\varphi$. In [6], a different, more general definition was proposed: $K(\varphi \wedge A\varphi)$. We will now investigate under what conditions the two notions are equivalent. It is straightforward to see that $K(\varphi \wedge A\varphi)$ implies $K\varphi \wedge KA\varphi$, which in turn implies $K\varphi \wedge A\varphi$.[4] The other direction is slightly more demanding, and to prove it, we need an extra assumption.

[4] The last step requires that the accessibility relation is reflexive, i.e., we have that $\models K\varphi \rightarrow \varphi$. Since in our model uncertainty range is just a set of all possible worlds, the implied accessibility relation is reflexive (as well as symmetric and transitive).

Proposition 1. *If* $\models A\varphi \rightarrow KA\varphi$, *then* $\models (K\varphi \wedge A\varphi) \rightarrow K(\varphi \wedge A\varphi)$.[5]

In [6], the principle in the assumption of Proposition 1, expressed by the formula $A\varphi \rightarrow KA\varphi$, is called 'weak introspection' and deemed natural, though not adopted in their framework. As a matter of fact, assuming weak introspection would have strong consequences for our modeling: awareness sets would have to be identical for all possible worlds (valuations) across the epistemic causal awareness model. We will call this effect *uniform awareness.*

Proposition 2. *Let* $\mathcal{M} = \langle \mathcal{S}, \mathcal{F}, \mathcal{T}, \mathcal{A} \rangle$. $\mathcal{M} \models A\varphi \rightarrow KA\varphi$ *if and only if for all* $\mathtt{Val}, \mathtt{Val}' \in \mathcal{T}$, $\mathcal{A}(\mathtt{Val}) = \mathcal{A}(\mathtt{Val}')$.

Although the assumption of uniform awareness is consistent with all of our examples so far, we'd rather preserve the full generality. As we will see below, certain types of explicit intervention (specifically the one with what we call *differential awareness*) will have the power to break uniform awareness. We then keep to the interpretation from [6], and define explicit knowledge in the following way: $K^{Ex}\varphi := K(\varphi \wedge A\varphi)$.

3.2 Explicit Causal Intervention

We will now apply a similar methodology to the intervention operator, by introducing explicit intervention $[\overrightarrow{X} = \overrightarrow{x}]^{Ex}$, that stands for $[\overrightarrow{X} = \overrightarrow{x}][+\overrightarrow{X}]$. This operation corresponds to the expert performing an intervention while raising the awareness of the agent of the variables intervened upon (this could happen when the expert utters the counterfactual reasoning in the presence of the layperson). The order of the two operations does not matter. In general, we have $\models [\overrightarrow{X} = \overrightarrow{x}][+\overrightarrow{X}]\varphi \leftrightarrow [+\overrightarrow{X}][\overrightarrow{X} = \overrightarrow{x}]\varphi$, because the two operations change different components of the model.

Let us now take a closer look at how explicit intervention interacts with the awareness and knowledge operators. First recall that the regular intervention operator does not impact awareness (in general, we have: $\models [\overrightarrow{X} = \overrightarrow{x}]A\varphi \leftrightarrow A[\overrightarrow{X} = \overrightarrow{x}]\varphi$) or implicit knowledge (we have: $\models [\overrightarrow{X} = \overrightarrow{x}]K\varphi \leftrightarrow K[\overrightarrow{X} = \overrightarrow{x}]\varphi$). The same property persists for explicit knowledge:

Proposition 3. *The formula* $[\overrightarrow{X} = \overrightarrow{x}]K^{Ex}\varphi \leftrightarrow K^{Ex}[\overrightarrow{X} = \overrightarrow{x}]\varphi$ *is valid in the class of epistemic causal awareness models.*

Things change when we consider explicit intervention: $\not\models [\overrightarrow{X} = \overrightarrow{x}]^{Ex}A\varphi \rightarrow A[\overrightarrow{X} = \overrightarrow{x}]^{Ex}\varphi$. To see that, it is enough to consider a model $\mathcal{M}$ and a $\mathtt{Val}$ in $\mathcal{M}$ such that $set(\overrightarrow{X}) \not\subseteq \mathcal{A}(\mathtt{Val})$. For similar models, the explicit version of the 'no-learning' postulate also fails, formally:

Proposition 4. *The formula* $[\overrightarrow{X} = \overrightarrow{x}]^{Ex}K^{Ex}\varphi \rightarrow K^{Ex}[\overrightarrow{X} = \overrightarrow{x}]^{Ex}\varphi$ *is not valid in the class of epistemic causal awareness models.*

[5] The proofs of all propositions can be found in Appendix A.

3.3 Differential Awareness

Another version of explicit intervention stems from the intuition that only those variables that change due to a given intervention are added to the awareness set (a version of a counterfactual *noticing*). We call this *explicit intervention with differential awareness*, and denote it with the operator $[\overrightarrow{X} = \overrightarrow{x}]^{\Delta}$. Before we specify its semantics, we need to define the notion of a *difference set.*

Definition 10. *Let* $\mathcal{M} = \langle S, \mathcal{F}, \mathcal{T}, \mathcal{A} \rangle$, $\mathtt{Val}, \mathtt{Val}' \in \mathcal{T}$. *The* difference set *between* $\mathtt{Val}$ *and* $\mathtt{Val}'$ *is* $\Delta(\mathtt{Val}, \mathtt{Val}') = \{X \in U \cup V \mid \mathtt{Val}(X) \neq \mathtt{Val}'(X)\}$.

The semantics of our new explicit intervention with differential awareness is given in the following way:

$$(\mathcal{M}, \mathtt{Val}) \models [\overrightarrow{X} = \overrightarrow{x}]^{\Delta}\varphi \text{ iff } (\mathcal{M}_{+\Delta}, \mathtt{Val}) \models \varphi,$$

where $\mathcal{M}_{+\Delta} = \langle S, \mathcal{F}_{\overrightarrow{X}=\overrightarrow{x}}, \mathcal{T}^{\mathcal{F}}_{\overrightarrow{X}=\overrightarrow{x}}, \mathcal{A}^{\Delta} \rangle$, and for any $\mathtt{Val} \in \mathcal{T}$, $\mathcal{A}^{\Delta}(\mathtt{Val}) = \mathcal{A}(\mathtt{Val}) \cup \Delta(\mathtt{Val}, \mathtt{Val}^{\mathcal{F}}_{\overrightarrow{X}=\overrightarrow{x}})$.

Example 5. Let us now return to our Billie example, and this time let us assume that Billie is only aware of the button. The expert can counterfactually reason that had the button been pressed, the sprinkler would turn on, and would be noticed by Billie (she would become aware of the variable S). That would happen only in the possible world in which the circuit-breaker is closed, see Fig. 4.

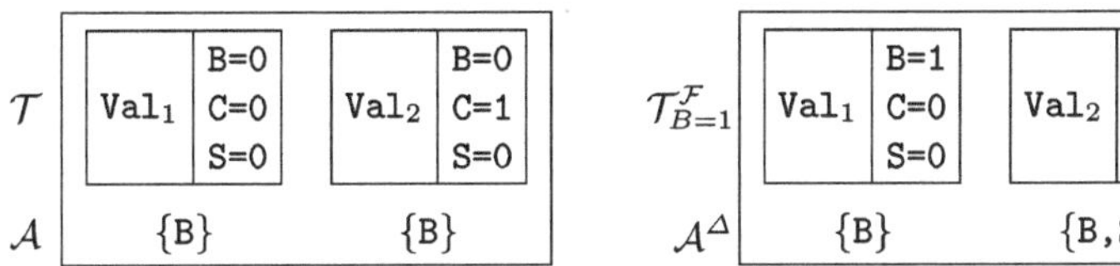

Fig. 4. Billie's valuations before (left) and after (right) explicit intervention with differential awareness $[B = 1]^{\Delta}$.

Example 5 illustrates how the operation of explicit intervention with differential awareness violates the principle of uniform awareness (used in Proposition 2). In the right-hand side of Fig. 4, $\mathcal{A}^{\Delta}$ gives *different* awareness sets for each of the two possible worlds: $\mathcal{A}^{\Delta}(\mathtt{Val}_1) = \{B\}$ and $\mathcal{A}^{\Delta}(\mathtt{Val}_2) = \{B, S\}$. We can then conclude that the differential awareness operator is incompatible with the weak introspection axiom discussed in Sect. 3.1 and in [6].

There are many further interesting questions about the proposed logic. For instance, our setting only allows reasoning about interventions on the level of implicit knowledge, i.e., only the expert performs counterfactual reasoning. A significant and potentially fruitful direction would be to consider including the learner's representation of structural functions, restricted to the awareness sets. Another, even more immediate open question is about a sound and complete axiomatization of our logics $\mathcal{L}_{\mathsf{PAKCA+}}$ and $\mathcal{L}_{\mathsf{PAKCA+}\Delta}$. As complete systems for $\mathcal{L}_{\mathsf{PAKC}}$ [1] and the dynamic logic of awareness [6] were already established, this task is attainable. We leave these topics for the extended version of this paper.

4 Hybrid Causal Graphs

In this section, we introduce the notion of a *hybrid causal graph*, and show how it can be used to graphically represent causal models that may enable an expert to reason and act proactively. The overall idea is to depict the causal graph (the DAG representing the direct causal influence between variables) on two levels: the complete representation of the expert's knowledge on the top (the t-layer) and the partial knowledge of layperson on the bottom (the b-layer, limited to only those variables that the layperson is aware of).

Example 6. Assume again that Billie is aware only of the button and the sprinkler (and knows of their causal connection). The gardener is aware of the complete functioning of the garden watering system, and of Billie's limited awareness. When Billie calls him on the phone complaining that the sprinkler does not work, he performs a counterfactual reasoning involving the variables she is aware of and the ones she does not consider at all (following the terminology of [13], we will call them *shadow* variables). He can then decide which of the variables can (or should) be revealed to Billie. Other variables, either because they are irrelevant for her immediate goals or because they require specialized tools, should remain unknown to her. The circuit-breaker can be easily manipulated by Billie, so it's a good candidate to be revealed, while a complicated sequence of valves and adapters in the pipe (which could be stuck or clogged) is not, perhaps because of a risk of injury or voiding the guarantee of the equipment.

In Fig. 5, the top (expert) and the bottom (layperson) layers are displayed. The two layers are connected by dashed arrows that map the elements known to the expert to some elements the layperson either is already aware of or could become aware of by expanding her current structure (we call these mappings *parent functions*). We highlight two important types of such mappings. Firstly, the expert might reveal 'shadow' variables. Some of them might be anticipated by the layperson because they directly connect to the component she considers. For that purpose, we introduce the *one-step closure* of the layperson's awareness (see Definition 12, point 3). Secondly, an edge in the layperson's level can be a *black-box*, i.e., can hide a bunch of variables that make the connection possible (we define such black-boxes in Definition 12, point 4). The parent mappings can be seen as a guidance system for the proactivity strategy of the expert, Fig. 5 depicts it for the case in Example 6. Below, we provide a formal definition of such a hybrid causal graph based on an epistemic causal awareness model.

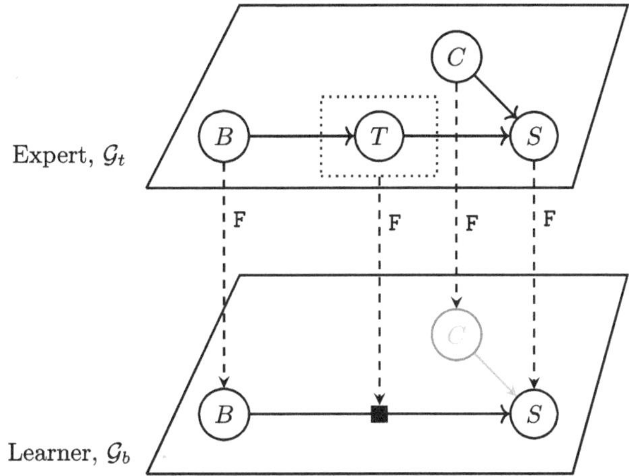

Fig. 5. Hybrid Causal Graph of the 'Billie-example'. Solid arrows: simple edges. Dotted rectangle: set of variables (here the set contains only a single variable, T, which stands for the tap) mapping to a black-box edge. Dashed arrows: parent functions. Circle: vertex, causal variable. Gray: vertices/edges that the learner is unaware of (shadow variables). The full hybrid causal graph is then $\mathcal{G}_{\mathbb{C}} = (\mathcal{G}_t, \mathcal{G}_b, \mathtt{F})$.

Definition 11. *Let $\mathcal{M} = \langle S, \mathcal{F}, \mathcal{T}, \mathcal{A} \rangle$ be an epistemic causal awareness model with uniform awareness. Then $\mathbf{A}_{\mathcal{M}} := \{X \mid X \in \mathcal{A}(\mathtt{Val})$ for some $\mathtt{Val} \in \mathcal{T}\}$.*[6]

Definition 12. *Let $\mathcal{M} = \langle S, \mathcal{F}, \mathcal{T}, \mathcal{A} \rangle$ be an epistemic causal awareness model with uniform awareness.*

1. *The* implicit causal graph *of $\mathcal{M}$ is $\mathcal{G}_{\mathcal{M}} = (\mathcal{X}, \mathcal{E})$, where $\mathcal{X} = U \cup V$, and $\mathcal{E} = \{(X, Y) \in (U \cup V)^2 \mid X \hookrightarrow_{\mathcal{F}} Y\}$.*
2. *The* explicit causal graph *of $\mathcal{M}$ is the implicit causal graph of $\mathcal{M}$ restricted to the awareness set, i.e., $\mathcal{G}^{Ex}_{\mathcal{M}} = (\mathcal{X}^{Ex}, \mathcal{E}^{Ex})$, where $\mathcal{X}^{Ex} = \mathbf{A}_{\mathcal{M}}$, and $\mathcal{E}^{Ex} = \{(X, Y) \in (\mathbf{A}_{\mathcal{M}})^2 \mid X \hookrightarrow^{+}_{\mathcal{F}} Y\}$.*
3. *Let $\mathcal{G}_{\mathcal{M}} = (\mathcal{X}, \mathcal{E})$ be the implicit causal graph of $\mathcal{M}$. The* one-step closure *of the explicit causal graph $\mathcal{G}^{Ex}_{\mathcal{M}} = (\mathcal{X}^{Ex}, \mathcal{E}^{Ex})$ is $\hat{\mathcal{G}}^{Ex}_{\mathcal{M}} = (\hat{\mathcal{X}}^{Ex}, \hat{\mathcal{E}}^{Ex})$, where:*
   - *$\hat{\mathcal{X}}^{Ex} = \mathcal{X}^{Ex} \cup \{Y \in \mathcal{X} \setminus \mathbf{A}_{\mathcal{M}} \mid (X, Y) \in \mathcal{E}$ or $(Y, X) \in \mathcal{E}$ for some $X \in \mathcal{X}^{Ex}\}$;*
   - *$\hat{\mathcal{E}}^{Ex} = \mathcal{E}^{Ex} \cup \{(X, Y) \mid X \in \mathcal{X}^{Ex}, Y \in \hat{\mathcal{X}}^{Ex}$ and $(X, Y) \in \mathcal{E}\}$.*
4. *An edge $(X, Y) \in \mathcal{E}^{Ex}$ is a* black box *if $X \hookrightarrow^{+}_{\mathcal{F}} Y$, but it is not the case that $X \hookrightarrow_{\mathcal{F}} Y$. We denote by $\blacksquare(\mathcal{G}^{Ex}_{\mathcal{M}})$ the set of all such black boxes in $\mathcal{G}^{Ex}_{\mathcal{M}}$.*

Definition 13. *Let $\mathcal{M} = \langle S, \mathcal{F}, \mathcal{T}, \mathcal{A} \rangle$ be an epistemic causal awareness model with uniform awareness. A* hybrid causal graph *of $\mathcal{M}$ is a tuple $\mathcal{G}_{\mathbb{C}} = (\mathcal{G}_t, \mathcal{G}_b, F)$, such that: $\mathcal{G}_t := \mathcal{G}_{\mathcal{M}}$, $\mathcal{G}_b := \hat{\mathcal{G}}^{Ex}_{\mathcal{M}}$, and $F := \{F_{sh}, F_{\blacksquare}\}$, where:*

[6] For simplicity, we include the assumption of uniform awareness from Proposition 2. In such cases, the expert knows which causal variables Billie is aware of, regardless of which possible world is the actual one.

- $F_{sh} : \mathcal{X}_t \to \hat{\mathcal{X}}_b$ *is a partial identity function, with* $F_{sh}(X) = X$*, if* $X \in \hat{\mathcal{X}}_b$*;*
- $F_{\blacksquare} : \mathcal{P}(\mathcal{X}_t) \to \blacksquare(\mathcal{G}_b)$ *is a partial function, with* $F_{\blacksquare}(\mathbf{Z}) = (X, Y)$*, if* $\mathbf{Z} = \{Z \mid X \hookrightarrow^{+}_{\mathcal{F}} Z \hookrightarrow^{+}_{\mathcal{F}} Y\}$.

We call the structure a hybrid causal graph, because it combines elements of simple DAGs, hierarchical graphs and hypergraphs; the black-box parent function is a hyper-edge mapping a set of vertices in the t-layer to a pair of vertices in the b-layer.[7]

5 Learning Procedures

All of the dynamic operators discussed in Sect. 3 can be rendered as algorithmic procedures, which allows reflecting on the possible variations of their implementation and analyzing their complexity. In this section we will focus on the two new intervention operators we proposed, $[\overrightarrow{X} = \overrightarrow{x}]^{Ex}$ and $[\overrightarrow{X} = \overrightarrow{x}]^{\Delta}$.

The explicit intervention procedure for the $[\overrightarrow{X} = \overrightarrow{x}]^{Ex}$ operator introduced in Sect. 3.2 can be seen as a sequential combination of two basic procedures of Consider and Intervene, corresponding to $[+\overrightarrow{X}]$ and $[\overrightarrow{X} = \overrightarrow{x}]$, respectively (see Algorithms 3 and 4 in Appendix B). In Algorithm 1, we display the pseudocode of the corresponding InterveneExplicit procedure.

Algorithm 1. InterveneExplicit($\mathcal{M}$,$\overrightarrow{X} = \overrightarrow{x}$)

Input
$\mathcal{M} = \langle S, \mathcal{F}, \mathcal{T}, \mathcal{A} \rangle$: an epistemic causal awareness model
$\overrightarrow{X} = \overrightarrow{x}$: an assignment
Output $\mathcal{M}'$: the updated model after explicit intervention

1: $\mathcal{M}_{+} \leftarrow$ Consider($\mathcal{M}, \overrightarrow{X}$)
2: $\mathcal{M}' \leftarrow$ Intervene($\mathcal{M}_{+}, \overrightarrow{X} = \overrightarrow{x}$)
3: **return** $\mathcal{M}'$

The regular intervention (procedure Intervene, Algorithm 4 in Appendix B) corresponds to the standard intervention operator $[\overrightarrow{X} = \overrightarrow{x}]$, which in combination with implicit knowledge operator K results in the no-learning effect (see Sect. 3). In InterveneExplicit, apart from Intervene, in line 1 we also execute the Consider procedure that augments the awareness function (see Algorithm 3 in Appendix B). Since the explicit knowledge operator K^{Ex} takes into account the awareness set, explicit intervention allows circumventing the no-learning effect with respect to K^{Ex} (but not with respect to K).

[7] A more specific definition of a black-box parent function, that maps a set of subgraphs (instead of a set of vertices) in the t-layer to an edge in the b-layer is a possible variant. But since we here assume that agents become aware of (sets) of variables, and get the causal links 'for free', the solution presented here suffices.

Example 7. Let us see how the change in the no-learning principle can be illustrated on an example. Consider again a sprinkler and a button. On the level of implicit knowledge and regular intervention, we have that if $[B = 1]K\ S = 1$ then $K[B = 1]S = 1$. Intuitively speaking, assume that after the gardener has performed a counterfactual reasoning about pressing the button, he ended up knowing that the sprinkler would be on. Just by the virtue of the fact that he was able to perform such a reasoning in the first place, and because he has full access to the epistemic causal awareness model, we also have that he actually could not have any uncertainty about the formula $[B = 1]S = 1$ in the first place—he knew that pressing the button would turn on the sprinkler. On the algorithmic level, invoking INTERVENE($\mathcal{M}, B = 1$) results in a model $\mathcal{M}'$ in which $K\ S = 1$, i.e., $S = 1$ in all `Val` $\in \mathcal{T}'$. If that's the case, then there is no world in $\mathcal{M}$, for which the execution of INTERVENE($\mathcal{M}, B = 1$) could result in a situation where it is not the case that $S = 1$, so $K[B = 1]S = 1$.

Enter Billie. Let's say she is aware of the sprinkler but unaware of the button and hence she cannot turn the sprinkler on or off. The gardener knows all that. He can infer that Billie does not know at this time that he could tell her that pressing the button would turn on the sprinkler. That is so, simply because B is not in Billie's awareness, so she can't even conceive such a statement. This means that $K^{Ex}[B = 1]^{Ex}S = 1$ is false (recall that K^{Ex} codes Billie's knowledge). However, it is the case that $K^{Ex}S = 1$, since Billie is aware of the sprinkler and can see that it is on. It would also be the case that had the gardener called Billie on the phone saying: 'if only you pressed the button ...', hence $[B = 1]^{Ex}K^{Ex}S = 1$ is true. On the algorithmic level, it is so because the INTERVENEEXPLICIT($\mathcal{M}, B = 1$) that codes $[B = 1]^{Ex}$, augments the awareness function $\mathcal{A}$.

Note that the order of execution of CONSIDER and INTERVENE in Algorithm 1 does not matter with respect to the outcome. This is because updating the agent's awareness is done independently from intervention (see Sect. 3.2). In contrast, the update of the awareness function in the case of differential awareness requires executing the intervention first. This is visible in the pseudo-code of the procedure INTERVENEDIFFAWARE, displayed in Algorithm 2. Lines 3–5 in the algorithm constitute the difference to INTERVENEEXPLICIT: for each possible world the difference set is computed according to Definition 10 (the variables that changed values as a result of the intervention), and the awareness function is augmented accordingly for each possible world.

Similar considerations as the above can be applied to (combinations of) the remaining dynamic operators. We will address this issue in future work.

Our account of learning about causality based on the hybrid causal graph representation in Sect. 4 opens up an interesting approach for investigating other learning procedures. The top (expert's) and bottom (layperson's) layers come with their separate representations of the set of structural functions. While the top layer gives an accurate account of the causal graph, in the bottom the layperson's structural functions are only an approximation of the true state of affairs. When the layperson considers a newly discovered set of shadow variables, we

Algorithm 2. $\textsc{InterveneDiffAware}(\mathcal{M}, \overrightarrow{X} = \overrightarrow{x})$

Input
$\mathcal{M} = \langle S, \mathcal{F}, \mathcal{T}, \mathcal{A} \rangle$: an epistemic causal awareness model
$\overrightarrow{X} = \overrightarrow{x}$: an assignment
Output
$\mathcal{M}^{\Delta}$: the updated model after differential awareness intervention

$\mathcal{A}' = null$
$\mathcal{M}_{\overrightarrow{X}=\overrightarrow{x}} = \langle S, \mathcal{F}_{\overrightarrow{X}=\overrightarrow{x}}, \mathcal{T}^{\mathcal{F}}_{\overrightarrow{X}=\overrightarrow{x}}, \mathcal{A} \rangle \leftarrow \textsc{Intervene}(\mathcal{M}, \overrightarrow{X} = \overrightarrow{x})$
for all $\mathtt{Val} \in \mathcal{T}$ **do**
    $\mathcal{A}'(\mathtt{Val}^{\mathcal{F}}_{\overrightarrow{X}=\overrightarrow{x}}) = \mathcal{A}(\mathtt{Val}) \cup \Delta(\mathtt{Val}, \mathtt{Val}^{\mathcal{F}}_{\overrightarrow{X}=\overrightarrow{x}})$
end for
$\mathcal{M}^{\Delta} = (S, \mathcal{F}_{\overrightarrow{X}=\overrightarrow{x}}, \mathcal{T}^{\mathcal{F}}_{\overrightarrow{X}=\overrightarrow{x}}, \mathcal{A}')$
return $\mathcal{M}^{\Delta}$

can require that they come with (some of) their causal links (in the form of the structural functions). These functional specifications might have to be partial, especially if they refer to variables the layperson is not aware of. One practical possibility is to limit the causal connections to those variables that were previously foreshadowed (as in the one-step closure in Sect. 4). Continuing the story of Example 7, the procedure of explicit intervention in the hybrid causal graph could code the gardener's 'minimal' help by reasoning: The parent function applied to C in $\mathcal{G}_t$ outputs a shadow variable as a child in $\mathcal{G}_b$. Given Billie's goal of turning on the sprinkler, she can be made aware of the circuit C. This information is enough as she can further act herself on the new variable to find out how it relates to other variables, and if any configuration allows her to achieve her goal. For now, such cases of gradual discovery of variables and their structural functions remain a topic for future work.

6 Conclusion and Discussion

In this paper, we proposed a new dynamic epistemic logic with public announcements, causal interventions and dynamic awareness, $\mathcal{L}_{\mathtt{PAKCA+}}$ (and $\mathcal{L}_{\mathtt{PAKCA+}\Delta}$), by combining the dynamic epistemic logic of causality proposed in [1] and the dynamic logic of awareness from [6]. We discuss the interactions of the different elements: the implications of various definitions of explicit knowledge, the connection between uniform awareness and the principle of weak (awareness) introspection from [6], and different definitions of explicit interventions (in particular, we proposed a new notion of explicit interventions with differential awareness). We show that the inclusion of explicit knowledge and explicit awareness invalidates the principle of *no-learning*, which troubled the authors in [1]. Even though our logic is single-agent in the traditional sense, we propose an interpretation in which the level of implicit knowledge represents the expert who reasons about a layperson's explicit knowledge (restricted to an awareness set) in a causal epistemic model. We propose to bring these logical considerations closer to practice by introducing a hybrid causal graph based on which an expert agent can make proactive decisions about which hidden causal variables to reveal to the layperson.

The importance of research on causality in AI, and its combination with epistemic reasoning to inform proactive human-AI systems, has been recently pointed out in [12]. Our original approach to reasoning about causality that supports dynamic awareness can arguably contribute to developing proactive AI systems. A tighter integration between our logics and proactive AI agents is desirable in the future. Hybrid causal graphs with appropriate algorithms could help determine which variables, in the Billie example, she should be made aware of to achieve a short-term goal (sprinkler is on) or maintenance goals (grass stays green). If the 'AI-gardener' *knows* about Billie's *awareness* of the button and the sprinkler and her *unawareness* of the circuit, then it can proactively inform her about the circuit and its *causal* link to the sprinkler when it anticipates that Billie aims to turn on the sprinkler. If the learner (Billie) is supposed to learn things by herself (e.g., as a pedagogic method), the expert might compute the *minimal relevant* announcement for updating the learner's awareness and knowledge. How to do this is an open question. In related work[8] a DEL-based approach is used to compute a minimal relevant announcement to inform the agent of a false expectation which would lead to an undesirable outcome.

There are numerous directions for future work. So far, we have considered one expert agent and one learner agent. Multiple learners might manifest in multiple layers in a hybrid causal graph. An apprentice ℓ_1 might need to be aware of more of the actual causal structure than a complete novice ℓ_2. Also, several expert agents with different areas of expertise might be useful. In another extension, a learner could query the experts when needed.

To dive further into the meaning of the shadow variables, inspired by the approach of [13], our framework could be extended with *speculative knowledge* [21], which concerns anything an agent believes might be possible based on their awareness. This would allow our learner to perform counterfactual interventions on the shadow variables, i.e., Billie would be able to speculatively know about the existence of the circuit without relying on the gardener when she discovers that the sprinkler is not on after pressing the button. Another interesting extension of our proposed logic is to incorporate the operator of losing awareness of causal variables: an agent might choose to ignore a variable, or she might forget it due to a lack of cognitive resources. In [6], in addition to $[+\chi]$, a *drop* operator $[-\chi]$ is introduced. By adding it to our logic, we can reason about the learner forgetting variables by becoming unaware of them. Finally, the issue of a sound and complete axiomatization of our logic, $\mathcal{L}_{\mathsf{PAKCA+}}$ (and $\mathcal{L}_{\mathsf{PAKCA+}\triangle}$) is left for future work. The system will combine those proposed in [6] and [1].

Related Work. So far, the research on logical approaches to learning about causality is scarce, with a recent example of [20]. There are, however, studies

[8] Bolander, T. and Grosinger, H.J., *Reasoning about Beliefs and Expectations Using Plausibility Models for Epistemic Proactivity* (2025), under review in Journal of Logic, Language and Information.

in optimal experiment/intervention design (e.g., [8]), but they do not provide a logical account. On the other hand, logical accounts of awareness, especially syntactic accounts, are numerous and were mentioned throughout the paper. For instance, in a recent work outlined in [13], the authors explore dynamic awareness for both single- and multi-agent probabilistic belief models on propositions, where the model is updated using a transition rule. The agents insert shadow propositions as placeholders when they suspect they are unaware of certain propositions. Apart from awareness, agents can have other cognitive stances towards information, such as: knowledge, belief, awareness, attention, implicit knowledge, implicit belief, suspension of judgment, or unawareness [2]. In this paper, we focus on awareness, but a very similar notion of attention also concerns selective focus on certain stimuli or information at a given time. Attention can be a scarce resource and acts as a filter on the information available [4]. More information may make the filter even stricter. Awareness, on the other hand, is defined as a conscious access to or representation of information [2]. A general logic of attention was defined in [3], where a generalization of edge-conditioned event models and attention of arbitrary formulas is extended to multi-agent interactions, where agents can be attentive to others' beliefs. There are also studies of attention and causal reasoning, for instance [5], where eye-tracking experiments demonstrate how causal reasoning is done via attention and mental models, on a counterfactual simulation model of human causal judgment.

Acknowledgements. This research is funded by the Swedish Research Council (Vetenskapsrådet), by projects 2021-05542 and 2023-04349, on H. J. Grosingers part.

Appendix

Appendix A: Proofs from Sect. 3

Proposition 1. If $\models A\varphi \rightarrow KA\varphi$, then $\models (K\varphi \wedge A\varphi) \rightarrow K(\varphi \wedge A\varphi)$.

Proof. Assume that $\models A\varphi \rightarrow KA\varphi$ $(*)$. Take an arbitrary epistemic causal awareness model $\mathcal{M}$ and a $\mathtt{Val}$ in $\mathcal{M}$. Assume that $\mathcal{M}, \mathtt{Val} \models K\varphi \wedge A\varphi$, and so $\mathcal{M}, \mathtt{Val} \models K\varphi$ and $\mathcal{M}, \mathtt{Val} \models A\varphi$. By the latter and $(*)$, we get $\mathcal{M}, \mathtt{Val} \models KA\varphi$. Hence, $\mathcal{M}, \mathtt{Val} \models K\varphi \wedge KA\varphi$, and finally $\mathcal{M}, \mathtt{Val} \models K(\varphi \wedge A\varphi)$.

Proposition 2. Let $\mathcal{M} = \langle S, \mathcal{F}, \mathcal{T}, \mathcal{A} \rangle$. $\mathcal{M} \models A\varphi \rightarrow KA\varphi$ if and only if $\mathtt{Val}, \mathtt{Val}' \in \mathcal{T}$, $\mathcal{A}(\mathtt{Val}) = \mathcal{A}(\mathtt{Val}')$.

Proof. $(\Rightarrow)$ Take an arbitrary epistemic causal awareness model $\mathcal{M}$ and assume that $\mathcal{M} \models A\varphi \rightarrow KA\varphi$. Take any $\mathtt{Val}$ in $\mathcal{M}$ and assume that $X \in \mathcal{A}(\mathtt{Val})$ (and that $x \in \mathcal{R}(X)$), and so $\mathcal{M}, \mathtt{Val} \models A(X = x)$. By weak introspection, we obtain: $\mathcal{M}, \mathtt{Val} \models KA(X = x)$. Finally, by the semantics of K, we get that for all $\mathtt{Val}'$ in $\mathcal{M}$: $\mathcal{M}, \mathtt{Val}' \models A(X = x)$, and so $X \in \mathcal{A}(\mathtt{Val}')$ for all $\mathtt{Val}' \in \mathcal{T}$.
$(\Leftarrow)$ Assume that $\mathcal{M}$ has uniform awareness. Take any $\mathtt{Val} \in \mathcal{T}$ and assume that $\mathcal{M}, \mathtt{Val} \models A\varphi$. Then $v(\varphi) \subseteq \mathcal{A}(\mathtt{Val})$. By uniform awareness, $v(\varphi) \subseteq \mathcal{A}(\mathtt{Val}')$ for all $\mathtt{Val}' \in \mathcal{T}$, so, by the semantics of K: $\mathcal{M}, \mathtt{Val} \models KA\varphi$. Since $\mathtt{Val}$ was arbitrary, we get $\mathcal{M} \models A\varphi \rightarrow KA\varphi$.

Proposition 3. The formula $[\overrightarrow{X} = \overrightarrow{x}]K^{Ex}\varphi \leftrightarrow K^{Ex}[\overrightarrow{X} = \overrightarrow{x}]\varphi$ is valid in the class of epistemic causal awareness models.

Proof. The following are equivalent:

- $[\overrightarrow{X} = \overrightarrow{x}]K^{Ex}\varphi$
- $[\overrightarrow{X} = \overrightarrow{x}]K(\varphi \wedge A\varphi)$
- $[\overrightarrow{X} = \overrightarrow{x}](K\varphi \wedge KA\varphi)$
- $[\overrightarrow{X} = \overrightarrow{x}]K\varphi \wedge [\overrightarrow{X} = \overrightarrow{x}]KA\varphi$
- $K[\overrightarrow{X} = \overrightarrow{x}]\varphi \wedge KA[\overrightarrow{X} = \overrightarrow{x}]\varphi$
- $K([\overrightarrow{X} = \overrightarrow{x}]\varphi \wedge A[\overrightarrow{X} = \overrightarrow{x}]\varphi)$
- $K^{Ex}[\overrightarrow{X} = \overrightarrow{x}]\varphi$

Proposition 4. The formula $[\overrightarrow{X} = \overrightarrow{x}]^{Ex}K^{Ex}\varphi \rightarrow K^{Ex}[\overrightarrow{X} = \overrightarrow{x}]^{Ex}\varphi$ is not valid in the class of epistemic causal awareness models.

Proof. Consider a model $\mathcal{M}$ and a $\mathtt{Val}$ in $\mathcal{M}$ such that $set(\overrightarrow{X}) \not\subseteq \mathcal{A}(\mathtt{Val})$ and $\mathcal{M}, \mathtt{Val} \models [\overrightarrow{X} = \overrightarrow{x}]^{Ex}K^{Ex}\varphi$. Since $set(\overrightarrow{X}) \subseteq v(K^{Ex}[\overrightarrow{X} = \overrightarrow{x}]^{Ex}\varphi)$, we have that $v(K^{Ex}[\overrightarrow{X} = \overrightarrow{x}]^{Ex}\varphi) \not\subseteq \mathcal{A}(\mathtt{Val})$, and so $\mathcal{M}, \mathtt{Val} \not\models K^{Ex}[\overrightarrow{X} = \overrightarrow{x}]^{Ex}\varphi$.

Appendix B: Algorithms from Sect. 5

Algorithm 3. CONSIDER($\mathcal{M}$,$\overrightarrow{X}$)

Input $\overrightarrow{X}$: a sequence of variables
$\mathcal{M} = \langle S, \mathcal{F}, \mathcal{T}, \mathcal{A} \rangle$: an epistemic causal awareness model
Output $\mathcal{M}'$: an updated model after awareness expansion

$\mathcal{A}' = null$
for each $\texttt{Val} \in \mathcal{T}$ **do**
    $\mathcal{A}'(\texttt{Val}) = \mathcal{A}(\texttt{Val}) \cup set(\overrightarrow{X})$
end for
$\mathcal{M}' = \langle S, \mathcal{F}, \mathcal{T}, \mathcal{A}' \rangle$
return $\mathcal{M}'$

Algorithm 4. Intervene($\mathcal{M}, \overrightarrow{X} = \overrightarrow{x}$)

Input $\mathcal{M} = \langle \langle U, V, \mathcal{R} \rangle, \mathcal{F}, \mathcal{T}, \mathcal{A} \rangle$: an epistemic causal awareness model, where
$\mathcal{F} = \{f_{X_i} \mid X_i \in V\}$ and $\mathcal{T} = \{\texttt{Val} \mid \texttt{Val}$ complies with $\mathcal{F}\}$
$\overrightarrow{X} = \overrightarrow{x}$: an assignment
Output $\mathcal{M}'$: the updated model after intervention

$\mathcal{F}' = \mathcal{F}$, $\mathcal{T}' = \mathcal{T}$
for each X_i in $\overrightarrow{X}$ **do**
    if $X_i \in V$ **then**
        f'_{X_i} = constant function x_i
        for each $\texttt{Val}' \in \mathcal{T}'$ **do**
            $\texttt{Val}'(X_i) = x_i$
        end for
    else
        for each $\texttt{Val}' \in \mathcal{T}'$ **do**
            $\texttt{Val}'(X_i) = x_i$
        end for
    end if
end for
for each $X_j \in V \setminus set(\overrightarrow{X})$ **do**
    for each $\texttt{Val}' \in \mathcal{T}'$ **do**
        $\texttt{Val}'(X_j)$ = value complying with $\mathcal{F}'$
    end for
end for
$\mathcal{M}' = \langle \langle U, V, \mathcal{R} \rangle, \mathcal{F}', \mathcal{T}', \mathcal{A} \rangle$
return $\mathcal{M}'$

References

1. Barbero, F., Schulz, K., Smets, S., Velázquez-Quesada, F.R., Xie, K.: Thinking about causation: a causal language with epistemic operators. In: Martins, M.A., Sedlár, I. (eds.) Dynamic Logic. New Trends and Applications, pp. 17–32. Springer, Cham (2020). https://doi.org/10.1007/978-3-030-65840-3_2

2. Belardinelli, G.: Epistemic Logic Models of Awareness, Attention, and Implicit Reasoning. Ph.D. thesis, Københavns Universitet (2023)
3. Belardinelli, G., Bolander, T., Watzl, S.: A logic of general attention using edge-conditioned event models. In: Kwok, J. (ed.) Proceedings of the Thirty-Fourth International Joint Conference on Artificial Intelligence, IJCAI 2025, pp. 4347–4355. International Joint Conferences on Artificial Intelligence Organization (2025). https://doi.org/10.24963/ijcai.2025/484
4. Belardinelli, G., Rendsvig, R.K.: Epistemic planning with attention as a bounded resource. In: Ghosh, S., Icard, T. (eds.) LORI 2021. LNCS, vol. 13039, pp. 14–30. Springer, Cham (2021). https://doi.org/10.1007/978-3-030-88708-7_2
5. Bello, P., Lovett, A., Briggs, G., O'Neill, K.: An attention-driven computational model of human causal reasoning. In: Proceedings of the Annual Meeting of the Cognitive Science Society (2018). https://escholarship.org/uc/item/7bw8d284
6. van Benthem, J., Velázquez-Quesada, F.R.: The dynamics of awareness. Synthese **177**(1), 5–27 (2010). https://doi.org/10.1007/s11229-010-9764-9
7. Charniak, E.: Bayesian networks without tears. AI Mag. **12**(4), 50 (1991). https://doi.org/10.1609/aimag.v12i4.918
8. Cooper, G.F., Yoo, C.: Causal discovery from a mixture of experimental and observational data. In: Proceedings of the Fifteenth Conference on Uncertainty in Artificial Intelligence, pp. 116–125. UAI 1999, Morgan Kaufmann Publishers Inc., San Francisco, CA, USA (1999). https://dl.acm.org/doi/10.5555/2073796.2073810
9. Endsley, M.R.: Toward a theory of situation awareness in dynamic systems. Hum. Factors J. Hum. Factors Ergon. Soc. **37**, 32–64 (1995). https://doi.org/10.1518/001872095779049543
10. Fagin, R., Halpern, J.Y.: Belief, awareness, and limited reasoning. Artif. Intell. **34**(1), 39–76 (1987). https://doi.org/10.1016/0004-3702(87)90003-8
11. Grant, A.M., Ashford, S.J.: The dynamics of proactivity at work. Res. Organ. Behav. **28**, 3–34 (2008). https://doi.org/10.1016/j.riob.2008.04.002
12. Grosinger, H.J.: The next level of long-term agent autonomy–proactively acquiring knowledge and abilities. In: Proceedings of the 24th International Conference on Autonomous Agents and Multiagent Systems, pp. 2859–2864 (2025). https://dl.acm.org/doi/10.5555/3709347.3744035
13. Halpern, J.Y., Piermont, E.: Dynamic awareness (2020). https://arxiv.org/abs/2007.02823
14. Lorini, E.: Designing artificial reasoners for communication. In: Proceedings of the 23rd International Conference on Autonomous Agents and Multiagent Systems, pp. 2690–2695 (2024). https://dl.acm.org/doi/abs/10.5555/3635637.3663259
15. Moreno, A.: Avoiding logical omniscience and perfect reasoning: a survey. AI Commun. **11**(2), 101–122 (1998). https://journals.sagepub.com/doi/abs/10.3233/EAI-1998-150
16. Pearl, J., Mackenzie, D.: The Book of Why: The New Science of Cause and Effect. Basic Books (2018). https://doi.org/10.1126/science.aau9731
17. Shanahan, M.: The frame problem. In: Zalta, E.N. (ed.) The Stanford Encyclopedia of Philosophy. Metaphysics Research Lab, Stanford University, Spring 2016 edn. (2016). https://plato.stanford.edu/archives/spr2016/entries/frame-problem/
18. Shvo, M., Hari, R., O'Reilly, Z., Abolore, S., Wang, S.Y.N., McIlraith, S.A.: Proactive robotic assistance via theory of mind. In: 2022 IEEE/RSJ International Conference on Intelligent Robots and Systems (IROS), pp. 9148–9155 (2022). https://doi.org/10.1109/IROS47612.2022.9981627

19. Sim, K.M.: Epistemic logic and logical omniscience: a survey. Int. J. Intell. Syst. **12**(1), 57–81 (1997). https://doi.org/10.1002/(SICI)1098-111X(199701)12:1⟨57::AID-INT3⟩3.0.CO;2-X
20. Thoft, K.B.P., Gierasimczuk, N.: Learning by intervention in simple causal domains. In: Gierasimczuk, N., Velázquez-Quesada, F.R. (eds.) Dynamic Logic. New Trends and Applications, pp. 104–118. Springer, Cham (2024). https://doi.org/10.1007/978-3-031-51777-8_7
21. van Ditmarsch, H., French, T., Velázquez-Quesada, F.R., Wáng, Y.N.: Implicit, explicit and speculative knowledge. Artif. Intell. **256**, 35–67 (2018). https://doi.org/10.1016/j.artint.2017.11.004

Hybrid Logics with Moving Names

Yiwen Ding[1], Krishna Manoorkar[1], Ni Wayan Switrayni[1,2(✉)], and Apostolos Tzimoulis[3]

[1] Vrije Universiteit Amsterdam, Amsterdam, The Netherlands
switrayniniwayan@gmail.com
[2] Department of Mathematics, University of Mataram, Mataram, Indonesia
[3] University of Luxembourg, Esch-sur-Alzette, Luxembourg
apostolos@tzimoulis.eu

Abstract. We introduce a family of hybrid logics extended with a dynamic "move" operator that reassigns nominals to accessible states. For each nominal i, the formula $\langle \mathrm{i} \rangle \varphi$ holds exactly when φ is true in the model obtained by moving i to some successors of the state it currently names. We embed this operator into three baseline systems—basic hybrid logic with only nominals, hybrid logic with the satisfaction operator @, and hybrid logic with binders—to obtain three corresponding hybrid logics with moving names. We investigate their expressive power, develop reduction axioms that internalize the dynamics back into the static languages, and demonstrate how some real-world scenarios can be formalized succinctly in these logics.

Keywords: Hybrid logic · Dynamic logic · Moving names · Expressive power · Translations · Bisimulations

1 Introduction

Hybrid logics [1,9,18] are extensions of classical modal logic that allow for reference to specific states in a Kripke model. Compared to classical modal logic, they offer a more favorable balance between expressive power and computational complexity. The basic hybrid logic extends classical modal logic with a new class of propositional variables called *nominals*, which are interpreted as singletons on

This project has received funding from the European Union's Horizon 2020 research and innovation programme under the Marie Skłodowska-Curie grant agreement No 101007627. The research of the first author is supported by the China Scholarship Council. The research of the second author is supported by the NWO grant KIVI.2019.001 awarded to Alessandra Palmigiano. The research of the third author is supported by the Indonesian Education Scholarship, Center for Higher Education Funding and Assessment, and Indonesian Endowment Fund for Education (No.3189/BPPT/BPI.LG/IV/2024). The research of the fourth author was supported by the Luxembourg National Research Fund (FNR) (INTER/DFG/23/17415164/LODEX) and by the Key Project of the Chinese Ministry of Education (22JJD720021).

J. Wang et al. (Eds.): DaLí 2025, LNCS 16472, pp. 23–39, 2026.
https://doi.org/10.1007/978-3-032-22626-6_2

Kripke model. Intuitively, those nominals serve as names for specific states in the model. To further enhance expressive power, various extensions of the basic hybrid logic have been proposed. Among the most well-studied are hybrid logic with the satisfaction operator @ and hybrid logic with the binder $\downarrow$. In those logics, formula $@_{\mathrm{i}}\varphi$ means "φ is true at the state which the nominal i names", and formula $\downarrow x.\varphi$ means "φ is true when the current state is bound to the variable x".

Originally introduced by Arthur Prior in the context of tense logic [24,25], hybrid logics have since evolved significantly, with contemporary developments shifting in focus and scope [10]. They are closely related to description logics [5,22], and have proven highly effective in applications involving knowledge representation and reasoning [14,17,22]. Their model theory is also well developed, with key notions such as standard translation, bisimulation, and foundational results including the van Benthem and GoldblattThomason theorems established [12,15]. Recently, hybrid logics have been applied to various kinds of philosophical logic, particularly to logics reasoning about different types of games [8,26].

Dynamic logics are a vast family of modal logics that aim to formally express changes in dynamic systems, through actions and processes. In particular, the study of dynamic modal logics that capture changes in the Kripke models, were initiated by the introduction of Public Announcement logic [19,23] and led to the development of Dynamic Epistemic Logics [6,7,29,30], which have been proved useful in modeling epistemic actions. In the past two decades, many different dynamic logics have been developed and studied such as sabotage logic [28], swap logic [3], and arrow update logic [21].

In this article, we introduce a family of dynamic hybrid logics by extending the three basic hybrid logics with a name-changing operator $\langle \mathrm{i} \rangle$ for each nominal i. A formula of the form $\langle \mathrm{i} \rangle\varphi$ holds precisely when φ is true in the model resulting from reassigning the nominal i to some successors of the state it currently denotes. These logics provide a natural framework for modeling various dynamic aspects of real-world scenarios. We also investigate the expressive power and finite model property of these dynamic hybrid logics by means of translations and bisimulations. In addition, we introduce a notion of bisimulations for the dynamic hybrid logics with and without satisfaction operator @ and obtain a bisimulation invariance result.

Structure of the Paper. Section 2 gathers preliminaries for the paper. Section 3 introduces hybrid logics with moving names and illustrates their usefulness through examples that model real-world scenarios. Section 4 investigates the reduction and expressive power of these logics. Section 5 defines bisimulations for hybrid logics with moving names and establishes several bisimulation invariance results. Section 6 concludes the paper and outlines some future research.

2 Preliminaries

In this section, we gather some basic preliminaries on semantics and expressive powers of hybrid logics. For more details about hybrid logics, we refer to [9,11, 15].

Let Prop, Nom, Svar be countably infinite sets of propositional variables, nominals, and state variables, respectively. We define the hybrid languages $\mathcal{H}(@)$ and $\mathcal{H}(@,\downarrow)$ by the following recursions:

$$\varphi ::= \top \mid p \mid \mathrm{i} \mid \varphi \vee \varphi \mid \neg\varphi \mid \Diamond\varphi \mid @_{\mathrm{i}}\varphi,$$

$$\psi ::= \top \mid p \mid \mathrm{t} \mid \psi \vee \psi \mid \neg\psi \mid \Diamond\psi \mid @_{\mathrm{t}}\psi \mid \downarrow \mathrm{x}.\psi,$$

where $\mathrm{i} \in$ Nom, $p \in$ Prop, $x \in$ Svar and $t \in$ Nom $\cup$ Svar. We use $\mathcal{H}$ to denote @-free fragment of $\mathcal{H}(@)$. In any of those languages, for any formula φ, let $\Box\varphi := \neg\Diamond\neg\varphi$, and other Boolean connectives defined as usual. Throughout this paper, we refer to all these languages collectively as languages for hybrid logics.

A *model* for hybrid logics is a tuple $\mathbb{M} = (W, R, V)$, where W is a non-empty set, $R \subseteq W \times W$ is a binary relation and V : Prop$\cup$Nom $\to \mathcal{P}(W)$ is a valuation, such that for each nominal $\mathrm{i} \in$ Nom, $V(\mathrm{i})$ is a singleton. We say i denotes the state in $V(\mathrm{i})$, or i is the name of the state in $V(\mathrm{i})$. Nominals are usually used for both the names and the states which they denote when the context is clear. Given any model $\mathbb{M} = (W, R, V)$, an *interpretation* for the language $\mathcal{H}(@,\downarrow)$ is a map g : Svar $\to W$. For any $x \in$ Svar and $w \in W$, $g[x := w]$ denotes the map different from g by assigning w to x.

Definition 1. *Given any model* $\mathbb{M} = (W, R, V)$*, an interpretation* g : Svar $\to$ W*, a state* $w \in W$*, and an* $\mathcal{H}(@,\downarrow)$*-formula* φ*,* $\mathbb{M}$ satisfies φ at w with g (written $\mathbb{M}, g, w \models \varphi$) *is defined inductively as following:*

- $\mathbb{M}, g, w \models \top$ *always.*
- $\mathbb{M}, g, w \models p \Leftrightarrow w \in V(p)$.
- $\mathbb{M}, g, w \models \mathrm{x} \Leftrightarrow w = g(\mathrm{x})$.
- $\mathbb{M}, g, w \models \mathrm{i} \Leftrightarrow w \in V(\mathrm{i})$.
- $\mathbb{M}, g, w \models \neg\varphi \Leftrightarrow \mathbb{M}, g, w \not\models \varphi$.
- $\mathbb{M}, g, w \models \varphi \vee \psi \Leftrightarrow \mathbb{M}, g, w \models \varphi$ *or* $\mathbb{M}, g, w \models \psi$.
- $\mathbb{M}, g, w \models \Diamond\varphi \Leftrightarrow \exists w'(wRw'$ *and* $\mathbb{M}, g, w' \models \varphi)$.
- $\mathbb{M}, g, w \models @_{\mathrm{i}}\varphi \Leftrightarrow \mathbb{M}, g, v \models \varphi$ *where* $V(\mathrm{i}) = \{v\}$.
- $\mathbb{M}, g, w \models @_{\mathrm{x}}\varphi \Leftrightarrow \mathbb{M}, g, v \models \varphi$ *where* $g(x) = v$.
- $\mathbb{M}, g, w \models \downarrow \mathrm{x}.\varphi \Leftrightarrow \mathbb{M}, g[\mathrm{x} := w], w \models \varphi$.

φ is valid in $\mathbb{M}$ with g (written $\mathbb{M}, g \models \varphi$)*, if* $\mathbb{M}, g, w \models \varphi$ *for any* $w \in W$*.* φ is valid (written $\models \varphi$)*, if* $\mathbb{M}, g \models \varphi$ *for any model* $\mathbb{M}$ *and interpretation* g*.*

For any $\mathcal{H}(@)$-formula φ, $\mathbb{M}$ *satisfies* φ *at* w *(written* $\mathbb{M}, w \models \varphi$*)* is defined similarly to $\mathcal{H}(@,\downarrow)$, but without reference to any interpretation[1]. A formula

[1] Note that for $\mathcal{H}(@)$-formulas, interpretations do not play any role in the definition of satisfaction relations.

φ is valid in $\mathbb{M}$ *(written* $\mathbb{M} \models \varphi$*)*, if for any $w \in W$, $\mathbb{M}, w \models \varphi$. A formula φ *is valid (written* $\models \varphi$*)*, if for all $\mathbb{M}$, $\mathbb{M} \models \varphi$. Finally, the satisfaction relation for $\mathcal{H}$-formulas can be defined as the restriction of the satisfaction relation for $\mathcal{H}(@)$. The validity of $\mathcal{H}$-formulas is also defined similarly.

Let $\mathcal{TH}$ (resp. $\mathcal{TH}(@)$, (resp. $\mathcal{TH}(@, \downarrow)$) be the tense hybrid logic language which is extension of $\mathcal{H}$ (resp. $\mathcal{H}(@)$, $\mathcal{H}(@, \downarrow)$) with the formulas of the form $\blacklozenge\varphi$, for $\varphi \in \mathcal{H}$ (resp.$\varphi \in \mathcal{H}(@)$, $\varphi \in \mathcal{H}(@, \downarrow)$) whose satisfaction on a model $(\mathbb{M}, g, w)$ for hybrid logics is defined as follows: $\mathbb{M}, g, w \models \blacklozenge\varphi \Leftrightarrow \exists w'(w'Rw \textit{ and } \mathbb{M}, g, w' \models \varphi)$[2].

We now recall some results about the expressive powers of hybrid logics.

Definition 2 (Definition 4.1.1, [15]). *Given two models* $\mathbb{M} = (W, R, V)$ *and* $\mathbb{M}' = (W', R', V')$, *an* $\mathcal{H}$-bisimulation *between* $\mathbb{M}$ *and* $\mathbb{M}'$ *is a binary relation* $Z \subseteq W \times W'$ *satisfying the following conditions:*

Atom *If* wZw', *then* $w \in V(p) \Leftrightarrow w' \in V'(p)$, *for any* $p \in \mathrm{Prop} \cup \mathrm{Nom}$.
Zig *If* wZw' *and* $w'R'v'$, *then there exists* $v \in W$, *such that* vZv' *and* wRv.
Zag *If* wZw' *and* wRv, *then there exists* $v' \in W'$, *such that* vZv' *and* $w'R'v'$.

An $\mathcal{H}(@)$-bisimulation *is an* $\mathcal{H}$*-bisimulation satisfying, in addition,*

Nom *If there exists nominal* $\mathrm{i} \in \mathrm{Nom}$ *s.t.* $w \in V(\mathrm{i})$ *and* $w' \in V'(\mathrm{i})$, *then* wZw'.

An $\mathcal{H}$*-bisimulation (resp.* $\mathcal{H}(@)$*-bisimulation) is* $\mathcal{TH}$*-bisimulation (resp.* $\mathcal{TH}(@)$*-bisimulation) if conditions* **Zig** *and* **Zag** *are also true for relations* R^{-1} *and* R'^{-1}.

The following theorem shows that $\mathcal{H}$ (resp. $\mathcal{H}(@)$, resp. $\mathcal{TH}$, resp. $\mathcal{TH}(@)$) formulas are invariant under $\mathcal{H}$ (resp. $\mathcal{H}(@)$, resp. $\mathcal{TH}$, resp. $\mathcal{TH}(@)$) bisimulations.

Theorem 1 (Theorem 4.1.2, [15]). *Let* $\mathbb{M}$, $\mathbb{N}$ *be two models, and* Z *is an* $\mathcal{H}$ *(resp.* $\mathcal{H}(@)$*, resp.* $\mathcal{TH}$*, resp.* $\mathcal{TH}(@)$*) bisimulation between* $\mathbb{M}$ *and* $\mathbb{N}$. *For any points* w, v *in* $\mathbb{M}$, $\mathbb{N}$ *respectively, if* wZv, *then* w *and* v *satisfy the same* $\mathcal{H}$ *(resp.* $\mathcal{H}(@)$*, resp.* $\mathcal{TH}$*, resp.* $\mathcal{TH}(@)$*) formulas.*

3 Hybrid Logics with Moving Names

In this section, we extend those hybrid logics in Sect. 2 with dynamic operators which allow for moving the nominals (names) along the accessibility relation in a model. We also give several examples to show that those dynamic operators are useful in modeling real-world scenarios.

For the language $\mathcal{H}(@, \downarrow)$ (resp. $\mathcal{H}(@)$, resp. $\mathcal{H}$), we denote its extension with the dynamic operator $\langle \mathrm{i} \rangle$ for each nominal $\mathrm{i} \in \mathrm{Nom}$ by $\mathcal{DH}(@, \downarrow)$ (resp. $\mathcal{DH}(@)$, resp. $\mathcal{DH}$). Throughout this paper, we refer to all these languages collectively as languages for hybrid logics with moving names.

[2] As with other formulas we can drop the interpretation g for $\mathcal{TH}$ and $\mathcal{TH}(@)$.

Given any model $\mathbb{M} = (W, R, V)$, nominal $\mathrm{i} \in \mathrm{Nom}$ and $w \in W$, we define a new model $\mathbb{M}_{[\mathrm{i} \to w]} = (W, R, V_{[\mathrm{i} \to w]})$, where $V_{[\mathrm{i} \to w]}$ agrees with V on all variables except i, and $V_{[\mathrm{i} \to w]}(\mathrm{i}) = \{w\}$. Intuitively, the new valuation $V_{[\mathrm{i} \to w]}$ changes the world referred by name i to w.

Definition 3. *Given any model* $\mathbb{M} = (W, R, V)$*, interpretation* g *on* $\mathbb{M}$*, nominal* $\mathrm{i} \in \mathrm{Nom}$*, states* $w, w' \in W$*, and* $\mathcal{DH}(@, \downarrow)$*-formula* φ*, the satisfaction relation for* $\mathcal{DH}(@, \downarrow)$ *is defined step-by-step as below:*

- $\mathbb{M}, g, w \models \varphi$ *is defined as same if the main connective of* φ *is in* $\mathcal{H}(@, \downarrow)$.
- $\mathbb{M}, g, w \models \langle \mathrm{i} \rangle \varphi \Leftrightarrow \exists w'(V(\mathrm{i})Rw'$ *and* $\mathbb{M}_{[\mathrm{i} \to w']}, g, w \models \varphi)$.

As satisfaction relations for language $\mathcal{DH}(@)$ and $\mathcal{DH}$, we can drop the reference to interpretations as in the case of $\mathcal{H}(@)$ and $\mathcal{H}$. For any formula φ in languages for hybrid logics with moving names, let $[\mathrm{i}]\varphi := \neg\langle \mathrm{i} \rangle \neg\varphi$. The validity of formulas in languages for hybrid logics with moving names is defined as same as in the case of hybrid logics.

The following example shows that these logics can be used to formalize various real-life scenarios.

Example 1. Consider a scenario where a robot is moving through a space with set W be the set of possible worlds representing different regions in the space. Let $R \subseteq W \times W$ be the relation such that $w_1 R w_2$ iff it is possible for the robot at location w_1 to move to location w_2 at the next instant. Let i be a nominal representing a particular region in the space. Then a formula $\langle \mathrm{i} \rangle \Diamond \mathrm{i}$ is true at some world $w \in W$ iff there is a location to which the robot can move from the worlds named by either i or w in the next instant.

Consider the formula $\langle \mathrm{i} \rangle @_{\mathrm{i}} \Box \mathrm{i}$. This formula is true at some world $w \in W$ if it is possible to move in one step from i to some place at which the robot can get stuck i.e. reach a location w_0 such that only the place reachable from it is itself.

Consider a slight variation of the above scenario with two robots whose present locations are given by nominals i and j. Consider the formula $[\mathrm{i}]\langle \mathrm{j} \rangle @_{\mathrm{i}} \Box \mathrm{i}$. This formula is true at some world $w \in W$ iff for any place one robot can move from world named i, another robot can move from world named j to a place which is not reachable from the new location of the first robot in a single step.

4 Reduction and Expressivity

In this section, we discuss the reduction and expressivity of hybrid logics with moving names.

Lemma 1. *Given any* $\mathrm{i} \in \mathrm{Nom}$*, and any formulas* $\varphi, \psi \in \mathcal{H}(@, \downarrow)$*,*

1. $\langle \mathrm{i} \rangle p \leftrightarrow p \wedge @_{\mathrm{i}} \Diamond \top$.
2. $\langle \mathrm{i} \rangle \mathrm{i} \leftrightarrow \blacklozenge \mathrm{i}$.

3. $\langle \mathrm{i}\rangle \mathrm{t} \leftrightarrow @_{\mathrm{i}}\Diamond\top \wedge \mathrm{t}$, *if* $\mathrm{t} \neq \mathrm{i}$.
4. $\langle \mathrm{i}\rangle(\varphi \vee \psi) \leftrightarrow \langle \mathrm{i}\rangle\varphi \vee \langle \mathrm{i}\rangle\psi$.
5. $\langle \mathrm{i}\rangle\Diamond\varphi \leftrightarrow \Diamond\langle \mathrm{i}\rangle\varphi$.
6. $\langle \mathrm{i}\rangle @_{\mathrm{t}}\varphi \leftrightarrow @_{\mathrm{t}}\langle \mathrm{i}\rangle\varphi$, *if* $\mathrm{t} \neq \mathrm{i}$.
7. *For any formula* φ *that does not contain* i, $\langle \mathrm{i}\rangle\varphi \leftrightarrow \varphi \wedge @_{\mathrm{i}}\Diamond\top$.

Proof. Let $\mathbb{M} = (W, R, V)$ be any model for hybrid logics, g be any interpretation on it, and $w \in W$ be any world.

1.
$$\begin{aligned}\mathbb{M}, g, w \models \langle \mathrm{i}\rangle p &\text{ iff } \exists w'(V(\mathrm{i})Rw' \wedge \mathbb{M}_{[\mathrm{i}\to w']}, g, w \models p)\\ &\text{ iff } \exists w'(V(\mathrm{i})Rw' \wedge w \in V_{[\mathrm{i}\to w']}(p))\\ &\text{ iff } \exists w'(V(\mathrm{i})Rw' \wedge w \in V(p))\\ &\text{ iff } \mathbb{M}, g, V(\mathrm{i}) \models \Diamond\top \text{ and } \mathbb{M}, g, w \models p\\ &\text{ iff } \mathbb{M}, g, w \models @_{\mathrm{i}}\Diamond\top \text{ and } \mathbb{M}, g, w \models p\\ &\text{ iff } \mathbb{M}, g, w \models @_{\mathrm{i}}\Diamond\top \wedge p\end{aligned}$$

2.
$$\begin{aligned}\mathbb{M}, g, w \models \langle \mathrm{i}\rangle \mathrm{i} &\text{ iff } \exists w'(V(\mathrm{i})Rw' \wedge \mathbb{M}_{[\mathrm{i}\to w']}, g, w \models \mathrm{i})\\ &\text{ iff } \exists w'(V(\mathrm{i})Rw' \wedge w \in V_{[\mathrm{i}\to w']}(\mathrm{i}))\\ &\text{ iff } \exists w'(V(\mathrm{i})Rw' \wedge w = w')\\ &\text{ iff } V(\mathrm{i})Rw\\ &\text{ iff } \mathbb{M}, g, w \models \blacklozenge \mathrm{i}\end{aligned}$$

3. We only give proof for $\mathrm{t} \in Nom$. The proof for $t \in Svar$ is similar.
$$\begin{aligned}\mathbb{M}, g, w \models \langle \mathrm{i}\rangle \mathrm{t} &\text{ iff } \exists w'(V(\mathrm{i})Rw' \wedge \mathbb{M}_{[\mathrm{i}\to w']}, g, w \models \mathrm{t})\\ &\text{ iff } \exists w'(V(\mathrm{i})Rw' \wedge w \in V_{[\mathrm{i}\to w']}(\mathrm{t}))\\ &\text{ iff } \exists w'(V(\mathrm{i})Rw' \wedge w \in V(\mathrm{t}))\\ &\text{ iff } \mathbb{M}, g, V(\mathrm{i}) \models \Diamond\top \text{ and } \mathbb{M}, g, w \models \mathrm{t}\\ &\text{ iff } \mathbb{M}, g, w \models @_{\mathrm{i}}\Diamond\top \text{ and } \mathbb{M}, g, w \models \mathrm{t}\\ &\text{ iff } \mathbb{M}, g, w \models @_{\mathrm{i}}\Diamond\top \wedge \mathrm{t}\end{aligned}$$

4.
$$\begin{aligned}\mathbb{M}, g, w \models \langle \mathrm{i}\rangle(\varphi \vee \psi) &\text{ iff } \exists w'(V(\mathrm{i})Rw' \wedge \mathbb{M}_{[\mathrm{i}\to w']}, g, w \models \varphi \vee \psi)\\ &\text{ iff } \exists w'(V(\mathrm{i})Rw' \wedge (\mathbb{M}_{[\mathrm{i}\to w']}, g, w \models \varphi \vee \mathbb{M}_{[\mathrm{i}\to w']}, g, w \models \psi))\\ &\text{ iff } \exists w'((V(\mathrm{i})Rw' \wedge \mathbb{M}_{[\mathrm{i}\to w']}, g, w \models \varphi) \vee (V(\mathrm{i})Rw' \wedge \mathbb{M}_{[\mathrm{i}\to w']}, g, w \models \psi))\\ &\text{ iff } \exists w'((V(\mathrm{i})Rw' \wedge \mathbb{M}_{[\mathrm{i}\to w']}, g, w \models \varphi)) \vee \exists w'((V(\mathrm{i})Rw' \wedge \mathbb{M}_{[\mathrm{i}\to w']}, g, w \models \psi))\\ &\text{ iff } \mathbb{M}, g, w \models \langle \mathrm{i}\rangle\varphi \vee \mathbb{M}, g, w \models \langle \mathrm{i}\rangle\psi\\ &\text{ iff } \mathbb{M}, g, w \models \langle \mathrm{i}\rangle\varphi \vee \langle \mathrm{i}\rangle\psi\end{aligned}$$

5.
$$\begin{aligned}\mathbb{M}, g, w \models \langle \mathrm{i}\rangle\Diamond\varphi &\text{ iff } \exists w'(V(\mathrm{i})Rw' \wedge \mathbb{M}_{[\mathrm{i}\to w']}, g, w \models \Diamond\varphi)\\ &\text{ iff } \exists w'\exists w''(V(\mathrm{i})Rw' \wedge wRw'' \wedge \mathbb{M}_{[\mathrm{i}\to w']}, g, w'' \models \varphi)\\ &\text{ iff } \exists w''\exists w'(wRw'' \wedge V(\mathrm{i})Rw' \wedge \mathbb{M}_{[\mathrm{i}\to w']}, g, w'' \models \varphi)\\ &\text{ iff } \mathbb{M}, g, w \models \Diamond\langle \mathrm{i}\rangle\ \varphi\end{aligned}$$

6. We only give proof for $\mathrm{t} \in Nom$. The proof for $t \in Svar$ is similar.
$$\begin{aligned}\mathbb{M}, g, w \models \langle \mathrm{i}\rangle @_{\mathrm{t}}\varphi &\text{ iff } \exists w'(V(\mathrm{i})Rw' \wedge \mathbb{M}_{[\mathrm{i}\to w']}, g, w \models @_{\mathrm{t}}\varphi)\\ &\text{ iff } \exists w'(V(\mathrm{i})Rw' \wedge \mathbb{M}_{[\mathrm{i}\to w']}, g, V_{[\mathrm{i}\to w']}(\mathrm{t}) \models \varphi)\\ &\text{ iff } \exists w'(V(\mathrm{i})Rw' \wedge \mathbb{M}_{[\mathrm{i}\to w']}, g, V(\mathrm{t}) \models \varphi)\\ &\text{ iff } \mathbb{M}, g, V(\mathrm{t}) \models \langle \mathrm{i}\rangle\varphi\\ &\text{ iff } \mathbb{M}, g, w \models @_{\mathrm{t}}\langle \mathrm{i}\rangle\varphi\end{aligned}$$

7. The proof follows from the following fact: Let $(\mathbb{M}, g, w)$ be a model for hybrid logic with $\mathbb{M} = (W, R, V)$. Let V' be any valuation s.t. $V'(p) = V(p)$ for any

$p \neq \mathrm{i} \in Prop \cup Nom$ and $\mathbb{M}' = (W, R, V')$. Then, for any formula $\varphi \in \mathcal{H}(@, \downarrow)$ not containing i, $\mathbb{M}, g, w \models \varphi$ iff $\mathbb{M}', g, w \models \varphi$. The proof for this fact is by straight-forward induction on the complexity of the formula φ.

The above lemma implies that the $\{\wedge, \neg\}$-free fragment of $\mathcal{DH}$ can be translated into $\mathcal{TH}(@)$, while the fragment of $\mathcal{DH}(@)$ that is $\{\wedge, \neg\}$-free and does not contain $@_\mathrm{i}$ in the scope of $\langle \mathrm{i} \rangle$ for any $\mathrm{i} \in \mathrm{Nom}$ can be translated into $\mathcal{TH}(@)$[3]. This observation immediately implies many properties, such as the finite model property (with the bound on the size of the model in terms of the length of the formula to be satisfied) and decidability, hold for these fragments of hybrid logics with moving names.

The following example shows how we can use equivalences in Lemma 1 to eliminate dynamic operators in favor of the tense operator $\blacklozenge$.

Example 2. Consider the formula $\varphi := \langle \mathrm{j} \rangle @_\mathrm{i} \Diamond (\mathrm{i} \vee \mathrm{j})$. For any model $(\mathbb{M}, g, w)$, $\mathbb{M}, g, w \models \varphi$ iff there is a world accessible from $V(\mathrm{j})$ and there is a reflexive arrow at $V(\mathrm{i})$, i.e. $V(\mathrm{i}) R V(\mathrm{i})$ or there exists a world w_0 which is accessible from both $V(\mathrm{i})$ and $V(\mathrm{j})$ (i.e. $V(\mathrm{i}) R w_0$ and $V(\mathrm{j}) R w_0$). Applying the series of equivalences in Lemma 1, we have

$$\begin{array}{ll} & \mathbb{M}, g, w \models \varphi \\ \text{iff} & \mathbb{M}, g, w \models \langle \mathrm{j} \rangle @_\mathrm{i} \Diamond (\mathrm{i} \vee \mathrm{j}) \\ \text{iff} & \mathbb{M}, g, w \models @_\mathrm{i} \Diamond (\langle \mathrm{j} \rangle \mathrm{i} \vee \langle \mathrm{j} \rangle \mathrm{j}) \\ \text{iff} & \mathbb{M}, g, w \models @_\mathrm{i} \Diamond ((\mathrm{i} \wedge @_\mathrm{j} \Diamond \top) \vee \blacklozenge \mathrm{j}) \\ \text{iff} & \mathbb{M}, g, w \models @_\mathrm{i} \Diamond (\mathrm{i} \wedge @_\mathrm{j} \Diamond \top) \vee @_\mathrm{i} \Diamond \blacklozenge \mathrm{j} \end{array}$$

By the definition of satisfaction relation, it is easy to see that this is equivalent to $\mathbb{M}, g, w \models @_\mathrm{i} \Diamond (\mathrm{i} \wedge @_\mathrm{j} \Diamond \top) \vee @_\mathrm{i} \Diamond \blacklozenge \mathrm{j}$ which has the exact same interpretation as the interpretation for φ discussed above.

Example 3. In item 6, the condition $\mathrm{t} \neq \mathrm{i}$ is necessary. To see this, consider the formula $\langle \mathrm{i} \rangle @_\mathrm{i} \mathrm{i}$. This formula is equivalent to $@_\mathrm{i} \Diamond \top$. Consider the model $\mathbb{M} = (W, R, V)$ with $W = \{w_1, w_2\}$, $R = \{(w_1, w_2)\}$, where $V(\mathrm{i}) = w_1$. Clearly, for any g, $\mathbb{M}, g, w_1 \models \langle \mathrm{i} \rangle @_\mathrm{i} \mathrm{i}$, whereas $\mathbb{M}, g, w_1 \not\models @_\mathrm{i} \langle \mathrm{i} \rangle \mathrm{i}$.

Example 4. The identity $\langle \mathrm{i} \rangle \varphi \wedge \langle \mathrm{i} \rangle \psi \rightarrow \langle \mathrm{i} \rangle (\varphi \wedge \psi)$ is not valid. Consider the model $\mathbb{M} = (W, R, V)$ with $W = \{w_1, w_2, w_3\}$, $R = \{(w_1, w_2), (w_1, w_3)\}$, where $V(\mathrm{i}) = w_1$, $V(\mathrm{j}) = w_2$, and $V(\mathrm{k}) = w_3$. Clearly, for any g, $\mathbb{M}, g, w_1 \models \langle \mathrm{i} \rangle @_\mathrm{i} \mathrm{j} \wedge \langle \mathrm{i} \rangle @_\mathrm{i} \mathrm{k}$, whereas $\mathbb{M}, g, w_1 \not\models \langle \mathrm{i} \rangle (@_\mathrm{i} \mathrm{j} \wedge @_\mathrm{i} \mathrm{k})$. Note that as $\langle \mathrm{i} \rangle$ preserves joins and not meets it must be true that it does not preserve negations, i.e. the identity $\langle \mathrm{i} \rangle \neg \varphi \leftrightarrow \neg \langle \mathrm{i} \rangle \varphi$ is not valid. We in fact show something stronger in the following parts.

We show that if we extend the $\{\wedge, \neg\}$-free fragment of $\mathcal{DH}$ with either $\wedge$ or $\neg$, the resulting logic can express formulas not expressible in $\mathcal{TH}(@)$.

[3] In fact, in the translated formulas, the operator $\blacklozenge$ is applied only to a nominal or to a formula consisting of nominals followed by n $\blacklozenge$ operators.

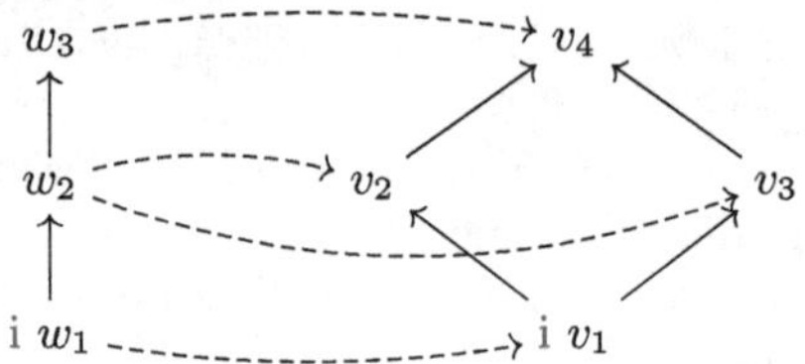

Fig. 1. The $\mathcal{H}(@)$-bisimulation between models $\mathbb{M}$ and $\mathbb{M}'$.

Example 5. Consider the modal language with only one nominal i and no propositional variables. Let $\mathbb{M} = (W, R, V)$ be a model, such that $W = \{w_1, w_2, w_3\}$, $R = \{(w_1, w_2), (w_2, w_3)\}$ and $V(\mathrm{i}) = \{w_1\}$; $\mathbb{M}' = (W', R', V')$ is another model, such that $W' = \{v_1, v_2, v_3, v_4\}$, $R' = \{(v_1, v_2), (v_1, v_3), (v_2, v_4), (v_3, v_4)\}$ and $V'(\mathrm{i}) = \{v_1\}$. Let $Z \subseteq W \times W'$ be a binary relation, such that $Z = \{(w_1, v_1), (w_2, v_2), (w_2, v_3), (w_3, v_4)\}$ (see Fig. 1). It is straightforward to check that Z is an $\mathcal{H}$-bisimulation and a $\mathcal{TH}(@)$-bisimulation, and formula $\langle \mathrm{i} \rangle \Box \mathrm{i} = \langle \mathrm{i} \rangle \neg \Diamond \neg \mathrm{i}$ is true at w_1 but not v_1. Therefore, an extension of $\{\wedge\}$-free fragment of $\mathcal{DH}$ cannot be translated into $\mathcal{TH}(@)$ and hence $\mathcal{H}$, $\mathcal{H}(@)$, and $\mathcal{TH}$.

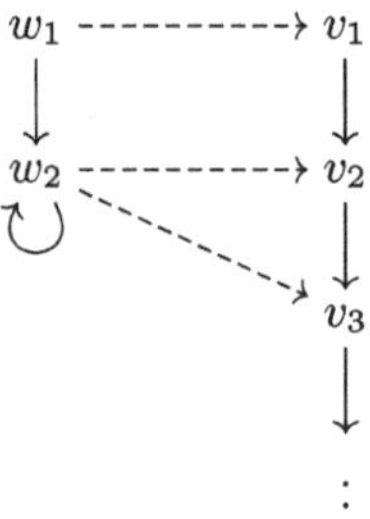

Fig. 2. The $\mathcal{H}(@)$-bisimulation between models $\mathbb{M}$ and $\mathbb{M}'$.

Example 6. Consider the modal language with only one nominal i and no propositional variables. Let $\mathbb{M} = (W, R, V)$ be a model, such that $W = \{w_1, w_2\}$, $R = \{(w_1, w_2), (w_2, w_2)\}$ and $V(\mathrm{i}) = \{w_1\}$; $\mathbb{M}' = (W', R', V')$ is another model, such that $W' = \{v_i \mid i \in \mathbb{N}\}$, $R' = \{(v_i, v_{i+1}) \mid i \in \mathbb{N}\}$ and $V'(\mathrm{i}) = \{v_1\}$. Let $Z \subseteq W \times W'$ be a binary relation, such that $Z = \{(w_1, v_1)\} \cup \{(w_2, v_i) \mid i \neq 1, i \in \mathbb{N}\}$ (see Fig. 2). It is straightforward to check that Z is an $\mathcal{H}$-bisimulation and in fact $\mathcal{TH}(@)$-bisimulation, and formula $\langle \mathrm{i} \rangle (\mathrm{i} \wedge \Diamond \mathrm{i})$ is true at w_2 but not v_2. Therefore, an extension of $\{\neg\}$-free fragment of $\mathcal{DH}$ cannot be translated into $\mathcal{TH}(@)$ and hence $\mathcal{H}$, $\mathcal{H}(@)$, and $\mathcal{TH}$.

We now show that the operator $\langle \mathrm{i} \rangle$ is definable in $\mathcal{H}(@, \downarrow)$.

Proposition 1. *Let $\mathbb{M} = (W, R, V)$ be any model for hybrid logics, g be any interpretation on it, and $w \in W$ be any world. For any formula $\varphi \in \mathcal{H}(@, \downarrow)$, and any nominal $\mathrm{i} \in \mathrm{Nom}$, $\mathbb{M}, g, w \models \langle \mathrm{i} \rangle \varphi \leftrightarrow \downarrow \mathrm{x}.(@_{\mathrm{i}} \Diamond \downarrow \mathrm{y}.(@_{\mathrm{x}}(\varphi[\mathrm{y}/\mathrm{i}]))$, where $\mathrm{x}, \mathrm{y} \in \mathit{Svar}$ do not appear in φ.*

Proof. The proof follows from the following series of equivalences.
$\mathbb{M}, g, w \models \downarrow \mathrm{x}.(@_{\mathrm{i}}\Diamond \downarrow \mathrm{y}.(@_{\mathrm{x}}(\varphi[\mathrm{y}/\mathrm{i}]))$
iff $\mathbb{M}, g[x := w], w \models @_{\mathrm{i}}\Diamond \downarrow \mathrm{y}.(@_{\mathrm{x}}(\varphi[\mathrm{y}/\mathrm{i}])$
iff $\mathbb{M}, g[x := w], V(\mathrm{i}) \models \Diamond \downarrow \mathrm{y}.(@_{\mathrm{x}}(\varphi[\mathrm{y}/\mathrm{i}])$
iff $\exists w'(V(\mathrm{i})Rw' \wedge \mathbb{M}, g[x := w], w' \models \downarrow \mathrm{y}.(@_{\mathrm{x}}\varphi[\mathrm{y}/\mathrm{i}]))$
iff $\exists w'(V(\mathrm{i})Rw' \wedge \mathbb{M}, g[x := w][y := w'], w' \models @_{\mathrm{x}}\varphi[\mathrm{y}/\mathrm{i}])$
iff $\exists w'(V(\mathrm{i})Rw' \wedge \mathbb{M}, g[x := w][y := w'], g[x := w][y := w'](\mathrm{x}) \models \varphi[\mathrm{y}/\mathrm{i}])$
iff $\exists w'(V(\mathrm{i})Rw' \wedge \mathbb{M}, g[x := w][y := w'], w \models \varphi[\mathrm{y}/\mathrm{i}])$
iff $\exists w'(V(\mathrm{i})Rw' \wedge \mathbb{M}_{[\mathrm{i} \to g[x:=w][y:=w'](y)]}, g[x := w][y := w'], w \models \varphi)$
iff $\exists w'(V(\mathrm{i})Rw' \wedge \mathbb{M}_{[\mathrm{i} \to w']}, g[x := w][y := w'], w \models \varphi)$
iff $\exists w'(V(\mathrm{i})Rw' \wedge \mathbb{M}_{[\mathrm{i} \to w']}, g, w \models \varphi)$
iff $\mathbb{M}, g, w \models \langle \mathrm{i} \rangle \varphi$

Therefore, the languages $\mathcal{DH}$, $\mathcal{DH}(@)$, and $\mathcal{DH}(@, \downarrow)$ all have a translation into $\mathcal{H}(@, \downarrow)$. In the next section, we will show that $\mathcal{DH}$ and $\mathcal{DH}(@)$ have strictly lower expressive power than $\mathcal{H}(@, \downarrow)$.

4.1 Expressing Local Frame Properties

In this section, we show how $\mathcal{DH}$ and $\mathcal{DH}(@)$ can be used to express the local versions (i.e. at some particular world) of commonly studied properties of Kripke frames such as reflexivity, symmetry, transitivity, functionality, etc. while also showing that many of these local properties[4] can not be expressed in $\mathcal{H}$ and $\mathcal{H}(@)$.

Table 1. Local properties and their dynamic correspondents

Local Property	Defining first-order expression for i	Correspondent
Reflexivity	$V(\mathrm{i})RV(\mathrm{i})$	$\Diamond \mathrm{i}$
Irreflexivity	$\neg(V(\mathrm{i})RV(\mathrm{i}))$	$\neg\Diamond \mathrm{i}$
Symmetry	$\forall w(V(\mathrm{i})Rw \implies wRV(\mathrm{i}))$	$\Box\Diamond \mathrm{i}$
Antisymmetry	$\forall w(V(\mathrm{i})Rw \implies \neg(wRV(\mathrm{i})))$	$\Box\neg\Diamond \mathrm{i}$
Transitivity	$\forall w_1 \forall w_2(V(\mathrm{i})Rw_1 \wedge w_1 R w_2 \implies V(\mathrm{i})Rw_2)$	$[\mathrm{i}][\mathrm{i}]\Diamond \mathrm{i}$
Antitransitivity	$\forall w_1 \forall w_2(V(\mathrm{i})Rw_1 \wedge w_1 R w_2 \implies \neg(V(\mathrm{i})Rw_2))$	$[\mathrm{i}][\mathrm{i}]\neg\Diamond \mathrm{i}$
Seriality	$\exists w(V(\mathrm{i})Rw)$	$\Diamond \top$
Functionality	$\forall w_1 \forall w_2(V(\mathrm{i})Rw_1 \wedge V(\mathrm{i})Rw_2 \implies w_1 = w_2)$	$[\mathrm{i}]\Box \mathrm{i}$
Euclideanness	$\forall w_1 \forall w_2(V(\mathrm{i})Rw_1 \wedge V(\mathrm{i})Rw_2 \implies w_1 R w_2)$	$[\mathrm{i}]\Box\Diamond \mathrm{i}$
Density	$\forall w(V(\mathrm{i})Rw \implies \exists w_1(V(\mathrm{i})Rw_1 \wedge w_1 R w))$	$[\mathrm{i}]\Diamond\Diamond \mathrm{i}$
Confluence	$\forall w_1 \forall w_2(V(\mathrm{i})Rw_1 \wedge V(\mathrm{i})Rw_2 \implies \exists w_3(w_1 R w_3 \wedge w_2 R w_3))$	$[\mathrm{i}]\Box\Diamond\langle \mathrm{i} \rangle \mathrm{i}$

For any property P in Table 1, we denote its dynamic correspondent by $\mathcal{D}(P)$. When the context is clear, we identify a property with its defining expression. For

[4] We call these properties local as they describe the local picture of the frame surrounding the world where they are being evaluated.

any property P, and any world w, we denote by $P(w)$ the first-order expression obtained by uniformly replacing $V(\mathrm{i})$ with w in P. The following proposition relates the properties and their dynamic correspondents in $\mathcal{DH}$ and $\mathcal{DH}(@)$.

Proposition 2. *Let* $\mathbb{M} = (W, R, V)$ *be any model for hybrid logics. For any property* P *in Table 1, and any world* $w \in W$,

1. $\mathbb{M}, w \models \mathrm{i} \wedge \mathcal{D}(P)$ *iff* $V(\mathrm{i}) = w \wedge P$ *holds in* $\mathbb{M}$.
2. $\mathbb{M}, w \models @_{\mathrm{i}}\mathcal{D}(P)$ *iff* P *holds in* $\mathbb{M}$.
3. $\mathbb{M}, w \models \mathrm{i} \wedge [\mathrm{i}]\Diamond(\mathrm{i} \wedge \mathcal{D}(P))$ *iff* $V(\mathrm{i}) = w \wedge \forall w'(V(\mathrm{i})Rw' \implies P(w'))$ *holds in* $\mathbb{M}$.
4. $\mathbb{M}, w \models @_{\mathrm{i}}[\mathrm{i}]@_{\mathrm{i}}\mathcal{D}(P)$ *iff* $\forall w'(V(\mathrm{i})Rw' \implies P(w'))$ *holds in* $\mathbb{M}$.
5. $\mathbb{M}, w \models \mathrm{i} \wedge \langle \mathrm{i} \rangle \Diamond(\mathrm{i} \wedge \mathcal{D}(P))$ *iff* $V(\mathrm{i}) = w \wedge \exists w'(V(\mathrm{i})Rw' \wedge P(w'))$ *holds in* $\mathbb{M}$.
6. $\mathbb{M}, w \models @_{\mathrm{i}}\langle \mathrm{i} \rangle @_{\mathrm{i}}\mathcal{D}(P)$ *iff* $\exists w'(V(\mathrm{i})Rw' \wedge P(w'))$ *holds in* $\mathbb{M}$.

Proof. We only give proofs for item 1. The proofs for all other items are analogous.

1. Reflexivity

$\mathbb{M}, w \models \mathrm{i} \wedge \Diamond \mathrm{i}$
iff $V(\mathrm{i}) = w \wedge \exists w'(wRw' \wedge V(\mathrm{i}) = w')$
iff $V(\mathrm{i}) = w \wedge V(\mathrm{i})RV(\mathrm{i})$

2. Irreflxivity

$\mathbb{M}, w \models \mathrm{i} \wedge \neg\Diamond \mathrm{i}$
iff $V(\mathrm{i}) = w \wedge \neg\exists w'(wRw' \wedge V(\mathrm{i}) = w')$
iff $V(\mathrm{i}) = w \wedge \forall w'(wRw' \implies V(\mathrm{i}) \neq w')$
iff $V(\mathrm{i}) = w \wedge \neg(V(\mathrm{i})RV(\mathrm{i}))$

3. Symmetry

$\mathbb{M}, w \models \mathrm{i} \wedge \Box\Diamond \mathrm{i}$
iff $V(\mathrm{i}) = w \wedge \forall w'(wRw' \implies \exists w''(w'Rw'' \wedge V(\mathrm{i}) = w''))$
iff $V(\mathrm{i}) = w \wedge \forall w'(wRw' \implies w'Rw))$
iff $V(\mathrm{i}) = w \wedge \forall w'(V(\mathrm{i})Rw' \implies w'RV(\mathrm{i}))$

4. Antisymmetry

$\mathbb{M}, w \models \mathrm{i} \wedge \Box\neg\Diamond \mathrm{i}$
iff $V(\mathrm{i}) = w \wedge \forall w'(wRw' \implies \neg\exists w''(w'Rw'' \wedge V(\mathrm{i}) = w''))$
iff $V(\mathrm{i}) = w \wedge \forall w'(wRw' \implies \forall w''(w'Rw'' \implies V(\mathrm{i}) \neq w''))$
iff $V(\mathrm{i}) = w \wedge \forall w'\forall w''(V(\mathrm{i})Rw' \wedge w'Rw'' \implies w'' \neq V(\mathrm{i}))$
iff $V(\mathrm{i}) = w \wedge \forall w'(V(\mathrm{i})Rw' \implies \neg(w'RV(\mathrm{i})))$

5. Transitivity

$\mathbb{M}, w \models \mathrm{i} \wedge [\mathrm{i}][\mathrm{i}]\Diamond \mathrm{i}$
iff $V(\mathrm{i}) = w \wedge \forall w'\forall w''(wRw' \wedge w'Rw'' \implies \mathbb{M}_{[\mathrm{i}\to w'']}, w \models \Diamond \mathrm{i})$
iff $V(\mathrm{i}) = w \wedge \forall w'\forall w''(wRw' \wedge w'Rw'' \implies \exists v(wRv \wedge \mathbb{M}_{[\mathrm{i}\to w'']}, v \models \mathrm{i}))$
iff $V(\mathrm{i}) = w \wedge \forall w'\forall w''(wRw' \wedge w'Rw'' \implies \exists v(wRv \wedge V_{[\mathrm{i}\to w'']}(\mathrm{i}) = v))$
iff $V(\mathrm{i}) = w \wedge \forall w'\forall w''(wRw' \wedge w'Rw'' \implies \exists v(wRv \wedge v = w'')$
iff $V(\mathrm{i}) = w \wedge \forall w'\forall w''(wRw' \wedge w'Rw'' \implies wRw'')$
iff $V(\mathrm{i}) = w \wedge \forall w'\forall w''(V(\mathrm{i})Rw' \wedge w'Rw'' \implies V(\mathrm{i})Rw'')$

6. AntiTransitivity

$\mathbb{M}, w \models \mathrm{i} \wedge [\mathrm{i}][\mathrm{i}]\neg\Diamond\mathrm{i}$
iff $V(\mathrm{i}) = w \wedge \forall w'\forall w''(wRw' \wedge w'Rw'' \implies \mathbb{M}_{[\mathrm{i}\to w'']}, w \models \neg\Diamond\mathrm{i}$
iff $V(\mathrm{i}) = w \wedge \forall w'\forall w''(wRw' \wedge w'Rw'' \implies \neg\exists v(wRv \wedge \mathbb{M}_{[\mathrm{i}\to w'']}, v \models \mathrm{i}))$
iff $V(\mathrm{i}) = w \wedge \forall w'\forall w''(wRw' \wedge w'Rw'' \implies \forall v(wRv \implies \mathbb{M}_{[\mathrm{i}\to w'']}, v \not\models \mathrm{i}))$
iff $V(\mathrm{i}) = w \wedge \forall w'\forall w''(wRw' \wedge w'Rw'' \implies \forall v(wRv \implies V_{[\mathrm{i}\to w'']}(\mathrm{i}) \neq v))$
iff $V(\mathrm{i}) = w \wedge \forall w'\forall w''(wRw' \wedge w'Rw'' \implies \forall v(wRv \implies w'' \neq v))$
iff $V(\mathrm{i}) = w \wedge \forall w'\forall w''(wRw' \wedge w'Rw'' \implies \neg(wRw''))$
iff $V(\mathrm{i}) = w \wedge \forall w'\forall w''(V(\mathrm{i})Rw' \wedge w'Rw'' \implies \neg(V(\mathrm{i})Rw''))$

7. Seriality

$\mathbb{M}, w \models \mathrm{i} \wedge \Diamond\top$
iff $V(\mathrm{i}) = w \wedge \exists w'(wRw')$
iff $V(\mathrm{i}) = w \wedge \exists w'(V(\mathrm{i})Rw')$

8. Functionality

$\mathbb{M}, w \models \mathrm{i} \wedge [\mathrm{i}]\Box\mathrm{i}$
iff $V(\mathrm{i}) = w \wedge \forall w'\forall w''((wRw' \wedge V(\mathrm{i})Rw'') \implies \mathbb{M}_{[\mathrm{i}\to w'']}, w' \models \mathrm{i})$
iff $V(\mathrm{i}) = w \wedge \forall w'\forall w''((wRw' \wedge V(\mathrm{i})Rw'') \implies V_{[\mathrm{i}\to w'']}(\mathrm{i}) = w')$
iff $V(\mathrm{i}) = w \wedge \forall w'\forall w''((wRw' \wedge V(\mathrm{i})Rw'') \implies w' = w'')$
iff $V(\mathrm{i}) = w \wedge \forall w'\forall w''((V(\mathrm{i})Rw' \wedge V(\mathrm{i})Rw'') \implies w' = w'')$

9. Euclideannes

$\mathbb{M}, w \models \mathrm{i} \wedge [\mathrm{i}]\Box\Diamond\mathrm{i}$
iff $V(\mathrm{i}) = w \wedge \forall w'\forall w''((wRw' \wedge V(\mathrm{i})Rw'') \implies \exists v(w'Rv \wedge \mathbb{M}_{[\mathrm{i}\to w'']}, v \models \mathrm{i}))$
iff $V(\mathrm{i}) = w \wedge \forall w'\forall w''((wRw' \wedge V(\mathrm{i})Rw'') \implies \exists v(w'Rv \wedge w'' = v))$
iff $V(\mathrm{i}) = w \wedge \forall w'\forall w''((V(\mathrm{i})Rw' \wedge V(\mathrm{i})Rw'') \implies w'Rw'')$

10. Density

$\mathbb{M}, w \models \mathrm{i} \wedge [\mathrm{i}]\Diamond\Diamond\mathrm{i}$
iff $V(\mathrm{i}) = w \wedge \forall w'(wRw' \implies \mathbb{M}_{[\mathrm{i}\to w']}, w \models \Diamond\Diamond\mathrm{i})$
iff $V(\mathrm{i}) = w \wedge \forall w'(wRw' \implies \exists v\exists v'(wRv \wedge vRv' \wedge \mathbb{M}_{[\mathrm{i}\to w']}, v' \models \mathrm{i}))$
iff $V(\mathrm{i}) = w \wedge \forall w'(wRw' \implies \exists v\exists v'(wRv \wedge vRv' \wedge w' = v'))$
iff $V(\mathrm{i}) = w \wedge \forall w'(wRw' \implies \exists v(wRv \wedge vRw'))$
iff $V(\mathrm{i}) = w \wedge \forall w'(V(\mathrm{i})Rw' \implies \exists v(V(\mathrm{i})Rv \wedge vRw'))$

11. Confluence

$\mathbb{M}, w \models \mathrm{i} \wedge [\mathrm{i}]\Box\Diamond\langle\mathrm{i}\rangle\mathrm{i}$
iff $V(\mathrm{i}) = w \wedge \forall w'\forall w''((wRw' \wedge V(\mathrm{i})Rw'') \implies \exists v\exists v'(w'Rv \wedge w''Rv' \wedge \mathbb{M}_{[\mathrm{i}\to v']}, v \models \mathrm{i}))$
iff $V(\mathrm{i}) = w \wedge \forall w'\forall w''((wRw' \wedge V(\mathrm{i})Rw'') \implies \exists v\exists v'(w'Rv \wedge w''Rv' \wedge v' = v))$
iff $V(\mathrm{i}) = w \wedge \forall w'\forall w''((wRw' \wedge V(\mathrm{i})Rw'') \implies \exists v(w'Rv \wedge w''Rv))$
iff $V(\mathrm{i}) = w \wedge \forall w'\forall w''((V(\mathrm{i})Rw' \wedge V(\mathrm{i})Rw'') \implies \exists v(w'Rv \wedge w''Rv))$

An important consequence of the ability of $\mathcal{DH}$ to express these local properties is that $\mathcal{DH}$ does not have the finite model property.

Proposition 3. *The logic $\mathcal{DH}$ does not have the finite model property.*

Proof. Consider the formula ϕ which is conjunction of the formulas of the form $\mathrm{i} \wedge \mathcal{D}(P) \wedge [\mathrm{i}]\Diamond(\mathrm{i} \wedge \mathcal{D}(P))$ for P being irreflexivity, asymmetry, transitivity, and seriality. We claim that ϕ has an infinite model but no finite model. A model for ϕ is given by $\mathbb{M} = (\mathbb{Z}, <, V)$, where $V(\mathrm{i}) = 0$. To see that ϕ does not have a finite model, note that every model of ϕ must contain a strict partial order with no maximal elements. However, every finite strict partial order must have a maximal element. Therefore, ϕ can not have a finite model.

Proposition 4. *The formula* i $\wedge \mathcal{D}(P)$ *(and hence* $\mathcal{D}(P)$*), where* P *is any of the properties transitivity, antitransitivity, functionality, Euclideanness, density, and confluence, is not definable in* $\mathcal{H}(@)$.

Proof. In all the proofs, we would use the language consisting of no propositional variables and only one nominal i.

1. Transitivity: Let $\mathbb{M} = (W, R, V)$ and $\mathbb{N} = (W', R', V')$ be models s.t. $W = \{w_0, w_1, w_2, w_3\}$, $R = \{(w_0, w_1), (w_1, w_2), (w_1, w_3), (w_0, w_2)\}$, $V(\mathrm{i}) = w_0$,$W' = \{v_0, v_1, v_2\}$, $R' = \{(v_0, v_1), (v_1, v_2), (v_0, v_2)\}$, $V'(\mathrm{i}) = v_0$. Let $Z = \{(w_0, v_0), (w_1, v_1), (w_2, v_2), (w_3, v_2)\}$. It is easy to check that Z is a $\mathcal{H}(@)$-bisimulation, but $\mathbb{M}, w_0 \not\models \mathrm{i} \wedge [\mathrm{i}][\mathrm{i}]\Diamond\mathrm{i}$, while $\mathbb{N}, v_0 \models \mathrm{i} \wedge [\mathrm{i}][\mathrm{i}]\Diamond\mathrm{i}$.

2. Antitransitivity: Let $\mathbb{M} = (W, R, V)$ and $\mathbb{N} = (W', R', V')$ be models s.t. $W = \{w_0, w_1, w_2, w_3\}$, $R = \{(w_0, w_1), (w_1, w_2), (w_0, w_3)\}$, $V(\mathrm{i}) = w_0$, $W' = \{v_0, v_1, v_2\}$, $R' = \{(v_0, v_1), (v_1, v_2), (v_0, v_2)\}$, $V'(\mathrm{i}) = v_0$. Let $Z = \{(w_0, v_0), (w_1, v_1), (w_2, v_2), (w_3, v_2)\}$. It is easy to check that Z is a $\mathcal{H}(@)$-bisimulation, but $\mathbb{M}, w_0 \models \mathrm{i} \wedge [\mathrm{i}][\mathrm{i}]\neg\Diamond\mathrm{i}$, while $\mathbb{N}, v_0 \not\models \mathrm{i} \wedge [\mathrm{i}][\mathrm{i}]\neg\Diamond\mathrm{i}$.

3. Functionality: Let $\mathbb{M} = (W, R, V)$ and $\mathbb{N} = (W', R', V')$ be models s.t. $W = \{w_0, w_1, w_2\}$, $R = \{(w_0, w_1), (w_0, w_2)\}$, $V(\mathrm{i}) = w_0$, $W' = \{v_0, v_1\}$, $R' = \{(v_0, v_1)\}$, $V'(\mathrm{i}) = v_0$. Let $Z = \{(w_0, v_0), (w_1, v_1), (w_2, v_1)\}$. It is easy to check that Z is a $\mathcal{H}(@)$-bisimulation, but $\mathbb{M}, w_0 \not\models \mathrm{i} \wedge [\mathrm{i}]\Box\mathrm{i}$, while $\mathbb{N}, v_0 \models \mathrm{i} \wedge [\mathrm{i}]\Box\mathrm{i}$.

4. Euclideanness: Let $\mathbb{M} = (W, R, V)$ and $\mathbb{N} = (W', R', V')$ be models s.t. $W = \{w_0, w_1, w_2\}$, $R = \{(w_0, w_1), (w_0, w_2), (w_1, w_1), (w_2, w_2)\}$, $V(\mathrm{i}) = w_0$, $W' = \{v_0, v_1\}$, $R' = \{(v_0, v_1), (v_1, v_1)\}$, $V'(\mathrm{i}) = v_0$. Let $Z = \{(w_0, v_0), (w_1, v_1), (w_2, v_1)\}$. It is easy to check that Z is a $\mathcal{H}(@)$-bisimulation, but $\mathbb{M}, w_0 \not\models \mathrm{i} \wedge [\mathrm{i}]\Box\Diamond\mathrm{i}$, while $\mathbb{N}, v_0 \models \mathrm{i} \wedge [\mathrm{i}]\Box\Diamond\mathrm{i}$.

5. Density: Let $\mathbb{M} = (W, R, V)$ and $\mathbb{N} = (W', R', V')$ be models s.t. $W = \{w_i, \mid i \in \mathbb{Z}^+\}$[5], $R = \{(w_i, w_{i+1}) \mid i \in \mathbb{Z}^+\}$, $V(\mathrm{i}) = w_0$, $W' = \{v_0, v_1\}$, $R' = \{(v_0, v_1), (v_1, v_1)\}$, $V'(\mathrm{i}) = v_0$. Let $Z = \{(w_0, v_0)\} \cup \{(w_i, v_1) \mid i \neq 0, i \in \mathbb{Z}^+\}$. It is easy to check that Z is a $\mathcal{H}(@)$-bisimulation, but $\mathbb{M}, w_0 \not\models \mathrm{i} \wedge [\mathrm{i}]\Diamond\Diamond\mathrm{i}$, while $\mathbb{N}, v_0 \models \mathrm{i} \wedge [\mathrm{i}]\Diamond\Diamond\mathrm{i}$.

6. Confluence: Let $\mathbb{M} = (W, R, V)$ and $\mathbb{N} = (W', R', V')$ be models s.t. $W = \{w_0, w_1, w_2, w_3, w_4\}$, $R = \{(w_0, w_1), (w_0, w_2)(w_1, w_3), (w_2, w_4)\}$, $V(\mathrm{i}) = w_0$, $W' = \{v_0, v_1, v_2, v_3\}$, $R' = \{(v_0, v_1), (v_0, v_2), (v_1, v_3), (v_2, v_3)\}$, $V'(\mathrm{i}) = v_0$. Let $Z = \{(w_0, v_0), (w_1, v_1), (w_2, v_2), (w_3, v_3), (w_4, v_3)\}$. It is easy to check that Z is a $\mathcal{H}(@)$-bisimulation, but $\mathbb{M}, w_0 \not\models \mathrm{i} \wedge [\mathrm{i}]\Box\Diamond\langle\mathrm{i}\rangle\mathrm{i}$, while $\mathbb{N}, v_0 \models \mathrm{i} \wedge [\mathrm{i}]\Box\Diamond\langle\mathrm{i}\rangle\mathrm{i}$.

5 Bisimulations for Hybrid Logic with Moving Names

In this section, we use the ideas from bisimulations in hybrid logics and dynamic logics to define an appropriate notion of bisimulations for hybrid logics with moving names.

A *pointed model* is a tuple $(\mathbb{M}, w)$, such that $\mathbb{M} = (W, R, V)$ is a model and $w \in W$. For any nominal $\mathrm{i} \in \mathrm{Nom}$, R_i is a binary relation defined on the class of pointed models as following: For any pointed models $(\mathbb{M}_1, w_1)$ and $(\mathbb{M}_2, w_2)$

[5] We use $\mathbb{Z}^+$ to denote non-negative integers.

with $\mathbb{M}_1 = (W_1, R_1, V_1)$ and $\mathbb{M}_2 = (W_2, R_2, V_2)$, $(\mathbb{M}_1, w_1)R_{\mathrm{i}}(\mathbb{M}_2, w_2)$ iff $\mathbb{M}_2 = \mathbb{M}_{1[\mathrm{i}\to w']}$ for some $w' \in W_1$ such that $V_1(\mathrm{i})R_1 w'$ and $w_2 = w_1$. Intuitively, the relation R_{i} captures how a (pointed) model changes when the dynamic operator $\langle \mathrm{i} \rangle$ is applied to it.

Lemma 2. *Let* $\mathbb{M} = (W, R, V)$ *be any model,* g *be any interpretation on* $\mathbb{M}$*, and* $w \in W$ *be any state. For any* $\mathcal{DH}(@, \downarrow)$*-formula* φ, $\mathbb{M}, g, w \models \langle \mathrm{i} \rangle \varphi$ *iff there exists a pointed model* $(\mathbb{M}', w')$*, such that* $(\mathbb{M}, w)R_{\mathrm{i}}(\mathbb{M}', w')$ *and* $\mathbb{M}', g, w' \models \varphi$.

Proof. Let $\varphi \in \mathcal{DH}(@, \downarrow)$, then $\mathbb{M}, g, w \models \langle \mathrm{i} \rangle \varphi$ iff $\exists w'(V(\mathrm{i})Rw' \wedge M_{[\mathrm{i}\to w']}, g, w \models \varphi)$. Thus, $(M_{[\mathrm{i}\to w']}, w)$ is the required pointed model. The reverse direction follows straightforwardly from the definition of R_{i}.

Definition 4. *Given any pointed model* $(\mathbb{M}_1, w_1)$ *and* $(\mathbb{M}_2, w_2)$*, a* $\mathcal{DH}$-bisimulation between $(\mathbb{M}_1, w_1)$ and $(\mathbb{M}_2, w_2)$ *is a binary relation* Z *over the class of pointed models satisfying the following conditions:*

- **Atom** *If* $(\mathbb{M}_1, w_1)Z(\mathbb{M}_2, w_2)$*, then* $w_1 \in V_1(p) \Leftrightarrow w_2 \in V_2(p)$*, for any* $p \in$ Prop $\cup$ Nom.
- **Zig**$_\Diamond$ *If* $(\mathbb{M}_1, w_1)Z(\mathbb{M}_2, w_2)$ *and* $w_2 R_2 v_2$*, then there exists* $v_1 \in W_1$*, such that* $(\mathbb{M}_1, v_1)Z(\mathbb{M}_2, v_2)$ *and* $w_1 R_1 v_1$.
- **Zag**$_\Diamond$ *If* $(\mathbb{M}_1, w_1)Z(\mathbb{M}_2, w_2)$ *and* $w_1 R_1 v_1$*, then there exists* $v_2 \in W_2$*, such that* $(\mathbb{M}_1, v_1)Z(\mathbb{M}_2, v_2)$ *and* $w_2 R_2 v_2$.
- **Zig**$_{\langle \mathrm{i} \rangle}$ *For every* $\mathrm{i} \in Nom$*, if* $(\mathbb{M}_1, w_1)Z(\mathbb{M}_2, w_2)$ *and* $(\mathbb{M}_2, w_2)R_{\mathrm{i}}(\mathbb{M}'_2, w'_2)$ *for some pointed model* $(\mathbb{M}'_2, w'_2)$*, then there exists a pointed model* $(\mathbb{M}'_1, w'_1)$*, such that* $(\mathbb{M}'_1, w'_1)Z(\mathbb{M}'_2, w'_2)$ *and* $(\mathbb{M}_1, w_1)R_{\mathrm{i}}(\mathbb{M}'_1, w'_1)$.
- **Zag**$_{\langle \mathrm{i} \rangle}$ *For every* $\mathrm{i} \in Nom$*, if* $(\mathbb{M}_1, w_1)Z(\mathbb{M}_2, w_2)$ *and* $(\mathbb{M}_1, w_1)R_{\mathrm{i}}(\mathbb{M}'_1, w'_1)$ *for some pointed model* $(\mathbb{M}'_1, w'_1)$*, then there exists a pointed model* $(\mathbb{M}'_2, w'_2)$*, such that* $(\mathbb{M}'_1, w'_1)Z(\mathbb{M}'_2, w'_2)$ *and* $(\mathbb{M}_2, w_2)R_{\mathrm{i}}(\mathbb{M}'_2, w'_2)$.

A $\mathcal{DH}(@)$-bisimulation *is an* $\mathcal{DH}$*-bisimulation satisfying, in addition,*

- **Nom** *For any* $(\mathbb{M}_1, w_1)$ *and* $(\mathbb{M}_2, w_2)$*, if there exists* $\mathrm{i} \in$ Nom *s.t.* $w_1 \in V_1(\mathrm{i})$ *and* $w_2 \in V_2(\mathrm{i})$*, then* $(\mathbb{M}_1, w_1)Z(\mathbb{M}_2, w_2)$.

Theorem 2. *Let* $(\mathbb{M}_1, w_1)$*, and* $(\mathbb{M}_2, w_2)$ *be two pointed models, and* Z *be a* $\mathcal{DH}$*(resp.* $\mathcal{DH}(@)$*)-bisimulation between them. Then for any formula* $\varphi \in \mathcal{DH}$ *(resp.* $\varphi \in \mathcal{DH}(@)$*),* $\mathbb{M}_1, w_1 \models \varphi$ *iff* $\mathbb{M}_2, w_2 \models \varphi$.

Proof. The proof is by induction on the complexity of φ. The proof for the base case and the inductive cases for all the operators except $\langle \mathrm{i} \rangle$ are as in the proof of Theorem 1. Hence, we only prove the inductive case for operator $\langle \mathrm{i} \rangle$. Suppose $\varphi = \langle \mathrm{i} \rangle \psi$ for some $\mathrm{i} \in Nom$. Then by Lemma 2, $\mathbb{M}_1, w_1 \models \langle \mathrm{i} \rangle \psi$ iff there exists a pointed model $(\mathbb{M}'_1, w'_1)$ s.t. $(\mathbb{M}_1, w_1)R_{\mathrm{i}}(\mathbb{M}'_1, w'_1)$ and $\mathbb{M}'_1, w'_1 \models \psi$. By condition **Zag**$_{\langle \mathrm{i} \rangle}$, $(\mathbb{M}_1, w_1)R_{\mathrm{i}}(\mathbb{M}'_1, w'_1)$ and $(\mathbb{M}_1, w_1)Z(\mathbb{M}_2, w_2)$ imply there exists a pointed model $(\mathbb{M}'_2, w'_2)$ such that $(\mathbb{M}_2, w_2)R_{\mathrm{i}}(\mathbb{M}'_2, w'_2)$ and $(\mathbb{M}'_1, w'_1)Z(\mathbb{M}'_2, w'_2)$. From the induction hypothesis, we have $\mathbb{M}'_1, w'_1 \models \psi$ iff $\mathbb{M}'_2, w'_2 \models \psi$. Therefore, there exists a pointed model $(\mathbb{M}'_1, w'_1)$ such that $(\mathbb{M}_1, w_1)R_{\mathrm{i}}(\mathbb{M}'_1, w'_1)$ and $\mathbb{M}'_1, w'_1 \models \psi$ imply that there exists a pointed model $(\mathbb{M}'_2, w'_2)$ such that $(\mathbb{M}_2, w_2)R_{\mathrm{i}}(\mathbb{M}'_2, w'_2)$ and $\mathbb{M}'_2, w'_2 \models \psi$. Therefore, by Lemma 2, $\mathbb{M}_2, w_2 \models \langle \mathrm{i} \rangle \psi$.

We can similarly prove the reverse implication using the condition **Zig**$_{\langle \mathrm{i} \rangle}$.

By using the definition of bisimulation above, we can show that the binder is not definable in both $\mathcal{DH}(@)$ and $\mathcal{DH}$.

Proposition 5. *The binder* $\downarrow$ *is not definable in languages* $\mathcal{DH}(@)$ *and* $\mathcal{DH}$.

Proof. Consider the language $\mathcal{DH}(@)$ with only one nominal i and no propositional variables. Let $\mathbb{M} = (W, R, V)$ and $\mathbb{N} = (W', R', V')$ be two models, such that $W = \{w_1, w_2\}, R = \{(w_1, w_1)\}, V(\mathrm{i}) = \{w_2\}$; and $W' = \mathrm{N} \cup \{v_1\}, R' = S, V'(\mathrm{i}) = \{v_1\}$, such that N is the set of natural numbers and S is the successor relation on N. Note that $R_\mathrm{i} = \emptyset$. Now we consider the binary relation Z defined as $Z := \{((\mathbb{M}, w_1), (\mathbb{N}, n)) \mid n \in \mathrm{N}\} \cup \{((\mathbb{M}, w_2), (\mathbb{N}, v_1))\}$ (See Fig. 3). It is straightforward to check that Z is indeed a $\mathcal{DH}(@)$-bisimulation between $(\mathbb{M}, w_1)$ and $(\mathbb{N}, 1)$. However, the formula $\downarrow x.\Diamond x$ is true at $(\mathbb{M}, w_1)$ but not at $(\mathbb{N}, 1)$. Therefore, $\downarrow$ is not definable in $\mathcal{DH}(@)$, and hence not definable in $\mathcal{DH}$.

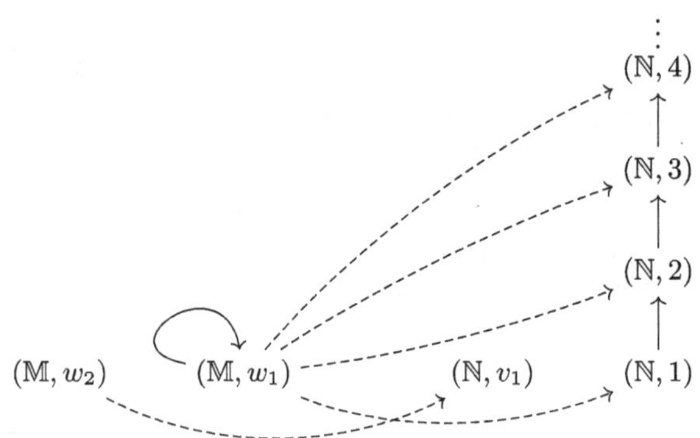

Fig. 3. The $\mathcal{DH}(@)$-bisimulation between $(\mathbb{M}, w_1)$ and $(\mathbb{N}, 1)$

6 Conclusions

We have introduced dynamic extensions of hybrid logics $\mathcal{H}$, $\mathcal{H}(@)$, and $\mathcal{H}(@, \downarrow)$ by adding operators that enable the movement of nominals along the accessibility relation. These extensions, denoted by $\mathcal{DH}$, $\mathcal{DH}(@)$, and $\mathcal{DH}(@, \downarrow)$ constitute hybrid logics with moving names. We have showed that these logics naturally capture scenarios involving agent movement on graphs.

We identified fragments of $\mathcal{DH}$ and $\mathcal{DH}(@)$ that can be translated into the static hybrid logic $\mathcal{TH}(@)$ using reduction axioms. However, we also showed that the full dynamic logics $\mathcal{DH}$ and $\mathcal{DH}(@)$ are strictly more expressive than $\mathcal{TH}(@)$. Additionally, we proved that $\mathcal{DH}(@, \downarrow)$ can be translated into $\mathcal{H}(@, \downarrow)$.

Furthermore, we showed that $\mathcal{DH}$ and $\mathcal{DH}(@)$ can express local versions of standard frame properties such as reflexivity, symmetry, transitivity, and seriality. We showed that we can in fact formalize the existence of an unbounded strict partial order within the model in $\mathcal{DH}$, establishing that $\mathcal{DH}$ and $\mathcal{DH}(@)$ lack the finite model property. We also defined an appropriate notion of bisimulation for $\mathcal{DH}$ and $\mathcal{DH}(@)$, using it to show that $\mathcal{H}(@, \downarrow)$ is strictly more expressive than $\mathcal{DH}(@)$.

This work opens several directions for future research:

Model Theory. We introduced a notion of bisimulation for $\mathcal{DH}$ and $\mathcal{DH}(@)$. It would be worthwhile to explore whether model-theoretic results such as the Hennessy-Milner theorem or van Benthem's characterization theorem can be established for these logics. Relatedly, Goldblatt-Thomason theorems have been developed for various hybrid logics [15, Chapter 4], and efforts have been made to extend Sahlqvist theory [16] to the hybrid setting. Extending these results to our dynamic hybrid logics would be an interesting direction for future research.

Axiomatization and Tableaux. In this paper, we showed that $\mathcal{DH}(@, \downarrow)$ can be translated into $\mathcal{H}(@, \downarrow)$. Thus, complete axiomatization for $\mathcal{DH}(@, \downarrow)$ is obtained by adding the reduction axioms of the form $\langle \mathrm{i} \rangle \varphi \leftrightarrow \downarrow \mathrm{x}.(@_{\mathrm{i}} \Diamond \downarrow \mathrm{y}.(@_{\mathrm{x}}(\varphi[\mathrm{y}/\mathrm{i}]))$ for every nominal i to the complete axiomatization of $\mathcal{H}(@, \downarrow)$. In the future, we aim to provide complete axiomatizations for $\mathcal{DH}$ and $\mathcal{DH}(@)$. Additionally, tableaux calculi have been developed for both hybrid logics [13,27] and dynamic logics [2,4,20]. Exploring tableaux calculi for $\mathcal{DH}$ and $\mathcal{DH}(@)$ is a natural next step.

Decidability. Since, $\mathcal{H}(@, \downarrow)$ is known to be undecidable [12, Section 5], $\mathcal{DH}(@, \downarrow)$ is also undecidable. We have showed in this paper that $\mathcal{DH}$ and $\mathcal{DH}(@)$ do not have the finite model property. A key open question is whether these logics are also undecidable.

Extension to more Expressive Logics. Our framework admits several meaningful extensions. First, hybrid logics with existential quantification over nominals, such as $\mathcal{H}(\exists)$ and $\mathcal{H}(@, \exists)$[6], have been studied in the literature [15]. Extending these logics with dynamic operators like $\langle \mathrm{i} \rangle$ is a natural direction.

Second, additional dynamic operators that capture different kinds of nominal movement can be introduced. Two particularly interesting cases are: (1) $\langle \mathrm{i} \rangle^{-1}$ which allows for the movement of nominals along R^{-1}.$\mathbb{M}, g, w \models \langle \mathrm{i} \rangle^{-1} \phi$ iff there exists w' such that $w' R V(\mathrm{i})$ and $\mathbb{M}_{[V(\mathrm{i}) \to w']}, g, w \models \phi$ and (2) $\langle \mathrm{i} \rangle^*$ which allows for the movement of nominals along any finite path in R, i.e., $\mathbb{M}, g, w \models \langle \mathrm{i}^* \rangle \phi$ iff there exists w' such that $V(\mathrm{i}) R^* w'$ and $\mathbb{M}_{[V(\mathrm{i}) \to w']}, g, w \models \phi$, where R^* is the transitive closure of R.

Lastly, we may consider generalizing the underlying modal framework itself. For instance, one could explore dynamic hybrid extensions of intuitionistic modal logic or other dynamic modal systems.

[6] The binder operator $\downarrow x.()$ is definable in $\mathcal{H}(\exists)$.

References

1. Areces, C., ten Cate, B.: 14 hybrid logics. In: Studies in Logic and Practical Reasoning, vol. 3, pp. 821–868. Elsevier (2007)
2. Areces, C., Fervari, R., Hoffmann, G.: Tableaux for relation-changing modal logics. In: Fontaine, P., Ringeissen, C., Schmidt, R.A. (eds.) Frontiers of Combining Systems, pp. 263–278. Springer, Heidelberg (2013)
3. Areces, C., Fervari, R., Hoffmann, G.: Swap logic. Log. J. IGPL **22**(2), 309–332 (2014)
4. Aucher, G., Benthem, J.V., Grossi, D.: Modal logics of sabotage revisited. J. Logic Comput. **28**(2), 269–303 (2017). https://doi.org/10.1093/logcom/exx034
5. Baader, F., Horrocks, I., Lutz, C., Sattler, U.: An Introduction to Description Logic. Cambridge University Press, Cambridge (2017)
6. Baltag, A., Moss, L.S.: Logics for epistemic programs. Synthese **139**, 165–224 (2004)
7. Baltag, A., Moss, L.S., Solecki, S.: The logic of public announcements, common knowledge, and private suspicions. In: Proceedings of the 7th Conference on Theoretical Aspects of Rationality and Knowledge (TARK 1998), pp. 43–56. Morgan Kaufmann Publishers, Los Altos, CA (1998)
8. van Benthem, J., Li, L., Shi, C., Yin, H.: Hybrid sabotage modal logic. J. Log. Comput. **33**(6), 1216–1242 (2023)
9. Blackburn, P.: Representation, reasoning, and relational structures: a hybrid logic manifesto. Logic J. IGPL **8**(3), 339–365 (2000)
10. Blackburn, P.: Arthur prior and hybrid logic. Synthese **150**(3), 329–372 (2006)
11. Blackburn, P., De Rijke, M., Venema, Y.: Modal Logic, vol. 53. Cambridge University Press, Cambridge (2001)
12. Blackburn, P., Seligman, J.: Hybrid languages. J. Logic Lang. Inform. **4**, 251–272 (1995)
13. Bolander, T., Braüner, T.: Tableau-based decision procedures for hybrid logic. J. Log. Comput. **16**(6), 737–763 (2006). https://doi.org/10.1093/logcom/exl008
14. Braüner, T.: Hybrid Logic and its Proof-theory, vol. 37. Springer (2010)
15. ten Cate, B.D.: Model theory for extended modal languages. University of Amsterdam (2004)
16. Conradie, W., Robinson, C.: On sahlqvist theory for hybrid logics. J. Log. Comput. **27**(3), 867–900 (2017)
17. Fensel, D., et al.: Knowledge Graphs. Springer (2020)
18. Gargov, G., Goranko, V.: Modal logic with names. J. Philos. Log. **22**, 607–636 (1993)
19. Gerbrandy, J., Groeneveld, W.: Reasoning about information change. J. Logic Lang. Inform. **6**, 147–169 (1997)
20. Hansen, J.U.: Terminating tableaux for dynamic epistemic logics. Electron. Notes Theor. Comput. Sci. **262**, 141–156 (2010)
21. Kooi, B., Renne, B.: Arrow update logic. Rev. Symb. Logic **4**(4), 536–559 (2011)
22. Nardi, D., Brachman, R.J., et al.: An introduction to description logics. In: Description Logic Handbook, vol. 1, p. 40 (2003)
23. Plaza, J.A.: Logics of public communications. In: Emrich, M.L., Pfeifer, M.S., Hadzikadic, M., Ras, Z.W. (eds.) Proceedings of the Fourth International Symposium on Methodologies for Intelligent Systems: Poster Session Program, pp. 201–216. Oak Ridge National Laboratory, ORNL/DSRD-24 (1989)
24. Prior, A.: Past, Present and Future. Oxford University Press (1967)

25. Prior, A.: Papers on Time and Tense. Oxford University Press (1968)
26. Sano, K., Liu, F., Li, D.: Hybrid logic of the hide and seek game. Studia Logica 1–33 (2024)
27. Tzakova, M.: Tableau calculi for hybrid logics. In: International Conference on Automated Reasoning with Analytic Tableaux and Related Methods, pp. 278–292. Springer (1999)
28. Van Benthem, J.: An essay on sabotage and obstruction. In: Mechanizing Mathematical Reasoning: Essays in Honor of Jörg H. Siekmann on the Occasion of His 60th Birthday, pp. 268–276. Springer (2005)
29. Van Benthem, J.: Logical Dynamics of Information and Interaction. Cambridge University Press (2011)
30. Van Ditmarsch, H., van Der Hoek, W., Kooi, B.: Dynamic Epistemic Logic, vol. 337. Springer (2007)

Deductive Systems for Dynamic Quantum Logic

Tomoaki Kawano(✉)

Kanagawa University, Yokohama, Japan
tomoakikawano.tk@gmail.com

Abstract. Dynamic quantum logic (DQL) is a modal logic for analyzing dynamic propositions of quantum physics. Although a Hilbert-style axiomatization of DQL has already been constructed, its completeness and the sequent calculus of DQL have not been well studied. In this paper, we construct a sequent calculus for DQL and discuss the soundness and completeness theorems with respect to dynamic quantum models. Through this, we also prove the completeness of the Hilbert-style system.

1 Introduction

Quantum logic (QL) is a non-classical logic which describes the propositional spaces of quantum physics. There are many different types of research on QL, ranging from relatively simple to complex. Among them, *orthomodular logic* (OML), which treats simple binary relational frames (Kripke frames) or lattices as semantics, has been studied as a basis of QL [6,8,23]. OML is based on the *closed subspaces* and orthogonal relation between states in a *Hilbert space*, the state space of quantum particles. *Quantum states* of a particle is represented by the one-dimensional closed subspaces in a Hilbert space.

The language and frames for OML, such as *OM-frames*, do not include fundamental *dynamical* concepts like *unitary evolutions* and *projections* (and composite *linear operators*) in a Hilbert space. *Dynamic quantum logic* (DQL) was constructed to express the dynamics of quantum physics and quantum computing [1–5]. DQL can express more detailed propositions than OML and also includes concepts from OML, which is intricately combined with dynamical notions. A Hilbert-style deductive system for DQL is given in [1,3,5]. Although the *completeness* of the system is mentioned in [1,3], the analysis is only outlined, and the proofs of the soundness and completeness theorems in the following sense have not been discussed: A formula ϕ is deducible in the deduction system iff ϕ is valid in all the frames of DQL. Therefore, in this study, we construct a new deduction system for DQL and prove the soundness and completeness theorem. We also use it to prove the completeness theorem of the Hilbert-style system.

Sequent calculus (Gentzen-style system) is more valuable than the Hilbert-style system in some situations. We adopt sequent calculus in this study because it is more convenient for proving completeness and was already developed in the

J. Wang et al. (Eds.): DaLí 2025, LNCS 16472, pp. 40–57, 2026.
https://doi.org/10.1007/978-3-032-22626-6_3

context of QL [10,16–18,23–25]. The references [1–5] discuss formulas and *frame conditions* of DQL, but the definitions used in these studies differ slightly across studies. In this study, we employ the basic logical symbols and frame conditions common to all of these. In DQL, almost the same language as the language used in *dynamic logic* (DL) [13] is adopted. DL describes the propositions of the flow of program executions in a classical computer. Although the languages are almost the same, their meaning in models are somewhat different. This difference is due to the difference between classical and quantum mechanics (quantum computers). In DL, the *action symbol* π represents programs in classical computers. The formula $[\pi]\phi$ represents the proposition "After the program π executes, ϕ is true." The basic components of an action π consist of a set of symbols that represent the basic programs with no particular conditions, and a set of *tests*, which are actions that check the truth of propositions. Since unitary evolutions and projections are fundamental to the change of state in a quantum computer, in DQL, the basic components of π are symbols that represent these concepts [1–5]. The set of possible worlds of the model of DQL is an abstract representation of the state space within a single Hilbert space. Unitary evolutions and projections are also abstractly represented by binary relations on the model. Although particularly complex conditions for actions are not imposed in the models of DL, we must impose many conditions on the models of DQL to mimic the Hilbert space and operators on it. *Hermitian conjugation* of action is also an essential factor in quantum computers. For example, the Hermitian conjugate of the unitary evolution corresponds to the inverse transform of the unitary evolution, which is essential when discussing program invertibility of quantum computing. In DQL, the symbol $\dagger$ corresponding to the Hermite conjugate is used. The projections appear as a movement of state when testing the truth of a proposition. In QL, fundamental propositions correspond to closed subspaces of a Hilbert space. When we try to observe whether proposition ϕ is true or false, if we observe that ϕ is true, the state is projected onto the closed subspace where ϕ is true. In this way, in quantum mechanics, the state changes when information is gained. In the frame of DQL, this property is expressed by binary relations representing the projections. This is another important difference between DL and DQL. In DL, the state does not change when the propositions are tested.

The sequent calculus **SDQL** constructed in this study is based on the sequent calculi for DL [14,27] and the sequent calculi for OML [15,23]. The *Kleene star*, which expresses iteration of a program, causes some problems in the sequent calculus for DL. The language for DQL does not include Kleene star because most quantum programming does not involve while-loops, it is not essential in quantum mechanics [3]. Therefore, in **SDQL**, problems caused by iteration will not occur. However, due to another issue, **SDQL** does not satisfy the cut-elimination theorem. Because the purpose of this study is to construct a system that satisfies the completeness theorem, this problem will be left as future work.

In Sect. 2, we review the basics of the DQL. In Sect. 3, we discuss the sequent calculus for DQL. In Sect. 4, we compare the Hilbert-style system with the new sequent calculus. Since this study is a result of mathematical logic, a detailed explanation of quantum physics and quantum information science is omitted. For more details and quantum theoretical aspects of DQL, see [1–5].

2 Dynamic Quantum Logic

In this section, we review the basics of DQL. The language of DQL consists of a countable infinite set of propositional variables $p, q, r, \ldots$, a countable infinite set $\mathcal{U}$ of variables for unitary evolutions $U, U_1, U_2, \ldots$, the logical connectives $\neg, \wedge, \Box$, action connectives $^\dagger, \cup, ;, ?$, and the brackets $[\,]$ for modal operators.

Definition 1. *The formulas and actions of DQL are defined as follows:*

$$\textit{Formula}\ \phi ::= p \mid \neg\phi \mid \phi \wedge \phi \mid \Box\phi \mid [\pi]\phi$$
$$\textit{Action}\ \pi ::= U \mid \pi^\dagger \mid \pi \cup \pi \mid \pi;\pi \mid \phi?$$

The symbols $\phi, \psi, \chi, \ldots$ denote formulas. $\pi, \pi_1, \pi_2, \ldots$ denote actions. $\Gamma, \Delta, \Sigma, \ldots$ denote sets of formulas. The set of all formulas of DQL is denoted by Ω. In addition, the following abbreviations are introduced: $\bot = p \wedge \neg p$, $\phi \vee \psi = \neg(\neg\phi \wedge \neg\psi)$, $\phi \to \psi = \neg\phi \vee \psi$, $\sim\phi = \Box\neg\phi$, $\Diamond\phi = \neg\Box\neg\phi$, $\langle\pi\rangle\phi = \neg[\pi]\neg\phi$, and $\phi \sqcup \psi = \sim(\sim\phi \wedge \sim\psi)$. Same as negation, $\sim$ takes precedence over $\wedge$ and $\sqcup$.

The symbol $\Box$ represents *non-orthogonality* of states. $\Box\phi$ means "The state is orthogonal to all state that ϕ is false." Intuitively, this symbol is used to represent OML concepts in DQL. The symbol $\dagger$ represents the Hermitian conjugate. The symbol $\cup$ represents a choice of action, and ; represents a sequence of actions. Although $\phi?$ has a similar interpretation as in DL, there is an essential difference between $\phi?$ in DL versus $\phi?$ in DQL. In both cases, $[\phi?]\psi$ is read as "After testing whether ϕ is true or not, if ϕ is true, then ψ is true." In DL, a state never changes after testing a proposition in contrast to DQL where they may change after testing a proposition (projection onto the closed subspace that ϕ is true). The model for DQL must reflect how the state changes with observation by means of a binary relation condition, and the truth of ψ is determined by the post-observation state.

Definition 2. *A* dynamic frame *is a tuple* $(W, \bot, \sigma, \to_{P?}, \to_u)$ *consists of:*

1. W is a non-empty set representing a set of quantum states.
2. $\bot$ is a binary relation on W which is irreflexive and symmetric. This relation represents an *orthogonal relation* between quantum states.
3. $\sigma \subseteq \mathfrak{P}(W)$. ($\mathfrak{P}(W)$ is the power set of W.) This set represents the collection of sets whose closures are intended to be treated as subspaces.
4. For each $X \in \sigma$, we define a binary relation $X?$ on W. This relation represents the projection onto X. Then, $\to_{P?}$ is defined as the set of all such binary relations.
5. For each $U \in \mathcal{U}$, we define a binary relation U on W. This relation represents the unitary transformation U. Then, $\to_u$ is defined as the set of all such binary relations. (We use the same symbols $U, U_1, U_2, \ldots$ for actions and binary relations.)

The subscript P stands for projection, and u stands for unitary evolution. We write xRy or $x(R)y$ if $(x, y) \in R$ for a relation R. For $x \in W$ and $X \subseteq W$, we write $x \perp X$ if for all $y \in X$, $x \perp y$. Given $X \subseteq W$, we define the set $X^{\perp} = \{x \in W | x \perp X\}$. We say that X is *$\perp$-closed* if $X^{\perp\perp} = X$. We write $x \not\perp y$ if not $x \perp y$. As $\perp$ is an irreflexive symmetric relation, we can regard $\not\perp$ as a reflexive symmetric relation on W. Note that for all $X \subseteq W$, $X \subseteq X^{\perp\perp}$. Therefore, if $X^{\perp\perp} \subseteq X$, then X is $\perp$-closed. A $\perp$-closed set represents a closed subspace of a Hilbert space.

Definition 3. Composite relations *in a dynamic quantum frame are inductively defined as follows:*

$$
\begin{aligned}
&x(\pi_1; \pi_2)y \Leftrightarrow \text{for some } z \in W,\ x(\pi_1)z \text{ and } z(\pi_2)y,\\
&x(\pi_1 \cup \pi_2)y \Leftrightarrow x(\pi_1)y \text{ or } x(\pi_2)y,\\
&x(X?^{\dagger})y \Leftrightarrow x(X?)y,\\
&x(U^{\dagger})y \Leftrightarrow y(U)x,\\
&x((\pi_1; \pi_2)^{\dagger})y \Leftrightarrow x(\pi_2^{\dagger}; \pi_1^{\dagger})y,\\
&x((\pi_1 \cup \pi_2)^{\dagger})y \Leftrightarrow x(\pi_1^{\dagger})y \text{ or } x(\pi_2^{\dagger})y,\\
&x((\pi)^{\dagger\dagger})y \Leftrightarrow x(\pi)y.
\end{aligned}
$$

Definition 4. *A* dynamic quantum frame $(W, \perp, \sigma, \rightarrow_{P?}, \rightarrow_u)$ *is a dynamic frame that satisfies the following conditions:*

F1 If for some $X \in \sigma$, x is related to y by $X?$, then $x \not\perp y$.
F2 For all $X \in \sigma$, $X^{\perp\perp} \in \sigma$. Furthermore, $x(X?)y$ iff $x((X^{\perp\perp})?)y$.
F3 (Partial functionality of $X?$) For all $x, y, z \in W$ and $X \in \sigma$, if $x(X?)y$ and $x(X?)z$, then $y = z$.
F4 (Adequacy) If $x \in X \in \sigma$, then $x(X?)x$.
F5 (Repeatability) For all $x, y \in W$, if $X \in \sigma$ is $\perp$-closed and $x(X?)y$, then $y \in X$.
F6 (Adjointness) For all relations R including composite relations, if $x(R)y$ and $y \not\perp z$, then there exists $w \in W$ such that $z(R^{\dagger})w$ and $w \not\perp x$.
F7 (Functionality for U) For all $x \in W$ and U, $\exists! y \in W$ such that $x(U)y$.
F8 (Functionality for $U^{\dagger}$) For all $x \in W$ and $U^{\dagger}$, $\exists! y \in W$ such that $x(U^{\dagger})y$.
F9 (Universal accessibility) For all $x, y \in W$, there exists $z \in W$ such that $x \not\perp z$ and $z \not\perp y$.

Each of these conditions represents an important property in Hilbert spaces. For example, F1 expresses that projection onto an orthogonal closed subspace is not possible. F2 expresses that a projection onto a subset that is not a closed subspace is defined as a projection onto its closure. F9 expresses that one can move to any state in the Hilbert space by following the non-orthogonal relation twice (Vectors $(1,0)$ and $(0,1)$ are orthogonal but $(1,0) \not\perp (1/\sqrt{2}, 1/\sqrt{2}) \not\perp (0,1)$).

We say that a binary relation R is *adjoint* if it satisfies the statements in F6. Intuitively, the following lemma represent important natures of linear operators in a Hilbert space, such as the product of conjugate operators are also conjugate.

Lemma 1. *If all relations $X?, U$, and $U^\dagger$ are adjoint in $(W, \perp, \sigma, \rightarrow_{P?}, \rightarrow_u)$, then all composite relations are adjoint in the frame.*

Proof. We prove this by induction on the construction of the relation R. We only prove the cases for $R = R_1; R_2$. For the other cases, the proof is more straightforward. Suppose $x(R)y$ and $y \not\perp z$. Then, there exists $w \in W$ such that $x(R_1)w$ and $w(R_2)y$. From the adjointness of R_2, there exists $v \in W$ such that $z(R_2^\dagger)v$ and $v \not\perp w$. Then, from the adjointness of R_1, there exists $u \in W$ such that $v(R_1^\dagger)u$ and $u \not\perp x$. Therefore, R is adjoint because $z(R^\dagger)u$. □

Definition 5. *A* dynamic model *is defined by $(W, \perp, \sigma, \rightarrow_{P?}, \rightarrow_u, V)$ consists of:*

1. $(W, \perp, \sigma, \rightarrow_{P?}, \rightarrow_u)$ is a dynamic frame.
2. V is a function that assigns each propositional variable p to an element of σ, and satisfies that for all $\phi \in \Omega$, following $\|\phi\|$ are included in σ.

Definition 6. *The set $\|\phi\|$ on a dynamic model are defined by induction on the composition of ϕ as follows:*

$$
\begin{aligned}
&\|p\| = V(p)\\
&\|\phi \wedge \psi\| = \|\phi\| \cap \|\psi\|\\
&\|\neg\phi\| = \|\phi\|^c\\
&\|\Box\phi\| = \{x \in W |\text{ for all } y \in W, \text{ if } x \not\perp y, \text{ then } y \in \|\phi\|\}\\
&\|[\phi?]\psi\| = \{x \in W |\text{ for all } y \in W, \text{ if } x(\phi?)y, \text{ then } y \in \|\psi\|\}\\
&\|[U]\phi\| = \{x \in W |\text{ for all } y \in W, \text{ if } x(U)y, \text{ then } y \in \|\phi\|\}\\
&\|[\pi_1;\pi_2]\phi\| = \|[\pi_1][\pi_2]\phi\|\\
&\|[\pi_1 \cup \pi_2]\phi\| = \|[\pi_1]\phi\| \cap \|[\pi_2]\phi\|\\
&\|[\psi?^\dagger]\phi\| = \|[\psi?]\phi\|\\
&\|[U^\dagger]\phi\| = \{x \in W |\text{ for all } y \in W, \text{ if } x(U^\dagger)y, \text{ then } y \in \|\phi\|\}\\
&\|[(\pi_1;\pi_2)^\dagger]\phi\| = \|[\pi_2^\dagger;\pi_1^\dagger]\phi\|\\
&\|[(\pi_1 \cup \pi_2)^\dagger]\phi\| = \|[\pi_1^\dagger \cup \pi_2^\dagger]\phi\|\\
&\|[\pi^{\dagger\dagger}]\phi\| = \|[\pi]\phi\|
\end{aligned}
$$

$x(\phi?)y$ denotes $\|\phi\| = X$ and $x(X?)y$.

Note that intersection appears in the definition of $\|[\pi_1 \cup \pi_2]\phi\|$. In the diamond definition, the union appears $\|\langle\pi_1 \cup \pi_2\rangle\phi\| = \|\langle\pi_1\rangle\phi\| \cup \|\langle\pi_2\rangle\phi\|$.

A formula ϕ is *true* at $x \in W$ if $x \in \|\phi\|$. We write $x \models \phi$ if ϕ is true at x. A formula ϕ is *false* at $x \in W$ if $x \notin \|\phi\|$. ϕ is *valid* in a DQM $(W, \perp, \sigma, \rightarrow_{P?}, \rightarrow_u, V)$ if ϕ is true at all $x \in W$.

Definition 7. *A dynamic model $(W, \perp, \sigma, \rightarrow_{P?}, \rightarrow_u, V)$ is called a* dynamic quantum model *(DQM) if $(W, \perp, \sigma, \rightarrow_{P?}, \rightarrow_u)$ is a dynamic quantum frame.*

From the definition of $\|\Box\phi\|$, we can confirm that the symbol $\sim$ represents the *quantum negation* of OML. That is, $\sim \phi$ is true at x iff $x \in \|\phi\|^\perp$. From the universal accessibility of a model, $\Box\Box\phi$ means that ϕ is valid in a model. Furthermore, the abbreviation $T(\phi) = \Box\Box(\sim\sim \phi \rightarrow \phi)$ is used for convenience:

Because $\phi \rightarrow \sim\sim \phi$ is always true by symmetry of $\not\perp$, if $\sim\sim \phi \rightarrow \phi$ is true at all $x \in W$, then $\|\phi\|$ is $\perp$-closed. Therefore, $T(\phi)$ implies that $\|\phi\|$ is $\perp$-closed.

In OML, *Sasaki implication* $\sim \phi \sqcup (\phi \wedge \psi)$ and *Sasaki projection* $\phi \wedge (\sim \phi \sqcup \psi)$ play an important role [8,12]. Sasaki implication is considered as one of the fundamental implications in OML because the *orthomodular law* $\phi \wedge (\sim \phi \sqcup (\phi \wedge \psi)) \rightarrow \psi$ is a form of modus ponens for Sasaki implication. Because $\phi \wedge [\phi]\psi \rightarrow \psi$ is valid in all DQM, as we see from the following lemma and theorem, the orthomodular law is partially valid in DQM. This fact is useful when applying the rules of sequent calculi for OML to the sequent calculus of DQL.

Lemma 2. *In any DQM $(W, \perp, \sigma, \rightarrow_{P?}, \rightarrow_u, V)$, if $x \models \psi$, $x(\phi?)y$ and $\|\phi\|$ is $\perp$-closed, then $y \models \phi \wedge (\sim \phi \sqcup \psi)$.*

Proof. $y \models \phi$ because of $x(\phi?)y$ and repeatability. Note that $\|(\sim \phi \sqcup \psi)\| = \|\Box\neg(\Box\neg\psi \wedge \phi)\|$. For the sake of contradiction, suppose $y \not\models \Box\neg(\Box\neg\psi \wedge \phi)$. Then, there exists $z \in W$ such that $y \not\perp z$ and $z \models \Box\neg\psi \wedge \phi$. From the adequacy, $z(\phi?)z$. From the adjointness and the partial functionality of $\phi?$, $x \not\perp z$. This condition is contradictory to $x \models \psi$ and $z \models \Box\neg\psi$. □

Theorem 1. *In any DQM $(W, \perp, \sigma, \rightarrow_{P?}, \rightarrow_u, V)$, if $\|\phi\|$ is $\perp$-closed, $\|[\phi?]\psi\| \subseteq \| \sim \phi \sqcup (\phi \wedge \psi)\|$. In particular, if $\|\phi\|$ and $\|\psi\|$ are $\perp$-closed, $\|[\phi?]\psi\| = \| \sim \phi \sqcup (\phi \wedge \psi)\|$.*

Proof. Since $\|\phi\|$ is $\perp$-closed, $\| \sim \phi \sqcup (\phi \wedge \psi)\| = \|\Box\neg(\sim\sim \phi \wedge \sim (\phi \wedge \psi)\| = \|\Box\neg(\phi \wedge \sim (\phi \wedge \psi)\|$. For the sake of contradiction, suppose $x \not\models \sim \phi \sqcup (\phi \wedge \psi)$ and $x \models [\phi?]\psi$. Then, there exists $y \in W$ such that $x \not\perp y$ and $y \models \phi \wedge \sim (\phi \wedge \psi)$. From the adequacy, $y(\phi?)y$. From the adjointness of a frame, there exists $z \in W$ such that $x(\phi?)z$ and $z \not\perp y$. Since $\|\phi\|$ is $\perp$-closed, from $x \models [\phi?]\psi$ and repeatability, $z \models \phi \wedge \psi$. This contradicts $y \models \sim (\phi \wedge \psi)$.

For the second part of the theorem, for the sake of contradiction, suppose $x \models \sim \phi \sqcup (\phi \wedge \psi)$ and $x \not\models [\phi?]\psi$. Then, there exists $y \in W$ such that $x(\phi?)y$ and $y \not\models \psi$. Suppose $\|\psi\|$ is $\perp$-closed. Then, $y \not\models \sim\sim \psi$. So there exists $z \in W$ such that $y \not\perp z$ and $z \models \sim \psi$. Using the adjointness property of the frame, there exists $w \in W$ such that $z(\phi?)w$ and $w \not\perp x$. From Lemma 2, $w \models \phi \wedge (\sim \phi \sqcup \sim \psi)$. This contradicts $x \models \sim \phi \sqcup (\phi \wedge \psi)$. □

3 Sequent Calculus for DQL

In this section, we discuss a sequent calculus for DQL. A *sequent* is defined by $\Gamma \Rightarrow \Delta$ where Γ and Δ are sets of formulas. We define that a sequent $\Gamma \Rightarrow \Delta$ is false at $x \in W$ of a DQM if all $\phi \in \Gamma$ is true at x and all $\phi \in \Delta$ is false at x. If $\Gamma \Rightarrow \Delta$ is not false at x, then it is true at x. If $\Gamma \Rightarrow \Delta$ is true at all $x \in W$ in a DQM, then $\Gamma \Rightarrow \Delta$ is valid in the DQM.

Definition 8 (Sequent calculus SDQL).

Γ, Δ, Π, and Σ are finite sets.
Axiom:

$$\frac{}{\phi \Rightarrow \phi}$$

Rules:

$$\frac{\Gamma \Rightarrow \Delta, \phi \quad \phi, \Pi \Rightarrow \Sigma}{\Gamma, \Pi \Rightarrow \Delta, \Sigma} \text{(cut)}$$

$$\frac{\Gamma \Rightarrow \Delta}{\phi, \Gamma \Rightarrow \Delta} \text{(w L)} \quad \frac{\Gamma \Rightarrow \Delta}{\Gamma \Rightarrow \Delta, \phi} \text{(w R)} \quad \frac{\Gamma \Rightarrow \Delta, \phi}{\neg\phi, \Gamma \Rightarrow \Delta} (\neg\text{L}) \quad \frac{\phi, \Gamma \Rightarrow \Delta}{\Gamma \Rightarrow \Delta, \neg\phi} (\neg\text{R})$$

$$\frac{\phi, \Gamma \Rightarrow \Delta}{\phi \wedge \psi, \Gamma \Rightarrow \Delta} (\wedge\text{L}) \quad \frac{\psi, \Gamma \Rightarrow \Delta}{\phi \wedge \psi, \Gamma \Rightarrow \Delta} (\wedge\text{L}) \quad \frac{\Gamma \Rightarrow \Delta, \phi \quad \Gamma \Rightarrow \Delta, \psi}{\Gamma \Rightarrow \Delta, \phi \wedge \psi} (\wedge\text{R})$$

$$\frac{\phi, \Gamma \Rightarrow \Delta}{\Box\phi, \Gamma \Rightarrow \Delta} (\Box 1) \quad \frac{\Box\Box\Gamma \Rightarrow \phi}{\Box\Box\Gamma \Rightarrow \Box\phi} (\Box 2) \quad \frac{\Gamma \Rightarrow \Box\Delta, \phi}{\Box\Gamma \Rightarrow \Delta, \Box\phi} (\Box 3)$$

$$\frac{\Box\Box\Gamma \Rightarrow \phi}{\Box\Box\Gamma \Rightarrow [\pi]\phi} (\pi) \quad \frac{[\psi?]\phi, \Gamma \Rightarrow \Delta}{\Box\phi, \Gamma \Rightarrow \Delta} (\Box) \quad \frac{\Gamma \Rightarrow [\pi]\Box\Delta, \phi}{[\pi^\dagger]\Box\Gamma \Rightarrow \Delta, [\pi^\dagger]\Box\phi} \text{(Ad)}$$

$$\frac{\Gamma \Rightarrow \Delta}{[U]\Gamma \Rightarrow [U]\Delta} \text{(U)} \quad \frac{\Gamma \Rightarrow \Delta}{[U^\dagger]\Gamma \Rightarrow [U^\dagger]\Delta} (\text{U}^\dagger) \quad \frac{\Gamma \Rightarrow \Delta}{[\phi?]\Gamma \Rightarrow [\phi?]\Delta} \text{(P)}^*$$

$$\frac{[\phi?]\psi, \Gamma \Rightarrow \Delta}{[\phi?^\dagger]\psi, \Gamma \Rightarrow \Delta} (\text{P}^\dagger\text{L}) \quad \frac{\Gamma \Rightarrow \Delta, [\phi?]\psi,}{\Gamma \Rightarrow \Delta, [\phi?^\dagger]\psi} (\text{P}^\dagger\text{R})$$

$$\frac{\Gamma \Rightarrow \Delta, \phi \quad \psi, \Gamma \Rightarrow \Delta}{[\phi?]\psi, \Gamma \Rightarrow \Delta} \text{(PL)} \quad \frac{T(\phi), < \phi? > \Gamma \Rightarrow \psi}{T(\phi), \Gamma \Rightarrow [\phi?]\psi} \text{(PR)}^{**}$$

$$\frac{\Gamma \Rightarrow [U]\Delta, \phi}{[U^\dagger]\Gamma \Rightarrow \Delta, [U^\dagger]\phi} \text{(UA1)} \quad \frac{\Gamma \Rightarrow [U^\dagger]\Delta, \phi}{[U]\Gamma \Rightarrow \Delta, [U]\phi} \text{(UA2)}$$

$$\frac{[\psi?]\phi, \Gamma \Rightarrow \Delta}{[\sim\sim \psi?]\phi, \Gamma \Rightarrow \Delta} (\sim\sim \text{L}) \quad \frac{\Gamma \Rightarrow \Delta, [\psi?]\phi}{\Gamma \Rightarrow \Delta, [\sim\sim \psi?]\phi} (\sim\sim \text{R})$$

$$\frac{[\pi_1]\phi, \Gamma \Rightarrow \Delta}{[\pi_1 \cup \pi_2]\phi, \Gamma \Rightarrow \Delta} (\cup \text{L}) \quad \frac{[\pi_2]\phi, \Gamma \Rightarrow \Delta}{[\pi_1 \cup \pi_2]\phi, \Gamma \Rightarrow \Delta} (\cup \text{L})$$

$$\frac{\Gamma \Rightarrow \Delta, [\pi_1]\phi \quad \Gamma \Rightarrow \Delta, [\pi_2]\phi}{\Gamma \Rightarrow \Delta, [\pi_1 \cup \pi_2]\phi} (\cup \text{R}) \quad \frac{\Gamma \Rightarrow \Delta, [\pi_1^\dagger]\phi \quad \Gamma \Rightarrow \Delta, [\pi_2^\dagger]\phi}{\Gamma \Rightarrow \Delta, [(\pi_1 \cup \pi_2)^\dagger]\phi} (\cup\dagger \text{R})$$

$$\frac{[\pi_1^\dagger]\phi, \Gamma \Rightarrow \Delta}{[(\pi_1 \cup \pi_2)^\dagger]\phi, \Gamma \Rightarrow \Delta} (\cup\dagger \text{L}) \quad \frac{[\pi_2^\dagger]\phi, \Gamma \Rightarrow \Delta}{[(\pi_1 \cup \pi_2)^\dagger]\phi, \Gamma \Rightarrow \Delta} (\cup\dagger \text{L})$$

$$\frac{[\pi_1][\pi_2]\phi, \Gamma \Rightarrow \Delta}{[\pi_1; \pi_2]\phi, \Gamma \Rightarrow \Delta} (; \text{L}) \quad \frac{\Gamma \Rightarrow \Delta, [\pi_1][\pi_2]\phi}{\Gamma \Rightarrow \Delta, [\pi_1; \pi_2]\phi} (; \text{R})$$

$$\frac{[\pi_2^\dagger; \pi_1^\dagger]\phi, \Gamma \Rightarrow \Delta}{[(\pi_1; \pi_2)^\dagger]\phi, \Gamma \Rightarrow \Delta} (;\dagger \text{L}) \quad \frac{\Gamma \Rightarrow \Delta, [\pi_2^\dagger; \pi_1^\dagger]\phi}{\Gamma \Rightarrow \Delta, [(\pi_1; \pi_2)^\dagger]\phi} (;\dagger \text{R})$$

$$\frac{[\pi]\phi, \Gamma \Rightarrow \Delta}{[\pi^{\dagger\dagger}]\phi, \Gamma \Rightarrow \Delta} (\dagger\dagger \text{ L}) \quad \frac{\Gamma \Rightarrow \Delta, [\pi]\phi}{\Gamma \Rightarrow \Delta, [\pi^{\dagger\dagger}]\phi} (\dagger\dagger \text{ R})$$

*Δ is required to be non-empty.

**$< \phi? > \chi = \phi \wedge (\sim \phi \sqcup \chi)$. $< \phi? > \Gamma = \{ < \phi? > \chi | \chi \in \Gamma\}$.
where $\Box\Gamma = \{\Box\phi|\phi \in \Gamma\}$, $\Box\Box\Gamma = \{\Box\Box\phi|\phi \in \Gamma\}$, $[U]\Gamma = \{[U]\phi|\phi \in \Gamma\}$, $[U^{\dagger}]\Gamma = \{[U^{\dagger}]\phi|\phi \in \Gamma\}$, $[\phi?]\Gamma = \{[\phi?]\psi|\psi \in \Gamma\}$, and $[\pi]\Box\Gamma = \{[\pi]\Box\phi|\phi \in \Gamma\}$.

We use a diamond form $< \phi? > \chi$ for $\phi \wedge (\sim \phi \sqcup \chi)$ because it is semantically equivalent to $\sim [\phi?] \sim \chi$ in OML in a certain way. $\langle \phi? \rangle$ is defined as $\neg[\phi?]\neg$. The shapes of rules (U) and (P) are the same, but Δ in (P) cannot be the empty set. This reflects the distinction between functionality and partial functionality.

In the sequent calculus for DL, the introduction rules for $[\phi?]\psi$ are identical to the rules for $\phi \to \psi$ in the sequent calculus of classical logic **LK** because $[\phi?]\psi$ and $\phi \to \psi$ are semantically equivalent in DL [27]. Although something similar to this notion is true in **SDQL**, we regard (PL) and (PR) as rules for conditions rather than introduction rules. The rule (PR) can be viewed as a generalisation of the following rule for orthomodular law (in the Sasaki implication form) in OML [15].

$$\frac{\phi \wedge (\sim \phi \sqcup \chi) \Rightarrow \psi}{\chi \Rightarrow \sim \phi \sqcup (\phi \wedge \psi)}$$

(PL) and (PR) are rules that represent adequacy and repeatability conditions, while (P) is the introduction rule for the formula $[\phi?]\psi$ and also represents the partial functionality condition. We will see these concepts in the proof of Lemma 4.

Similar to adequacy and repeatability, conditions and definitions of relations are represented by corresponding rules. For example, ($\Box$1) corresponds to the reflexivity of relation $\not\perp$, and ($\Box$3) corresponds to symmetry. (UA1) and (UA2) correspond to $x(U^{\dagger})y \Leftrightarrow y(U)x$. We review some conditions and corresponding axioms and rules of modal logic related to the dynamic quantum frame conditions in appendix. Comparing these rules with the rules in the **SDQL** will show the correspondence.

Theorem 2 (Soundness). *If $\Gamma \Rightarrow \Delta$ is provable in* **SDQL**, *then $\Gamma \Rightarrow \Delta$ is valid in all DQM.*

Proof. We prove this by induction on the construction of the proof of sequent $\Gamma \Rightarrow \Delta$. We only prove the cases for which the last rule used in the proof is (PL), (PR), ($\sim\sim$ L) and ($\sim\sim$ R). The proofs for the other cases can be done in the same way as these or as in the case of traditional modal logic.

Suppose the upper sequents of (PL) are valid, and suppose $x \models [\phi?]\psi$ and $x \models \chi$ for all $\chi \in \Gamma$. Then, from assumption, there exists $\theta \in \Delta \cup \{\phi\}$ such that $x \models \theta$. If $\theta \in \Delta$, the lower sequent is true at x. If $\theta = \phi$, from the adequacy of a model, $x(\phi?)x$. Then, from $x \models [\phi?]\psi$, $x \models \psi$. From the assumption, there exists η such that $x \models \eta$ and $\eta \in \Delta$. Therefore, the lower sequent is true at x.

For (PR), the proof is established by using almost the same argument in the proof of Lemma 2.

Suppose the upper sequent of (∼∼L) (or (∼∼R)) is valid. From F2, $\|[\phi?]\psi\| = \|[\sim\sim \phi?]\psi\|$. Consequently, the lower sequent is true. □

We now discuss the completeness theorem. In a DQM, we say that $x, y \in W$ are connected by a finite sequence of relations $R_1, R_2, \ldots, R_n$, if there exist $z_1, z_2, \ldots, z_{n-1} \in W$ such that $x(R_1)z_1$, $z_1(R_2)z_2, \ldots$ $z_{n-1}(R_n)y$, where each R_i is $\not\perp$, an element of $\rightarrow_{P?}$, an element of $\rightarrow_u$, or an element of $\{U^\dagger | U \in\rightarrow_u\}$. Note that the "finite sequence of relations" is an equivalence relation due to the F1 of dynamic quantum frame, the reflexivity and symmetry of $\not\perp$, and $x(U^\dagger)y \Leftrightarrow y(U)x$. An *infinite sequent* is defined as $\Gamma \Rightarrow \Delta$, where Γ or Δ is an infinite set. We define $\Gamma' \Rightarrow \Delta'$ to be a *subsequent* of $\Gamma \Rightarrow \Delta$ if $\Gamma' \subseteq \Gamma$ and $\Delta' \subseteq \Delta$. We say that an infinite sequent is not provable if all its finite subsequents are not provable. For the completeness theorem, a canonical model $(W_c, \perp_c, \sigma_c, \rightarrow_{P?c}, \rightarrow_{uc}, V_c)$ for an unprovable sequent $\Gamma \Rightarrow \Delta$ is constructed by using infinite sequents. There exist at least one unprovable infinite sequent $\Gamma' \Rightarrow \Delta'$ such that $\Gamma \subseteq \Gamma'$, $\Delta \subseteq \Delta'$, and $\Gamma' \cup \Delta' = \Omega$ because (cut) is included in **SDQL**. (That is, if $\Gamma \Rightarrow \Delta$ is unprovable, for all $\phi \in \Omega$, at least one of $\phi, \Gamma \Rightarrow \Delta$ or $\Gamma \Rightarrow \Delta, \phi$ is unprovable. By adding every $\phi \in \Omega$ to either the left or right side of the sequent in this way, we obtain an infinite unprovable sequent as limit.) We adopt one of them as $\Gamma' \Rightarrow \Delta'$. We define $\perp_c$ by defining $\not\perp_c$.

1. W_c is defined as the set of all infinite sequents $\Pi \Rightarrow \Sigma$ that satisfy the following conditions:
   $\Pi \Rightarrow \Sigma$ is not provable in **SDQL**.
   $\Pi \cup \Sigma = \Omega$.
   $\Pi \Rightarrow \Sigma$ could be connected by a finite sequence of relations from $\Gamma' \Rightarrow \Delta'$ defined by the following relations.
2. $\not\perp_c \stackrel{\text{def}}{=} \{(\Pi \Rightarrow \Sigma, \Pi' \Rightarrow \Sigma')|$ If $\Box\Box\phi \in \Pi$, then $\Box\Box\phi \in \Pi'$. If $\Box\phi \in \Pi$, then $\phi \in \Pi'$. If $\phi \in \Sigma$, then $\Box\phi \in \Sigma'$.$\}$
3. $\rightarrow_{P?c}$ is defined as the set of all relations $\|\phi\|?(\phi \in \Omega)$ defined as follows:
   $\|\phi\|? \stackrel{\text{def}}{=} \{(\Pi \Rightarrow \Sigma, \Pi' \Rightarrow \Sigma')|$ For all $[\phi?]\psi$, if $[\phi?]\psi \in \Pi$, then $\psi \in \Pi'$, and if $[\phi?]\psi \in \Sigma$, then $\psi \in \Sigma'\}$.
4. $\rightarrow_{uc}$ is defined as the set of all relations $U(U \in \mathcal{U})$ defined as follows:
   $U \stackrel{\text{def}}{=} \{(\Pi \Rightarrow \Sigma, \Pi' \Rightarrow \Sigma')|$If $[U]\psi \in \Pi$, then $\psi \in \Pi'$. If $[U]\psi \in \Sigma$, then $\psi \in \Sigma'\}$.
5. $V_c(p) \stackrel{\text{def}}{=} \{\Pi \Rightarrow \Sigma | p \in \Pi\}$.
6. $\sigma_c \stackrel{\text{def}}{=} \{\|\phi\| | \phi \in \Omega\}$

$(\Gamma' \Rightarrow \Delta') \in W_c$ because $(\Gamma' \Rightarrow \Delta') \not\perp_c (\Gamma' \Rightarrow \Delta')$.

Lemma 3. *In a canonical model* $(W_c, \perp_c, \sigma_c, \rightarrow_{P?c}, \rightarrow_{uc}, V_c)$, $(\Pi \Rightarrow \Sigma) \in \|\phi\|$ *iff* $\phi \in \Pi$.

Proof. By simultaneous induction on formulas and actions.

If $\phi = p$, it is obvious from the definition.

If $\phi = \psi \wedge \chi$, $(\Pi \Rightarrow \Sigma) \in \|\phi\|$ iff $(\Pi \Rightarrow \Sigma) \in \|\psi\| \cap \|\chi\|$ iff $(\Pi \Rightarrow \Sigma) \in \|\psi\|$ and $\Pi \Rightarrow \Sigma) \in \|\chi\|$ iff $\psi \in \Pi$ and $\chi \in \Pi$. Therefore, $\psi \wedge \chi \in \Pi$ because $\psi, \chi \in \Pi$ iff $\psi \wedge \chi \in \Pi$ from the unprovability of $\Pi \Rightarrow \Sigma$.

If $\phi = \neg\psi$, $(\Pi \Rightarrow \Sigma) \in \|\neg\psi\|$ iff $(\Pi \Rightarrow \Sigma) \notin \|\psi\|$ iff $\psi \in \Sigma$ iff $\neg\psi \in \Pi$.

If $\phi = \Box\psi$, $(\Pi \Rightarrow \Sigma) \in \|\Box\psi\|$ iff for all $\Pi' \Rightarrow \Sigma'$ such that $(\Pi \Rightarrow \Sigma) \not\perp (\Pi' \Rightarrow \Sigma')$, $(\Pi' \Rightarrow \Sigma') \in \|\psi\|$. The inductive hypothesis gives $\psi \in \Pi'$. For the sake of contradiction, suppose $\Box\psi \in \Sigma$. Then, there exists $(\Pi'' \Rightarrow \Sigma'') \in W_c$ such that, for all $\chi \in \Omega$, if $\Box\chi \in \Pi$, then $\chi \in \Pi''$, if $\chi \in \Sigma$, then $\Box\chi \in \Sigma''$, if $\Box\Box\chi \in \Pi$, then $\Box\Box\chi \in \Pi''$, and $\psi \in \Sigma''$ because if $\Pi'' \Rightarrow \Sigma''$ is provable, then $\Pi \Rightarrow \Sigma$ is also provable by (cut), ($\Box$2), and ($\Box$3). Because $(\Pi \Rightarrow \Sigma) \not\perp (\Pi'' \Rightarrow \Sigma'')$, this is a contradiction. Therefore, $\Box\psi \in \Pi$. The opposite direction follows from the definition of $\not\perp_c$ and the inductive hypothesis.

If $\phi = [\psi?]\chi$, $(\Pi \Rightarrow \Sigma) \in \|[\psi?]\chi\|$ iff for all $\Pi' \Rightarrow \Sigma'$ such that $(\Pi \Rightarrow \Sigma)(\|\psi\|?)(\Pi' \Rightarrow \Sigma')$ and $(\Pi' \Rightarrow \Sigma') \in \|\chi\|$. The inductive hypothesis gives $\chi \in \Pi'$. By the definition of $\rightarrow_{P?c}$, $[\psi?]\chi \in \Pi$. The opposite direction is obvious.

For the case that $\phi = [U]\psi$, the proof is nearly identical to that for $[\psi?]\chi$.

If $\phi = [\psi?^\dagger]\chi$, $[\psi?^\dagger]\chi \in \Pi$ iff $[\psi?]\chi \in \Pi$ because $[\psi?^\dagger]\chi \Rightarrow [\psi?]\chi$ and $[\psi?]\chi \Rightarrow [\psi?^\dagger]\chi$ are provable by using (P†L) and (P†R).

For the case that $\phi = [U^\dagger]\psi$, suppose that $[U^\dagger]\psi \in \Pi$ and $(\Pi \Rightarrow \Sigma)(U^\dagger)(\Pi' \Rightarrow \Sigma')$. From Definition 3, $(\Pi' \Rightarrow \Sigma')(U)(\Pi \Rightarrow \Sigma)$. From the definition of $\rightarrow_{uc}$, $[U][U^\dagger]\psi \in \Pi'$. Then, $\psi \in \Pi'$ because $[U][U^\dagger]\psi \Rightarrow \psi$ is provable as follows:

$$
\dfrac{
\dfrac{
\dfrac{[U^\dagger]\psi \Rightarrow [U^\dagger]\psi}{[U^\dagger]\psi \Rightarrow [U^\dagger]\psi, \bot}\,(\text{w R})
}{[U][U^\dagger]\psi \Rightarrow \psi, [U]\bot}\,(\text{UA2})
\qquad
\dfrac{\dfrac{\vdots}{p \wedge \neg p \Rightarrow}}{[U]\bot \Rightarrow}\,(\text{U})
}{[U][U^\dagger]\psi \Rightarrow \psi}\,(\text{cut})
$$

From the inductive hypothesis, $(\Pi' \Rightarrow \Sigma') \models \psi$. Therefore, $(\Pi \Rightarrow \Sigma) \models [U^\dagger]\psi$. Next, suppose that $[U^\dagger]\psi \in \Sigma$. Then, using the fact that $\psi \Rightarrow [U][U^\dagger]\psi$ can be proved as follows, we can prove $(\Pi \Rightarrow \Sigma) \not\models [U^\dagger]\psi$ in the same manner.

$$
\dfrac{
\dfrac{\vdots}{\psi \Rightarrow \neg\neg\psi}
\qquad
\dfrac{
\dfrac{
\dfrac{
\dfrac{\psi \Rightarrow \psi}{\Rightarrow \neg\psi, \psi}\,(\neg\text{R})
}{\Rightarrow [U^\dagger]\neg\psi, [U^\dagger]\psi}\,(\text{U}^\dagger)
}{\Rightarrow \neg\psi, [U][U^\dagger]\psi}\,(\text{UA2})
}{\neg\neg\psi \Rightarrow [U][U^\dagger]\psi}\,(\neg\ \text{L})
}{\psi \Rightarrow [U][U^\dagger]\psi}\,(\text{cut})
$$

Cases of composite $\dagger$, ;, and $\cup$ are omitted because they are routine or nearly identical to those in DL. $\square$

Lemma 4. *A canonical model* $(W_c, \perp_c, \sigma_c, \rightarrow_{P?c}, \rightarrow_{uc}, V_c)$ *satisfies all conditions of the dynamic quantum frame.*

Proof. The symmetry and the reflexivity of $\not\angle_c$ are confirmed by using ($\Box$1), ($\Box$3), and the traditional method for proving the completeness theorems for modal logic [21,22]. The proofs for the dynamic quantum frame conditions are as follows:

F1 Suppose $(\Pi \Rightarrow \Sigma)(\|\phi\|?)(\Pi' \Rightarrow \Sigma')$. Because $\Box\Box\psi \Rightarrow [\phi?]\Box\Box\psi$ is provable by using the rule (π), if $\Box\Box\psi \in \Pi$, then $[\phi?]\Box\Box\psi \in \Pi$ for all ϕ. Therefore, $\Box\Box\psi \in \Pi'$.
If $\Box\psi \in \Pi$, then $[\phi?]\psi \in \Pi$ for all ϕ because $\Box\psi \Rightarrow [\phi?]\psi$ is provable by using ($\Box$). Therefore, $\psi \in \Pi'$.
For the sake of contradiction, suppose $\psi \in \Sigma$ but $\Box\psi \notin \Sigma'$. Since $\Pi \cup \Sigma = \Omega$, we find that $\neg\Box\psi \in \Sigma'$ because $\neg\Box\psi, \Box\psi \Rightarrow$ is provable. Therefore, $[\phi?]\neg\Box\psi \in \Sigma$. However, $\Rightarrow \psi, [\phi?]\neg\Box\psi$ is provable by using (cut), ($\Box$), and ($\Box$3). This is a contradiction.

F2 For all $[\phi?]\psi$, $[\phi?]\psi \in \Pi$ iff $[\sim\sim \phi?]\psi \in \Pi$ because $[\phi?]\psi \Rightarrow [\sim\sim \phi?]\psi$ and $[\sim\sim \phi?]\psi \Rightarrow [\phi?]\psi$ are provable by using $(\sim\sim L)$ and $(\sim\sim R)$. Therefore, from the definition of $\rightarrow_{P?c}$, $(\|\phi\|?)$ and $(\| \sim\sim \phi\|?)$ are equivalent.

F3 For all $\phi, \psi \in \Omega$, $[\phi?]\psi \in \Pi$ or $[\phi?]\psi \in \Sigma$ because $\Pi \cup \Sigma = \Omega$. Therefore, from the definition of $\rightarrow_{P?c}$, how $[\phi?]\psi$ is included in Π or Σ for all $\psi \in \Omega$ decides the next sequent related by $\phi?$. Therefore, if $(\Pi \Rightarrow \Sigma)(\|\phi\|?)(\Pi' \Rightarrow \Sigma')$ and $(\Pi \Rightarrow \Sigma)(\|\phi\|?)(\Pi'' \Rightarrow \Sigma'')$, then $\Pi' = \Pi''$ and $\Sigma' = \Sigma''$.

F4 Suppose, $X = \|\phi\|$. From Lemma 3, if $(\Pi \Rightarrow \Sigma) \in \|\phi\|$, then $\phi \in \Pi$. Therefore, $(\Pi \Rightarrow \Sigma)(\|\phi\|?)(\Pi \Rightarrow \Sigma)$ because $\phi, \psi \Rightarrow [\phi?]\psi$ and $\phi, [\phi?]\psi \Rightarrow \psi$ are provable by using (P),(PL), and (cut).

F5 Suppose $\|\phi\|$ is $\perp$-closed. From the universal accessibility of this canonical model (which can be proven before this condition), $T(\phi)$ is true at all $x \in W_c$. Then, from Lemma 3, $T(\phi) \in \Pi$ for all $(\Pi \Rightarrow \Sigma) \in W_c$. Then, $[\phi?]\phi \in \Pi$ because $T(\phi) \Rightarrow [\phi?]\phi$ is provable by using (PR). Suppose $(\Pi \Rightarrow \Sigma)(\|\phi\|?)(\Pi' \Rightarrow \Sigma')$. Then, from the definition of $\rightarrow_{P?c}$, $\phi \in \Pi'$. From Lemma 3, $(\Pi' \Rightarrow \Sigma') \in \|\phi\|$.

F6 We only prove the cases for which the relation is $(\|\phi\|?)$. Suppose $(\Pi \Rightarrow \Sigma)(\|\phi\|?)(\Pi' \Rightarrow \Sigma')$ and $(\Pi' \Rightarrow \Sigma') \not\angle_c (\Pi'' \Rightarrow \Sigma'')$. For all $\psi \in \Pi$, $[\phi?]\Box\langle\phi?\rangle\Diamond\psi \in \Pi$ because $\psi \Rightarrow [\phi?]\Box\langle\phi?\rangle\Diamond\psi$ is provable by using (cut), (Ad), ($\mathrm{P}^\dagger$L), ($\mathrm{P}^\dagger$R), and the rules for $\Box$. From the definition of $\not\angle_c$ and $\rightarrow_{P?c}$, $\langle\phi?\rangle\Diamond\psi \in \Pi''$. From Lemma 3 and partial functionality of $\rightarrow_{P?c}$, there exists $\Pi''' \Rightarrow \Sigma'''$ such that $(\Pi'' \Rightarrow \Sigma'')(\|\phi\|?)(\Pi''' \Rightarrow \Sigma''')$ and $\Diamond\psi \in \Pi'''$ for all $\psi \in \Pi$. Then, $(\Pi''' \Rightarrow \Sigma''') \not\angle_c (\Pi \Rightarrow \Sigma)$. That is, for all $\chi \in \Omega$, if $\Box\chi \in \Pi$ then $\chi \in \Pi'''$ because $\Diamond\Box\chi \Rightarrow \chi$ is provable. If $\chi \in \Sigma$(which means $\neg\chi \in \Pi$), then $\Box\chi \in \Sigma'''$ because $\Diamond\neg\chi, \Box\chi \Rightarrow$ is provable. If $\Box\Box\chi \in \Pi$, then $\Box\Box\chi \in \Pi'''$ because of the universal accessibility of this canonical model (which can be proven before this condition) and the definition of $\not\angle_c$. Proofs for the other cases are similar to the cases that have been described. For composite relations, use Lemma 1.

F7 From the definition of $\rightarrow_{uc}$, each $(\Pi \Rightarrow \Sigma) \in W_c$ has at least and at most one sequent $(\Pi' \Rightarrow \Sigma')$ such that $(\Pi \Rightarrow \Sigma)(U)(\Pi' \Rightarrow \Sigma')$ because both Π and Σ can be empty in the rule (U).

F8 From Definition 3, relation $U^\dagger$ is defined as the inverse relation of relation U. Each $(\Pi \Rightarrow \Sigma) \in W_c$ has at least one sequent $\Pi' \Rightarrow \Sigma'$ such that $(\Pi' \Rightarrow \Sigma')(U)(\Pi \Rightarrow \Sigma)$ because $[U]\Pi \Rightarrow [U]\Sigma$ is not provable. That is, if $[U]\Pi \Rightarrow [U]\Sigma$ is provable, then $\Pi \Rightarrow \Sigma$ is probable by the following method. $[U]\Pi \Rightarrow [U]\Sigma, \bot$ is proven by (w R). Then, $[U^\dagger][U]\Pi \Rightarrow \Sigma, [U^\dagger]\bot$ is proven by (UA1). From $\bot \Rightarrow (= p \wedge \neg p \Rightarrow)$, $[U^\dagger]\bot \Rightarrow$ is proven by ($U^\dagger$). By (cut), $[U^\dagger][U]\Pi \Rightarrow \Sigma$ is proven. For each $\psi \in \Pi$, $\psi \Rightarrow [U^\dagger][U]\psi$ is proven by using (cut), ($\neg$ L), ($\neg$ R), (U), and (UA1). Then, by (cut), $\Pi \Rightarrow \Sigma$ is proven.
Furthermore, each $(\Pi \Rightarrow \Sigma) \in W_c$ has at most one sequent $(\Pi' \Rightarrow \Sigma')$ such that $(\Pi' \Rightarrow \Sigma')(U)(\Pi \Rightarrow \Sigma)$. For the sake of contradiction, suppose $(\Pi' \Rightarrow \Sigma')(U)(\Pi \Rightarrow \Sigma)$, $(\Pi'' \Rightarrow \Sigma'')(U)(\Pi \Rightarrow \Sigma)$, $\phi \in \Pi'$, and $\phi \in \Sigma''$. From Lemma 3, $[U^\dagger]\phi \in \Sigma$. Then, from $(\Pi' \Rightarrow \Sigma')(U)(\Pi \Rightarrow \Sigma)$, $[U][U^\dagger]\phi \in \Sigma'$. This is a contradiction because $\phi \Rightarrow [U][U^\dagger]\phi$ is provable by using (UA2).

F9 Suppose $(\Pi \Rightarrow \Sigma)(\pi_1)\ldots(\pi_n)(\Pi' \Rightarrow \Sigma')$. That is, $(\Pi \Rightarrow \Sigma)$ and $(\Pi' \Rightarrow \Sigma')$ are related by the finite sequence of relations $\pi_1, \pi_2, \ldots, \pi_n$. $\{\Box\Box\phi | \Box\Box\phi \in \Pi\} \cup \{\phi | \Box\phi \in \Pi\} \Rightarrow \{\Box\phi | \phi \in \Sigma\}$ is not provable because if it is provable, then $\Pi \Rightarrow \Sigma$ is provable by using (w R), (cut), ($\Box$2), ($\Box$3) and provability of $\Box\bot \Rightarrow$ by ($\Box$1). This sequent is represented by $\Pi'' \Rightarrow \Sigma''$.
$\Pi'' \Rightarrow \Sigma'' \cup \{\Box\phi | \phi \in \Sigma'\}$ is also not provable by the following reason. For the sake of contradiction, suppose $\Pi'' \Rightarrow \Sigma'' \cup \{\Box\phi | \phi \in \Sigma'\}$ is probable. Then, there exists $\{\Box\phi_1, \ldots, \Box\phi_m\} \subseteq \{\Box\phi | \phi \in \Sigma'\}$ such that $\Pi'' \Rightarrow \Sigma'' \cup \{\Box\phi_1, \ldots, \Box\phi_m\}$ is provable. Because $\Box\phi_1 \wedge \ldots \wedge \Box\phi_m \Rightarrow \Box(\phi_1 \wedge \ldots \wedge \phi_m)$ is provable, $\Pi'' \Rightarrow \Sigma'', \Box(\phi_1 \wedge \ldots \wedge \phi_m)$ is also provable by using ($\wedge$R) and (cut). $\Pi \Rightarrow \Sigma, \Box\Box(\phi_1 \wedge \ldots \wedge \phi_m)$ is provable from $\Pi'' \Rightarrow \Sigma'', \Box(\phi_1 \wedge \ldots \wedge \phi_m)$ because of ($\Box$2), ($\Box$3), (cut), and the construction method of $\Pi'' \Rightarrow \Sigma''$. Then, $\Pi \Rightarrow \Sigma, [\pi_1]\ldots[\pi_n](\phi_1 \wedge \ldots \wedge \phi_m)$ is provable by (cut) and (π) (which deduce $\Box\Box(\phi_1 \wedge \ldots \wedge \phi_m) \Rightarrow [\pi_1]\ldots[\pi_n](\phi_1 \wedge \ldots \wedge \phi_m)$). However, from $(\Pi \Rightarrow \Sigma)(\pi_1)\ldots(\pi_n)(\Pi' \Rightarrow \Sigma')$ and Lemma 3, $[\pi_1]\ldots[\pi_n](\phi_1 \wedge \ldots \wedge \phi_m)$ is already included in Σ, which means $\Pi \Rightarrow \Sigma$ is provable. This a contradiction.
Then, there exist at least one unprovable infinite sequent $\Pi''_M \Rightarrow \Sigma''_M$ such that $\Pi'' \subseteq \Pi''_M$, $\Sigma'' \cup \{\Box\phi | \phi \in \Sigma'\} \subseteq \Sigma''_M$, and $\Pi''_M \cup \Sigma''_M = \Omega$. $(\Pi \Rightarrow \Sigma) \not\perp_c (\Pi''_M \Rightarrow \Sigma''_M)$ is clear from the way $\Pi'' \Rightarrow \Sigma''$ is constructed. For all $\Box\phi \in \Pi''_M$, $\phi \in \Pi'$ because if $\phi \in \Sigma'$, from $\Sigma'' \cup \{\Box\phi | \phi \in \Sigma'\} \subseteq \Sigma''_M$, $\Box\phi$ is included in both Π''_M and Σ''_M. For all $\phi \in \Sigma''_M$, $\Box\phi \in \Sigma'$ because $\neg\Box\neg\Box\phi \Rightarrow \phi$ is provable by ($\Box$3), $\neg\Box\neg\Box\phi \in \Sigma''_M$, $\Box\neg\Box\phi \in \Pi''_M$, and $\neg\Box\phi \in \Pi'$. Therefore, $(\Pi''_M \Rightarrow \Sigma''_M) \not\perp_c (\Pi' \Rightarrow \Sigma')$. □

Theorem 3 (Completeness). *If $\Gamma \Rightarrow \Delta$ is valid in all DQM, then $\Gamma \Rightarrow \Delta$ is provable in* **SDQL**.

Proof. Suppose $\Gamma \Rightarrow \Delta$ is not provable. Then, a canonical model for $\Gamma \Rightarrow \Delta$ can be constructed. By Lemma 3, $\Gamma \Rightarrow \Delta$ is false at $(\Gamma' \Rightarrow \Delta') \in W_c$. □

4 Hilbert-Style System and Sequent Calculus

A Hilbert-style system **HDQL** for DQL with reference to the systems in [1,3,5] is given below. References [1,3,5] each discuss Hilbert-style systems that differ subtly from one another. **HDQL** is based on those common parts, with a few minor modifications.

Basic Axioms and Rules
All the axioms and rules of modal logic **B** (traditional modal logic which is sound and complete with respect to reflexive and symmetrical relational frames) for $\Box$.
All the axioms and rules of DL (regarding $\mathcal{U}$ as the basic programs), except for the axioms of the conditions for test $[\phi?]$ and Kleene star.
Axioms and rules for conditions
(Testability) $\Box\phi \rightarrow [\psi?]\phi$
(Partial functionality) $\neg[\phi?]\psi \rightarrow [\phi?]\neg\psi$
(Adequacy) $\phi \wedge \psi \rightarrow \langle\phi?\rangle\psi$
(Repeatability) $T(\phi) \rightarrow [\phi?]\phi$
(Unitary functionality) $\neg[U]\phi \leftrightarrow [U]\neg\phi$
(Unitary bijectivity 1) $\phi \leftrightarrow [U; U^\dagger]\phi$
(Unitary bijectivity 2) $\phi \leftrightarrow [U^\dagger; U]\phi$
(Adjointness) $\phi \rightarrow [\pi]\Box\langle\pi^\dagger\rangle\Diamond\phi$
(Self adjointness) $[\phi?^\dagger]\psi \leftrightarrow [\phi?]\psi$
(Proper superpositions) $\langle\pi\rangle\Box\Box\phi \rightarrow [\pi']\phi$ ($\langle\pi\rangle$, $[\pi']$, or both could be empty)
(Proper superposition for $\Box$) $\langle\pi\rangle\Box\Box\phi \rightarrow \Box\phi$ ($\langle\pi\rangle$ could be empty)
Axioms for †
$[(\pi_1; \pi_2)^\dagger]\phi \leftrightarrow [\pi_2^\dagger; \pi_1^\dagger]\phi$
$[(\pi_1 \cup \pi_2)^\dagger]\phi \leftrightarrow [\pi_1^\dagger \cup \pi_2^\dagger]\phi$
$[\pi^{\dagger\dagger}]\phi \leftrightarrow [\pi]\phi$

Lemma 5. *Let ϕ be an axiom in the above list. Then, $\Rightarrow \phi$ is provable in* **SDQL**.

Proof. In the following proofs, some parts that can be deduced by classical logic are used without explanation.

(Repeatability)

$$\dfrac{\Rightarrow \psi \vee \neg\psi \qquad \dfrac{\dfrac{\dfrac{\phi \Rightarrow \phi}{T(\phi), \phi \Rightarrow \phi}\,(\text{w L})}{T(\phi), \phi \wedge (\sim\phi \sqcup (\psi \vee \neg\psi)) \Rightarrow \phi}\,(\wedge\ \text{L})}{T(\phi), \psi \vee \neg\psi \Rightarrow [\phi?]\phi}\,(\text{PR})}{\dfrac{T(\phi) \Rightarrow [\phi?]\phi}{\Rightarrow T(\phi) \rightarrow [\phi?]\phi}}\,(\text{cut})$$

(Proper superpositions)

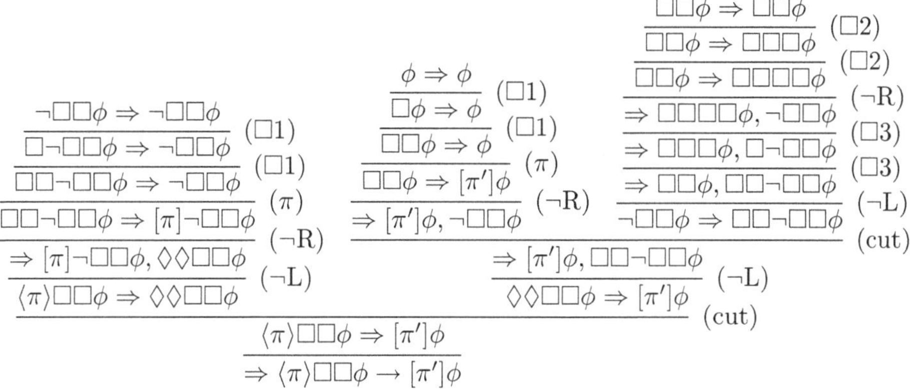

For the other cases, it is routine to prove by referring to these proofs or appendix and using the rule related to the condition. □

The *interpretation* τ of a sequent to a formula is defined as follows:

$$\tau(\Gamma \Rightarrow \Delta) = \bigwedge \Gamma \to \bigvee \Delta.$$

where $\bigwedge \Gamma$ is the conjunction of all formulas in Γ, $\bigvee \Delta$ is the disjunction of all formulas in Δ. Especially, $\tau(\Rightarrow \phi) = \phi$.

Lemma 6. *If $\Gamma \Rightarrow \Delta$ is provable in* **SDQL**, *then $\tau(\Gamma \Rightarrow \Delta)$ is provable in* **HDQL**.

Proof. We prove this by induction on the proof in **SDQL**. In the following proof, some explanations for deductions in classical logic and basic modal logic are omitted.

Suppose the last rule is (PR) and $\tau(T(\phi), <\phi?> \{\chi, \xi\} \Rightarrow \psi) = T(\phi) \wedge (\phi \wedge (\sim \phi \sqcup \chi)) \wedge (\phi \wedge (\sim \phi \sqcup \xi)) \to \psi$ is provable in **HDQL**. (Here the number of elements in Γ is assumed to be two, but the proof can be done in the general case as well). Then $T(\phi) \wedge \chi \wedge \xi \to [\phi?]\psi$ is proved as follows:

1. (Repeatability) $T(\phi) \to [\phi?]\phi$.
2. (Adjointness) and (Self adjointness) $(\chi \wedge \xi) \to [\phi?]\Box\langle\phi?\rangle\Diamond(\chi \wedge \xi)$.
   $(= (\chi \wedge \xi) \to [\phi?] \sim [\phi?] \sim (\chi \wedge \xi))$
3. (Partial functionality) $\langle\phi?\rangle \sim (\chi \wedge \xi) \to [\phi?] \sim (\chi \wedge \xi)$.
4. (Adequacy) $\phi\wedge \sim (\chi \wedge \xi) \to \langle\phi?\rangle \sim (\chi \wedge \xi)$.
5. From 3. and 4., $\phi\wedge \sim (\chi \wedge \xi) \to [\phi?] \sim (\chi \wedge \xi)$.
6. From 5., $\sim [\phi?] \sim (\chi \wedge \xi) \to\sim (\phi\wedge \sim (\chi \wedge \xi))$.
7. From 1., 2., and 6., $T(\phi) \wedge \chi \wedge \xi \to [\phi?](\phi\wedge \sim (\phi\wedge \sim (\chi \wedge \xi)))$.
8. From **B**, $\phi\wedge \sim (\phi\wedge \sim (\chi \wedge \xi)) \to (\phi \wedge (\sim \phi \sqcup \chi)) \wedge (\phi \wedge (\sim \phi \sqcup \xi))$.
9. From 8, $[\phi?](\phi\wedge \sim (\phi\wedge \sim (\chi \wedge \xi))) \to [\phi?]((\phi \wedge (\sim \phi \sqcup \chi)) \wedge (\phi \wedge (\sim \phi \sqcup \xi)))$.
10. From the inductive hypothesis, $[\phi?]T(\phi) \wedge [\phi?]((\phi \wedge (\sim \phi \sqcup \chi)) \wedge (\phi \wedge (\sim \phi \sqcup \xi))) \to [\phi?]\psi$.

11. From (Proper superpositions) and (Proper superpositions for $\Box$), $T(\phi) \to [\phi?]T(\phi)$.
12. From 7., 9., 10., and 11., $T(\phi) \wedge \chi \wedge \xi \to [\phi?]\psi$.

Suppose the last rule is (Ad) and $\tau(\chi, \xi \Rightarrow [\pi]\Box\theta, [\pi]\Box\eta, \phi) = \chi \wedge \xi \to [\pi]\Box\theta \vee [\pi]\Box\eta \vee \phi$ is provable in **HDQL**. Then $[\pi^\dagger]\Box\chi \wedge [\pi^\dagger]\Box\xi \to \theta \vee \eta \vee [\pi^\dagger]\Box\phi$ is proved as follows:

1. (Adjointness) $\neg\theta \to [\pi^\dagger]\Box\neg[\pi^{\dagger\dagger}]\Box\theta$.
2. From 1. and the axiom for actions, $\neg\theta \to [\pi^\dagger]\Box\neg[\pi]\Box\theta$.
3. (Adjointness) $\neg\eta \to [\pi^\dagger]\Box\neg[\pi^{\dagger\dagger}]\Box\eta$.
4. From 3. and the axiom for actions, $\neg\eta \to [\pi^\dagger]\Box\neg[\pi]\Box\eta$.
5. From the inductive hypothesis, $\neg[\pi]\Box\theta \wedge \neg[\pi]\Box\eta \wedge \chi \wedge \xi \to \phi$.
6. From 5., $\Box\neg[\pi]\Box\theta \wedge \Box\neg[\pi]\Box\eta \wedge \Box\chi \wedge \Box\xi \to \Box\phi$.
7. From 6., $[\pi^\dagger]\Box\neg[\pi]\Box\theta \wedge [\pi^\dagger]\Box\neg[\pi]\Box\eta \wedge [\pi^\dagger]\Box\chi \wedge [\pi^\dagger]\Box\xi \to [\pi^\dagger]\Box\phi$.
8. From 2., 4., and 7, $\neg\theta \wedge \neg\eta \wedge [\pi^\dagger]\Box\chi \wedge [\pi^\dagger]\Box\xi \to [\pi^\dagger]\Box\phi$.
9. From 8, $[\pi^\dagger]\Box\chi \wedge [\pi^\dagger]\Box\xi \to \theta \vee \eta \vee [\pi^\dagger]\Box\phi$.

For the other cases, it is routine to prove by referring to these proofs or appendix and using the basic properties of Hilbert-style system and sequent system of modal logic. $\Box$

Theorem 4. *The following three are equivalent.*

1. *$\Rightarrow \phi$ is provable in* **SDQL**.
2. *ϕ is provable in* **HDQL**.
3. *ϕ is valid in all DQM.*

Proof. 1. $\Leftrightarrow$ 2. Corollary of Lemma 5, Lemma 6, and basic properties of Hilbert-style system and sequent system of modal logic.
1. $\Leftrightarrow$ 3. Corollary of Theorem 2 and Theorem 3. $\Box$

Theorem 5 (Soundness and completeness for HDQL). *ϕ is valid in all DQM iff ϕ is provable in* **HDQL**.

Proof. Corollary of Theorem 4. $\Box$

5 Conclusion and Future Work

We construct a sequent calculus for DQL, by using both methods for a sequent calculus for DL and a sequent calculus for OML. The cut-elimination theorem and the *subformula property* do not hold in **SDQL**. As is generally known, if a model includes the symmetric relation, it is difficult to construct a sequent calculus that satisfies the cut-elimination theorem. As in the case of traditional modal logic, we cannot derive the sequent $p \Rightarrow \Box\Diamond p$ without (cut) in **SDQL**.

Hypersequent or *labeled sequent* systems [19,21,22,26] may solve this problem. The *tree-hypersequents calculus* for DL is constructed in [14]. However, this type of sequent calculus is not compatible with conditions for $\perp$-closed sets. That is, because $\perp$-closedness is the property of the whole frame (not a relation or single possible world), it is difficult to express it with a hyper sequent representing only part of the model, even if we use $T(\phi)$. This remains an open problem.

Acknowledgments. This work was supported by JSPS KAKENHI Grant Number JP20K19740.

Appendix: Correspondence Between Conditions, Axioms, and Rules

Suppose (W, R, V) is a Kripke model for traditional modal logic.
1. For all $x, y, z \in W$, if $(x, y) \in R$ and $(x, z) \in R$, then $y = z$.

$$\neg\Box\phi \to \Box\neg\phi \qquad \frac{\Gamma \Rightarrow \Delta}{\Box\Gamma \Rightarrow \Box\Delta}\ (\text{N1})$$

2. For all $x \in W$, there exists $y \in W$ such that $(x, y) \in R$.

$$\neg\Box\perp \qquad \frac{\Gamma \Rightarrow}{\Box\Gamma \Rightarrow}\ (\text{NE})$$

For the following conditions, a model is assumed to have multiple relations $R_1, R_2, R_3, \ldots$ We write $x(n)y$ if $x, y \in W$ have an n-th relationship.
3. For all $x, y \in W$, if there exist $x_1(= x), x_2, \ldots, x_n \in W$ and $\alpha_1, \alpha_2, \ldots, \alpha_n \in \{1, 2, , \ldots, n\}$ such that $x_1(\alpha_1)x_2, x_2(\alpha_2)x_3, \ldots, x_n(\alpha_n)y$, then there exists $z \in W$ such that $x(1)z, z(1)y$.

$$[1][1]\phi \to [\alpha][\beta]\ldots[\gamma]\phi \qquad \frac{[1][1]\Gamma \Rightarrow \phi}{[1][1]\Gamma \Rightarrow [\alpha]\phi}\ (\text{WT2})$$

where $\alpha, \beta, \ldots, \gamma$ is an arbitrary finite sequence of any relations contained in $\{1, 2, \ldots, n\}$. In (WT2), $\alpha \in \{1, 2, \ldots, n\}$. This condition is regarded as a weak version of the transitive condition. Intuitively, a transitive relation is interpreted as "If two elements of a model are connected by two steps, then these two elements are connected by one step." This condition is interpreted as "If two elements of a model are connected by some steps by arbitrary number relations, then these two elements are connected by a relation (1) within two steps."
4. If $x_1(\alpha_1)x_2, x_2(\alpha_2)x_3, \ldots, x_n(\alpha_n)x_{n+1}$, there exist $x_{n+2}, x_{n+3}, \ldots, x_{n+m} \in W$ such that $x_{n+1}(\alpha_{n+1})x_{n+2}, x_{n+2}(\alpha_{n+2})x_{n+3}, \ldots, x_{n+m}(\alpha_{n+m})x_1$.

$$\phi \to [\alpha_1][\alpha_2][\alpha_3]\langle\alpha_4\rangle\langle\alpha_5\rangle\langle\alpha_6\rangle\langle\alpha_7\rangle\phi \qquad \frac{\Gamma \Rightarrow [\alpha_4][\alpha_5][\alpha_6][\alpha_7]\Delta, \phi}{[\alpha_1][\alpha_2][\alpha_3]\Gamma \Rightarrow \Delta, [\alpha_1][\alpha_2][\alpha_3]\phi}\ (\text{GS})$$

($n = 3$ and $m = 4$ are assumed here.)

This condition can be regarded as a generalization for a symmetric relation condition. That is, if $n = m = 1$ and the language has only one modality symbol, this condition is identical to a condition for symmetric relation.

For details about frame conditions, see, for example, [7,11,20–22]. The following theorems are well known.

1. **LK** + (N1) is sound and complete with respect to Kripke models that satisfy condition 1.
2. **LK** + (NE) is sound and complete with respect to Kripke models that satisfy condition 2.
3. **LK** + (WT2) is sound and complete with respect to Kripke models that satisfy condition 3.
4. **LK** + (GS) is sound and complete with respect to Kripke models that satisfy condition 4.

References

1. Baltag, A., Smets, S.: The logic of quantum programs. QPL **2004**, 39–56 (2004)
2. Baltag, A., Smets, S.: Complete axiomatizations for quantum actions. Int. J. Theor. Phys. **44**(12), 2267–2282 (2005)
3. Baltag, A., Smets, S.: LQP: the dynamic logic of quantum information. Math. Struct. Comput. Sci. **16**(3), 491–525 (2006)
4. Baltag, A., Smets, S.: Quantum logic as a dynamic logic. Synthese **179**, 285–306 (2011)
5. Baltag, A., Smets, S.: The dynamic turn in quantum logic. Synthese **186**(3), 753–773 (2012)
6. Birkhoff, G., Von Neumann, J.: The logic of quantum mechanics. Ann. Math. **37**(4), 823–843 (1936)
7. Blackburn, P., de Rijke, M., Venema, Y.: Modal Logic. Cambridge University Press (2001)
8. Chiara, M.L.D., Giuntini, R.: Quantum logics. In: Gabbay, D.M., Guenthner, F. (ed.) Handbook Of Philosophical Logic, 2nd edn, vol. 6, no. 1, pp. 129–228 (2002)
9. Faggian, C., Sambin, G.: From basic logic to quantum logics with cut-elimination. Int. J. Theor. Phys. **37**(1), 31–37 (1998)
10. Fazio, D., Ledda, A., Paoli, F., John, G.S.: A substructural gentzen calculus for orthomodular quantum logic. Rev. Symb. Logic **16**(4), 1177–1198 (2023)
11. Fitting, M.: Cut-Free proof systems for Geach logics. IfCoLog J. Logics Appl. **2**(2), 17–64 (2015)
12. Hardegree, G.M.: Material implication in orthomodular (and boolean) lattices. Notre Dame J. Formal Logic **22**(2), 163–182 (1981)
13. Harel, D., Tiuryn, J., Kozen, D.: Dynamic Logic. MIT Press, Cambridge (2000)
14. Hill, B., Poggiolesi, F.: A contraction-free and cut-free sequent calculus for propositional dynamic logic. Stud. Logica **94**(1), 47–72 (2010)
15. Kawano, T.: Advanced kripke frame for quantum logic. In: Proceedings of WoLLIC, International Workshop on Logic, Language, Information, and Computation, pp. 237–249 (2018)

16. Kawano, T.: Labeled Sequent Calculus for Orthologic. To appear in Bulletin of the Section of Logic
17. Kawano, T.: Sequent calculi for orthologic with strict implication. Bull. Sect. Logic **51**(1), 73–89 (2022)
18. Kornell, A.: A natural deduction system for orthomodular logic. Rev. Symb. Logic **17**(3), 910–949 (2024)
19. Lellmann, B., Poggiolesi, F.: Nested sequent or tree-hypersequents: a survey. Saul Kripke on Modal logic (2023)
20. Lemmon, E.J., Scott, D.S.: An Introduction to Modal Logic: the Lemmon notes. B. Blackwell, Oxford (1977)
21. Negri, S.: Proof analysis in modal logic. J. Philos. Log. **34**, 507–544 (2005)
22. Negri, S.: Proof theory for modal logic. Philos Compass **6**(8), 523–538 (2011)
23. Nishimura, H.: Sequential method in quantum logic. J. Symb. Logic **45**(2), 339–352 (1980)
24. Nishimura, H.: Proof theory for minimal quantum logic I. Int. J. Theor. Phys. **33**(1), 103–113 (1994)
25. Nishimura, H.: Proof theory for minimal quantum logic II. Int. J. Theor. Phys. **33**(7), 1427–1443 (1994)
26. Poggiolesi, F.: Gentzen Calculi for Modal Propositional Logic. Springer (2010)
27. Rents, I.: Gentzen Type Sequent Calculus for Propositional Dynamic Logic. Institute of Computational Technologies, Russian Academy of Sciences, Technical Report (1998)

Satisfaction-Based Probabilistic Hoare Logic with While Loop: Weakest Precondition and Relative Completeness

Xin Sun[1(✉)], Xingchi Su[2], Xiaoning Bian[3], and Anran Cui[4]

[1] School of Mathematics and Statistics, Taishan University, Taian, China
xin.sun.logic@gmail.com
[2] Department of philosophy, Zhejiang University, Hangzhou, China
[3] Department of Computer Science and Technology, Tsinghua University, Beijing, China
[4] Faculty of Software Engineering, East China Normal University, Shanghai, China

Abstract. We establish the weakest precondition calculus for satisfaction-based probabilistic Hoare logic. In particular, we construct the weakest precondition for the While command. We use our weakest precondition calculus to solve the relative completeness problem of satisfaction-based probabilistic Hoare logic, which has been open since 1979.

Keywords: weakest precondition · Hoare logic · probabilistic program

1 Introduction

Hoare logic provides a formalization with logical rules for reasoning about the correctness of programs. It was originally designed by C. A. R. Hoare in 1969 in his seminal paper [15]. The underpinning idea captures the precondition and postcondition of executing a certain program. The precondition describes the property on which the command is based. The postcondition describes the property to which the command must lead after each correct execution. Hoare logic has become one of the most influential tools in formal verification of programs in the past decades. It has been successfully applied in the analysis of deterministic [15,16,26], non-deterministic [1,11,12], recursive [2,13,14], probabilistic [6,9,21,22] and quantum programs [10,18,25,27,28]. A comprehensive review of Hoare logic is referred to Apt, Boer, and Olderog [2,3].

Probabilistic Hoare logic (PHL) [6,9,17,20–22] is an extension of Hoare logic. It introduces probabilistic commands to handle programs with randomized behavior, providing tools to derive probabilistic assertions that guarantee that a program fulfills its intended behavior with certain probability. Ramshaw [21] developed the first PHL using a truth-functional assertion language. This type of PHL is called satisfaction-based PHL within the Hoare logic community. One limitation of Ramshaw's PHL is that it is not relatively complete and may not be able to prove some simple valid assertions. Relative completeness of Hoare logic means that if we had an oracle capable of checking the validity of assertions, then we could prove all valid Hoare triples. This notion was first

J. Wang et al. (Eds.): DaLí 2025, LNCS 16472, pp. 58–78, 2026.
https://doi.org/10.1007/978-3-032-22626-6_4

introduced by Cook in [7] and is now the most widely accepted notion of the completeness of Hoare logic. To avoid the relative incompleteness problem of satisfaction-based PHL, expectation-based PHL was introduced in a series of works [17,19,20]. This approach employs arithmetical assertions instead of truth-functional assertions. Although the recent work of Batz *et al.* [5] has proven that expectation-based PHL with the While loop is relatively complete, the work to date [6,8,9,22] has not proven the relative completeness of any satisfaction-based PHL with the While loop.

In Hoare logic, the weakest precondition calculus [11] plays an important role in program verification. By calculating the weakest precondition, we ensure that all necessary conditions are met before executing a program, thereby ensuring that the program behaves as intended. The weakest precondition calculus also plays a crucial role in proving the relative completeness of Hoare logic. In this work, we establish the weakest precondition calculus for satisfaction-based PHL. In particular, we construct the weakest precondition for the While command. We then use our weakest precondition calculus to solve the relative completeness problem of satisfaction-based PHL.

Our methodology is the following: We first establish the weakest precondition of probabilistic programs with deterministic assertions. Then we use it to construct the weakest preterm calculus of probabilistic expressions. Finally, we use the weakest preterm calculus to build the weakest precondition calculus of probabilistic assertions. The relative completeness of PHL is then obtained as a consequence of the weakest precondition calculus.

The structure of this paper is as follows. We first introduce our PHL with deterministic assertions in Sect. 2, in which we define the denotational semantics of deterministic assertions and construct its weakest precondition calculus. Then, Sect. 3 introduces the proof system for probabilistic assertions based on the weakest preconditions and proves that it is relatively complete. We conclude this paper with future work in Sect. 4. Due to the limitation of space, we put complicated proofs in the appendix.

2 Probabilistic Hoare Logic with Deterministic Assertion

Hoare logic is a formal system that reasons about *Hoare triples* of the form $\{\phi\}C\{\psi\}$. The intuitive reading of $\{\phi\}C\{\psi\}$ is the following: whenever the precondition ϕ is in the state before the execution of the command C, then after the execution and termination of C the postcondition ψ will hold. These assertions ϕ, ψ, also known as formulas, are based on deterministic and probabilistic expressions and will be defined in this section and in the following. The commands C are based on classical program statements such as assignment, conditional choice, while loop, etc.

2.1 Deterministic Expression and Assertion

Let $\mathbb{PV} = \{X, Y, Z, \ldots\}$ be a set of program variables denoted by capital letters. Let $\mathbb{LV} = \{x, y, z, \ldots\}$ be a set of logical variables. We assume that $\mathbb{LV}$ and $\mathbb{PV}$ are disjoint. Program variables are those variables that may occur in both programs and assertions, while logical variables may occur only in assertions. The program variables constitute deterministic expressions. Deterministic expressions are classified into arithmetic expression E and Boolean expression B. The arithmetic expression consists of integer constant $n \in \mathbb{Z}$ and variables from $\mathbb{PV}$. It also involves arithmetic operators between these components. The set of arithmetic operators (aop) is defined as $\{+,-,\times\} \subseteq \mathbb{Z} \times \mathbb{Z} \to \mathbb{Z}$.

Definition 1 (Arithmetic expressions). *Given a set of program variables $\mathbb{PV}$, we define the arithmetic expression E as follows:*

$$E := n \mid X \mid (E \; aop \; E).$$

The set of Boolean constants is $\mathbb{B} = \{\top, \bot\}$. We define relational operators (rop) to be performed on arithmetic expressions including $\{>,<,\geq,=,\leq\} \subseteq \mathbb{Z} \times \mathbb{Z} \to \mathbb{B}$.

Definition 2 (Boolean expressions). *The Boolean expression is defined as follows:*

$$B := \top \mid \bot \mid (E \; rop \; E) \mid \neg B \mid (B \wedge B).$$

The semantics of deterministic expressions is defined on deterministic states S which are denoted as mappings $S : \mathbb{PV} \to \mathbb{Z}$. Let $\mathbb{S}$ be the set of all deterministic states. Each state $S \in \mathbb{S}$ is a description of the value of all program variables. Consequently, the semantics of the arithmetic expressions is $[\![E]\!] : \mathbb{S} \to \mathbb{Z}$, which maps each deterministic state to an integer. Analogously, the semantics of Boolean expressions is $[\![B]\!] : \mathbb{S} \to \mathbb{B}$ that maps each state to a Boolean value.

Definition 3. *The semantics of arithmetic and Boolean expressions are defined inductively as follows:*

$$\begin{aligned}
[\![X]\!]S &= S(X) \\
[\![n]\!]S &= n \\
[\![E_1 \; aop \; E_2]\!]S &= [\![E_1]\!]S \; aop \; [\![E_2]\!]S \\
[\![\top]\!]S &= \top \\
[\![\bot]\!]S &= \bot \\
[\![E_1 \; rop \; E_2]\!]S &= [\![E_1]\!]S \; rop \; [\![E_2]\!]S \\
[\![\neg B]\!]S &= \neg [\![B]\!]S \\
[\![B_1 \wedge B_2]\!]S &= [\![B_1]\!]S \wedge [\![B_2]\!]S
\end{aligned}$$

Definition 4 (Syntax of deterministic assertion). *The deterministic assertions are defined by the following BNF:*

$$\phi := \top \mid \bot \mid (e \; rop \; e) \mid \neg\phi \mid (\phi \wedge \phi) \mid \forall x \phi$$

where e represents arithmetic expression build on $\mathbb{LV} \cup \mathbb{PV}$:

$$e := n \mid X \mid x \mid (e \; aop \; e).$$

The formula $\forall x\phi$ applies the universal quantifier to the logical variable x in formula ϕ.

An interpretation $I : \mathbb{LV} \mapsto \mathbb{Z}$ is a function that maps logical variables to integers. Given an interpretation I and a deterministic state S, the semantics of e is defined as follows.

$$\begin{aligned}
[\![n]\!]^I S &= n \\
[\![X]\!]^I S &= S(X) \\
[\![x]\!]^I S &= I(x) \\
[\![E_1 \ aop \ E_2]\!]^I S &= [\![E_1]\!]^I S \ aop \ [\![E_2]\!]^I S
\end{aligned}$$

The semantics of a deterministic assertion is denoted by $[\![\phi]\!]^I = \{S \mid S \models^I \phi\}$ which represents the set of all states that satisfy ϕ.

Definition 5 (Semantics of deterministic assertion). *The semantics of deterministic assertions is defined inductively as follows:*

$$\begin{array}{lcl}
[\![\top]\!]^I & = & \mathbb{S} \\
[\![\bot]\!]^I & = & \emptyset \\
[\![e_1 \ rop \ e_2]\!]^I & = & \{S \in \mathbb{S} \mid [\![e_1]\!]^I S \ rop \ [\![e_2]\!]^I S = \top\} \\
[\![\neg\phi]\!]^I & = & \mathbb{S} \backslash [\![\phi]\!]^I \\
[\![\phi_1 \wedge \phi_2]\!]^I & = & [\![\phi_1]\!]^I \cap [\![\phi_2]\!]^I \\
[\![\forall x\phi]\!]^I & = & \{S \mid for\ all\ integer\ n\ and\ I' = I[x \mapsto n],\ S \models^{I'} \phi\}
\end{array}$$

2.2 Commands

Commands are actions that we perform on program states. They change a deterministic state to a probabilistic distribution of deterministic states.

Definition 6 (Syntax of commands $\mathscr{L}_{PA}$). *The commands are defined inductively as follows:*

$$C := \texttt{skip} \mid X \leftarrow E \mid X \overset{\$}{\leftarrow} R \mid C_1; C_2 \mid \texttt{if}\, B \ \texttt{then}\, C_1 \ \texttt{else}\, C_2 \mid \texttt{while}\, B \ \texttt{do}\, C$$

where $R = \{a_1 : k_1, \cdots, a_n : k_n\}$ in which $\{k_1, \cdots, k_n\}$ is a set of integers and $a_1, \ldots, a_n$ are rational numbers such that $0 < a_i < 1$ and $a_1 + \ldots + a_n = 1$. B is a Boolean expression.

The command `skip` does nothing. $X \leftarrow E$ is the deterministic assignment. $X \overset{\$}{\leftarrow} R$ can be read as a value k_i is chosen with probability a_i and is assigned to X. $C_1; C_2$ is the sequential composition of C_1 and C_2 as usual. The last two expressions are the conditional choice and loop, respectively.

The semantics of commands are defined on probabilistic states. It shows how different commands *update* probabilistic states. A probabilistic state, denoted by μ, is a probability sub-distribution on deterministic states, which means that each $\mu : \mathbb{S} \rightarrow [0, 1]$ requires that $\Sigma_{S \in \mathbb{S}} \mu(S) \leq 1$. Let $D(\mathbb{S})$ refer to the set of probability sub-distributions on

$\mathbb{S}$. We use sub-distributions to take into account the situations where some programs may never terminate on certain states. For a deterministic state $S \in \mathbb{S}$, μ_S is a special probabilistic state that assigns the value of 1 to S and the value of 0 to any other state. A deterministic state S is a support for μ if $\mu(S) > 0$. The set of all supports of μ is denoted by $sp(\mu)$. Addition and scalar multiplication of sub-distributions means that they are added or scalar multiplied point-wisely.

Definition 7 (Semantics of command expressions). *The semantics of commands is a function* $[\![C]\!] : D(\mathbb{S}) \to D(\mathbb{S})$. *It is defined inductively as follows:*

- $[\![\texttt{skip}]\!](\mu) = \mu$
- $[\![X \leftarrow E]\!](\mu) = \sum_{S \in \mathbb{S}} \mu(S) \cdot \mu_{S[X \mapsto [\![E]\!]S]}$
- $[\![X \xleftarrow{\$} \{a_1 : k_1, ..., a_n : k_n\}]\!](\mu) = \sum_{i=1}^{n} a_i [\![X \leftarrow k_i]\!](\mu)$
- $[\![C_1;\, C_2]\!](\mu) = [\![C_2]\!]([\![C_1]\!](\mu))$
- $[\![\texttt{if}\, B\, \texttt{then}\, C_1\, \texttt{else}\, C_2]\!](\mu) = [\![C_1]\!](\downarrow_B(\mu)) + [\![C_2]\!](\downarrow_{\neg B}(\mu))$
- $[\![\texttt{while}\, B\, \texttt{do}\, C]\!](\mu) = \sum_{i=0}^{\infty} \downarrow_{\neg B}(([\![C]\!] \circ \downarrow_B)^i(\mu))$

We use $S[X \mapsto [\![E]\!]S]$ to denote the state that assigns variables the same values as S except that the variable X is assigned the value $[\![E]\!]S$. We use $\downarrow_B(\mu)$ to denote the distribution μ restricted to those states where B is true. Formally, $\downarrow_B(\mu) = \nu$ with $\nu(S) = \mu(S)$ if $[\![B]\!]S = \top$ and $\nu(S) = 0$ otherwise. We sometimes use $[\![C]\!]S$ to denote $[\![C]\!]\mu_S$. In general, if $\mu = [\![C]\!]S$, $S' \in sp(\mu)$ and $\mu(S') = a$. Then it means that the execution of command C from state S will end in state S' with probability a.

2.3 Weakest Precondition of Deterministic Assertions

A deterministic assertion describes some property of deterministic states. But how do we evaluate a deterministic assertion on probabilistic states? The semantics is given as follows:

$$\mu \models^I \phi \text{ iff for each support } S \text{ of } \mu, S \in [\![\phi]\!]^I.$$

We call it possibility semantics because the definition intuitively means that ϕ is true on a probabilistic state if and only if ϕ is true on all possible deterministic states indicated by the probabilistic state. A Hoare triple, written as $\{\phi\}C\{\psi\}$, is considered valid if, for every deterministic state that satisfies ϕ, executing command C results in a probabilistic state that satisfies ψ. Formally,

$$\models \{\phi\}C\{\psi\} \text{ iff for all interpretation } I \text{ and deterministic state } S,$$
$$\text{if } S \models^I \phi, \text{ then } [\![C]\!](\mu_S) \models^I \psi.$$

Let $wp(C,\psi)$ denote the weakest precondition of a command C and a postcondition ψ. Intuitively, $[\![wp(C,\psi)]\!]$ is the largest set of states starting from which if a program C is executed, the resulting states satisfy ψ. $[\![wp(C,\psi)]\!]$ is a precondition in the sense that $\models \{wp(C,\psi)\}C\{\psi\}$ and it is weakest since for each assertion ϕ, if $\models \{\phi\}C\{\psi\}$, then $\models \phi \to wp(C,\psi)$. In the strict sense, the weakest precondition is a semantic notion. That is, $[\![wp(C,\psi)]\!]$ is the weakest precondition of command C and postcondition ψ, while $wp(C,\psi)$ is a syntactic representative of the weakest precondition. In this article, however, we will treat the weakest precondition as a syntactic notion because this treatment is more convenient for our constructions and proofs.

Definition 8 (Weakest precondions). *The weakest precondition is defined inductively on the structure of commands as follows:*

1. $wp(\texttt{skip},\phi) = \phi$.
2. $wp(X \leftarrow E,\phi) = \phi[X/E]$.
3. $wp(X \xleftarrow{\$} \{a_1 : k_1,...,a_n : k_n\},\phi) = \phi[X/k_1] \wedge \ldots \wedge \phi[X/k_n]$.
4. $wp(C_1;C_2,\phi) = wp(C_1, wp(C_2,\phi))$.
5. $wp(\texttt{if}\, B\, \texttt{then}\, C_1\, \texttt{else}\, C_2,\phi) = (B \wedge wp(C_1,\phi)) \vee (\neg B \wedge wp(C_2,\phi))$.
6. $wp(\texttt{while}\, B\, \texttt{do}\, C,\phi) = \bigwedge_{k\geq 0} \psi_k$, *where* $\psi_0 = \top$ *and* $\psi_{i+1} = (B \wedge wp(C,\psi_i)) \vee (\neg B \wedge \phi)$.

We write $\phi[X/E]$ for the assertion obtained from ϕ by replacing the occurrence of X by E. Note that $\bigwedge_{k\geq 0} \psi_k$ is an assertion constructed by the infinite conjunction, which is not allowed in deterministic assertions. In fact, just like in classical Hoare logic, $wp(\texttt{while}\, B\, \texttt{do}\, C,\phi)$ can be equivalently expressed as an assertion without infinite conjunction by using Gödel's β predicate. For example, in the case where there is only one program variable X mentioned in C and ϕ, we have the following.

$$
\begin{gathered}
wp(\texttt{while}\, B\, \texttt{do}\, C,\phi) = \\
\forall k \forall m,n \geq 0((\beta^{\pm}(n,m,0,x) \wedge \\
\forall i(0 \leq i < k)(\forall x(\beta^{\pm}(n,m,i,X) \to B[X/x]) \wedge \\
\forall x,y(\beta^{\pm}(n,m,i,x) \wedge \beta^{\pm}(n,m,i+1,y) \to (wp(C,X=y) \wedge \\
\neg wp(C,\bot))[X/x]))) \to (\beta^{\pm}(n,m,k,X) \to (B \vee \phi)[X/x]))
\end{gathered}
$$

For more information, the interested readers may refer to [26]. Here we use the infinite conjunction for ease of proof.

The following propositions show the characterizing features of the weakest precondition. We omit the proof because of the limitation of space. The reader can find the missing proof in the full version of the paper [24].

Proposition 1. $\models \{wp(C,\phi)\}C\{\phi\}$.

Proposition 2. *If* $\models \{\phi\}C\{\psi\}$, *then* $\models \phi \to wp(C,\psi)$.

Proposition 3. $S \models wp(C,\phi)$ *iff* $[\![C]\!](S) \models \phi$.

In the next section, we will investigate probabilistic assertions and their corresponding proof system. The following lemma will be used in proofs in the next section.

Lemma 1. *For all* $i \geq 0$, *if* $S \models \neg wp(C^0, \neg B) \wedge \ldots \wedge \neg wp(C^{i-1}, \neg B) \wedge wp(C^i, \neg B)$, *then* $[\![\texttt{while}\, B\, \texttt{do}\, C]\!]\mu_S = [\![(\texttt{if}\, B\, \texttt{then}\, C\, \texttt{else}\, \texttt{skip})^i]\!]\mu_S$. *Here* C^0 *is* `skip` *and* C^{i+1} *is* $C;C^i$.

3 Probabilistic Hoare Logic with Probabilistic Assertion

3.1 Probabilistic Assertion

In order to describe the probabilistic aspects of probabilistic states, we need to expand the deterministic assertion to a probabilistic assertion. In this section, we first define extended deterministic assertion and rational expression (a.k.a. probabilistic expression) which are building blocks of probabilistic assertion.

The syntax of extended deterministic assertion is defined by the following BNF:

$$\phi := \top \mid \bot \mid (e\ rop\ e) \mid \phi[X/E] \mid wp(B,C,\infty) \mid \neg\phi \mid (\phi \wedge \phi) \mid \forall x \phi$$

Here e is an arithmetic expression built on $\mathbb{LV} \cup \mathbb{PV}$, E is an arithmetic expression built on $\mathbb{PV}$, B is a Boolean expression and C is a command. Other clauses are the same as in the definition of a deterministic assertion. The semantics of an extended deterministic assertion is an extension of the semantics of a deterministic assertion with the following clause.

- $S \models^I \phi[X/E]$ if $S[X \mapsto [\![E]\!]S] \models^I \phi$.
- $[\![wp(B,C,\infty)]\!]^I = \bigcap_{i=0}^{\infty}[\![\neg wp(C^i, \neg B)]\!]^I$.

It can be verified that the weakest precondition defined in the previous section extends naturally to an extended deterministic assertion.

Definition 9 (Rational expressions). *Let* $\mathbb{RV}$ *be a set of rational variables. The rational expression* r *is defined as follows:*

$$r := a \mid \mathfrak{r} \mid \mathbb{P}(\phi) \mid r\ aop\ r \mid \sum_{i=0}^{\infty} r_i$$

Here $a \in \mathbb{R}$ is a rational number, $\mathfrak{r} \in \mathbb{RV}$ is a rational variable and ϕ is an extended deterministic assertion. $\mathbb{P}(\phi)$ intuitively represents the probability that ϕ is true.

Given an interpretation I that maps logical variables to integers and maps rational variables to rational numbers, the semantics of rational expressions are defined in the standard way.

Definition 10 (Semantics of rational expressions). *Let* I *be an interpretation and* μ *be a probabilistic state, the semantics of rational expressions is defined inductively as follows.*

- $[\![a]\!]^I_\mu = a$.

- $[\![\mathfrak{r}]\!]^I_\mu = I(\mathfrak{r})$.
- $[\![\mathbb{P}(\phi)]\!]^I_\mu = \sum_{S \models^I \phi} \mu(S)$.
- $[\![r_1 \; aop \; r_2]\!]^I_\mu = [\![r_1]\!]^I_\mu \; aop \; [\![r_2]\!]^I_\mu$.
- $[\![\sum_{i=0}^{\infty} r_i]\!]^I_\mu = \sum_{i=0}^{\infty} [\![r_i]\!]^I_\mu$

The expression $\mathbb{P}(\phi)$ represents the probability that the deterministic formula ϕ holds in some probabilistic state. It can be computed by summing up all probabilities of these deterministic states that make ϕ true. $r_1 \; aop \; r_2$ characterizes the arithmetic calculation between two rational expressions. $\sum_{i=0}^{\infty} r_i$ is an infinite summation. Since an infinite addition of rational numbers may not converge, care should be taken when using a rational expression in this form. We will only use those rational expressions with infinite addition that appear in the weakest preterm calculus. The Lemma 3 will show that these rational expressions always converge. Moreover, in Sect. 3.3 we will give an alternative definition of rational expressions without using infinite addition. But in other parts of this article we will keep using infinite additions because they are convenient for proofs.

Now we define probabilistic assertions. The primitive type of probabilistic assertion is to show the relationship between two rational expressions. They are often used to compare two probabilities, for example $\mathbb{P}(X > 1) < \frac{1}{2}$.

Definition 11 (Syntax of probabilistic assertion). *Probabilistic assertions are defined inductively as follows.*

$$\Phi = (r \; rop \; r) \mid \neg\Phi \mid (\Phi \wedge \Phi)$$

where r is a rational expression.

For example, $(\mathbb{P}(X > 0) > \frac{1}{2}) \wedge \neg(\mathbb{P}(X > 1) < \mathbb{P}(X > 0))$ is a well-formed probabilistic assertion.

Definition 12 (Semantics of probabilistic assertion). *Given an interpretation I, the semantics of probabilistic assertion is defined on probabilistic states μ as follows:*

- $\mu \models^I r_1 \; rop \; r_2$ *if* $[\![r_1]\!]^I_\mu \; rop \; [\![r_2]\!]^I_\mu = \top$
- $\mu \models^I \neg\Phi$ *if not* $\mu \models^I \Phi$
- $\mu \models^I \Phi_1 \wedge \Phi_2$ *if* $\mu \models^I \Phi_1$ *and* $\mu \models^I \Phi_2$

3.2 Weakest Precondition of Probabilistic Assertion

We first construct the weakest preterms calculus, which are the main building blocks of the weakest precondition calculus. The terminology 'term' just refers to the rational expression in this paper. We inherit this word from Chadha *et al.* [6]. A weakest preterm with respect to a given command C and a term (rational expression) r intuitively denotes the term whose interpretation of the initial probabilistic state is the same as the interpretation of r on the resulting state after executing C. We also borrow the notion of the conditional term from Chadha *et al.* [6], which will be useful in defining the weakest preterm calculus.

Definition 13 (Conditional terms). *The conditional term r/B of a rational expression r with a Boolean formula B (a deterministic formula) is inductively defined as follows.*

- $a/B = a$
- $\mathfrak{x}/B = \mathfrak{x}$
- $\mathbb{P}(\phi)/B = \mathbb{P}(\phi \wedge B)$
- $(r_1 \; aop \; r_2)/B = r_1/B \; aop \; r_2/B$
- $(\sum_{i=0}^{\infty} r_i)/B = \sum_{i=0}^{\infty} (r_i/B)$

A conditional term intuitively denotes a probability under some condition described by a deterministic formula, which is shown by the next lemma.

Lemma 2. $[\![r/B]\!]^I_\mu = [\![r]\!]^I_{\downarrow B(\mu)}$.

Proof. The only nontrivial case is when the term r of the form $\mathbb{P}(\phi)$. In this case, we have the following.
$[\![\mathbb{P}(\phi)/B]\!]^I_\mu = [\![\mathbb{P}(\phi \wedge B)]\!]^I_\mu = [\![\mathbb{P}(\phi)]\!]^I_{\downarrow B(\mu)}$. □

Now we are fully prepared to define the weakest preterms. The weakest preterms for the rational expression a, $\mathfrak{x}$, and $r_1 \; aop \; r_2$ are straightforward. In terms of $\mathbb{P}(\phi)$, we need to split cases by different commands among which the WHILE command brings many subtleties.

Definition 14 (Weakest preterms). *The weakest preterm of a rational expression r with command C is inductively defined as follows.*

1. $pt(C,a) = a$
2. $pt(C,\mathfrak{x}) = \mathfrak{x}$
3. $pt(C, r_1 \; aop \; r_2) = pt(C,r_1) \; aop \; pt(C,r_2)$
4. $pt(C, \sum_{i=0}^{\infty} r_i) = \sum_{i=0}^{\infty} pt(C,r_i)$
5. $pt(\mathtt{skip}, \mathbb{P}(\phi)) = \mathbb{P}(\phi)$
6. $pt(X \leftarrow E, \mathbb{P}(\phi)) = \mathbb{P}(\phi[X/E])$
7. $pt(X \xleftarrow{\$} \{a_1 : k_1, \ldots, a_n : k_n\}, \mathbb{P}(\phi)) =$
$a_1\mathbb{P}(\phi[X/k_1] \wedge \neg\phi[X/k_2] \wedge \ldots \neg\phi[X/k_n]) +$
$a_2\mathbb{P}(\neg\phi[X/k_1] \wedge \phi[X/k_2] \wedge \neg\phi[X/k_3] \wedge \ldots \wedge \neg\phi[X/k_n]) + \ldots + a_n\mathbb{P}(\neg\phi[X/k_1] \wedge \ldots \wedge$
$\neg\phi[X/k_{n-1}] \wedge \phi[X/k_n]) +$
$(a_1 + a_2)\mathbb{P}(\phi[X/k_1] \wedge \phi[X/k_2] \wedge \neg\phi[X/k_3] \wedge \ldots \wedge \neg\phi[X/k_n]) + \ldots + (a_1 + \ldots +$
$a_n)\mathbb{P}(\phi[X/k_1] \wedge \ldots \wedge \phi[X/k_n])$
8. $pt(C_1;C_2, \mathbb{P}(\phi)) = pt(C_1, pt(C_2, \mathbb{P}(\phi)))$
9. $pt(\mathtt{if}\; B \;\mathtt{then}\; C_1 \;\mathtt{else}\; C_2, \mathbb{P}(\phi)) = pt(C_1, \mathbb{P}(\phi))/B + pt(C_2, \mathbb{P}(\phi))/(\neg B)$

10. $pt(\texttt{while}\,B\,\texttt{do}\,C, \mathbb{P}(\phi)) = \sum_{i=0}^{\infty} T_i$, *in which* T_i *is defined via the following procedure. We use the following abbreviation*

- $wp_B^C(i)$ *is short for* $\neg wp(C^0, \neg B) \wedge \ldots \wedge \neg wp(C^{i-1}, \neg B) \wedge wp(C^i, \neg B)$.
- WL *is short for* $\texttt{while}\,B\,\texttt{do}\,C$
- IF *is short for* $\texttt{if}\,B\,\texttt{then}\,C\,\texttt{else}\,\texttt{skip}$
- $SUM_{B,C,\phi}$ *is short for* $\sum_{i=0}^{\infty}(\mathbb{P}(wp_B^C(i))(pt((IF)^i, \mathbb{P}(\phi))/(wp_B^C(i))))$.

We will simply write $wp(i), wp(\infty), SUM$ *instead of* $wp_B^C(i), wp(B,C,\infty)$, $SUM_{B,C,\phi}$ *when* B, C, ϕ *are clear from context. For any probabilistic state* μ, *we let* $\mu_0 = \downarrow_{wp(\infty)} (\mu)$, $\mu_{i+1} = \downarrow_{wp(\infty)} ([\![C]\!]\mu_i)$.

- *Let* $T_0 = SUM$.
- *Let* $T_1 = \mathbb{P}(wp(\infty))(pt(C, SUM)/wp(\infty))$. *Equivalently,* T_1 *is the unique rational expression such that*

$$[\![T_1]\!]_\mu = [\![\mathbb{P}(wp(\infty))]\!]_\mu [\![SUM]\!]_{[\![C]\!]\mu_0}$$

for all probabilistic state μ.

- *Let* $T_2 = \mathbb{P}(wp(\infty))(pt(C, \mathbb{P}(wp(\infty)))/wp(\infty))(pt(C, pt(C, SUM)/wp(\infty))/wp(\infty))$. *Equivalently,* T_2 *is the unique rational expression such that*

$$[\![T_2]\!]_\mu = [\![\mathbb{P}(wp(\infty))]\!]_\mu [\![\mathbb{P}(wp(\infty))]\!]_{[\![C]\!]\mu_0} [\![SUM]\!]_{[\![C]\!]\mu_1}$$

for all probabilistic state μ.

- *In general, we let* T_i *be the unique rational expression such that*

$$[\![T_i]\!]_\mu = [\![\mathbb{P}(wp(\infty))]\!]_\mu [\![\mathbb{P}(wp(\infty))]\!]_{[\![C]\!]\mu_0} \cdots [\![\mathbb{P}(wp(\infty))]\!]_{[\![C]\!]\mu_{i-2}} [\![SUM]\!]_{[\![C]\!]\mu_{i-1}}$$

for all probabilistic states μ. *Equivalently, let* $f_{C,B}$ *be the function which maps a rational expression* r *to* $f_{C,B}(r) = pt(C,r)/wp(\infty)$. *Then*

$$T_i = f_{C,B}^i(SUM) \prod_{j=0}^{i-1} f_{C,B}^j(\mathbb{P}(wp(\infty))).$$

To sum up, we let

$$pt(\texttt{while}\,B\,\texttt{do}\,C, \mathbb{P}(\phi)) = \sum_{i=0}^{\infty} T_i = \sum_{i=0}^{\infty} f_{C,B}^i(SUM) \prod_{j=0}^{i-1} f_{C,B}^j(\mathbb{P}(wp(\infty))).$$

Let us explain the definition of $pt(\texttt{while}\,B\,\texttt{do}\,C, \mathbb{P}(\phi))$. Given a deterministic state S, $S \models wp_B^C(i)$ means that S will lead to a probabilistic state satisfying $\neg B$ (all its supports satisfy $\neg B$) after executing C for *exactly* i times. Thus, $SUM_{B,C,\phi}$ represents the probability of reaching a probabilistic state satisfying $\neg B$ by executing C in finitely many steps. In contrast, $S \models wp(B,C,\infty)$ means that S will never achieve a probabilistic state that satisfies $\neg B$. So, given a probabilistic state, some supports of it satisfy $wp_B^C(i)$

for some i (the probability of all of them is denoted by $SUM_{B,C,\phi}$), but others satisfy $wp(B,C,\infty)$. For a deterministic state that satisfies $wp(B,C,\infty)$, we execute C for 1 time to obtain a probabilistic state, the support being a set of deterministic states. Among these deterministic states, we use T_1 to represent those that will terminate at $\neg B$ after finite execution times. For those states which will not terminate on $\neg B$ in finite times, we execute C for the 2nd time. Then T_2 will appear. So on and so forth. In this way, we construct T_i for each $i \geq 1$.

Although the above definition of $pt(C,r)$ is a syntactic notion. In practice, we will always treat it as a semantic notion. That is, if $[\![pt(C,r_1)]\!]^I_\mu = [\![r_2]\!]^I_\mu$ for all probabilistic states μ and interpretation I, then we say that r_2 is a preterm of r_1 with command C. The above definition constitutes the weakest preterm calculus of probabilistic expression. It tells us how to calculate the weakest preterm for any given probabilistic expression and command. Now we use some examples to demonstrate the usage of the weakest preterm calculus.

• *Example 4.* Let us calculate $pt(\texttt{while}\top\texttt{ do skip},\mathbb{P}(\top))$. In this case, we have $wp(\texttt{skip}^i,\neg\top) = wp(\texttt{skip},\bot) = \bot$. (Here, we overload the symbol $=$ to represent semantic equivalence.) Then we know $wp(i) = \bot$ for all i and $wp(\infty) = \top$. Therefore, $SUM = 0$ because $\mathbb{P}(wp(i)) = \mathbb{P}(\bot) = 0$. Moreover, $T_1 = \mathbb{P}(wp(\infty))(pt(\texttt{skip}, SUM)/wp(\infty)) = \mathbb{P}(\top)(pt(\texttt{skip},0)/\top) = \mathbb{P}(\top)\cdot 0 = 0$. We also note that $f_{\texttt{skip},\top}(r) = pt(\texttt{skip},r)/\top = r$. Therefore, $T_i = 0$ for all i. We then know $pt(\texttt{while}\top\texttt{ do skip},\mathbb{P}(\top)) = 0$.

• *Example 5.* Let C be the following program:

$$X \xleftarrow{\$} \{\frac{1}{3}:0,\frac{2}{3}:1\};\texttt{if } X=0 \texttt{ then (while}\top\texttt{ do skip) else skip}$$

Let us calculate $pt(C,\mathbb{P}(\top))$. We have

$$pt(C,\mathbb{P}(\top)) = pt(X \xleftarrow{\$} \{\frac{1}{3}:0,\frac{2}{3}:1\},$$

$$pt(\texttt{if } X=0 \texttt{ then (while}\top\texttt{ do skip) else skip},\mathbb{P}(\top))).$$

$$pt(\texttt{if } X=0 \texttt{ then (while}\top\texttt{ do skip) else skip},\mathbb{P}(\top)) =$$

$$(pt(\texttt{while}\top\texttt{ do skip},\mathbb{P}(\top))/(X=0)) + (pt(\texttt{skip},\mathbb{P}(\top))/(\neg X=0)) =$$

$$(0/(X=0)) + (pt(\texttt{skip},\mathbb{P}(\top))/(\neg X=0)) =$$

$$0 + (\mathbb{P}(\top)/(\neg X=0)) = \mathbb{P}(X \neq 0).$$

Therefore, $pt(C,\mathbb{P}(\top)) = pt(X \xleftarrow{\$} \{\frac{1}{3}:0,\frac{2}{3}:1\},\mathbb{P}(X\neq 0)) =$
$\frac{1}{3}\mathbb{P}(0\neq 0 \wedge \neg 1 \neq 0) + \frac{2}{3}\mathbb{P}(\neg 0 \neq 0 \wedge 1 \neq 0) + (\frac{1}{3}+\frac{2}{3})\mathbb{P}(0\neq 0 \wedge 1 \neq 0) = \frac{1}{3}\mathbb{P}(\bot) + \frac{2}{3}\mathbb{P}(\top) + \mathbb{P}(\bot) = \frac{2}{3}\mathbb{P}(\top)$.

Now we prove the characterization lemma of the weakest preterms. This lemma shows that the value of the weakest preterm of a rational expression r with command C in a probabilistic state μ is the same as the value of r on the state obtained by executing C on μ.

Lemma 3 (characterization lemma of preterms). *Given a probabilistic state μ, an interpretation I, a command C and a rational expression r,*

$$[\![pt(C,r)]\!]^I_\mu = [\![r]\!]^I_{[\![C]\!]\mu}$$

The weakest preterm calculus and the above characterization lemma of the weakest preterms calculus are the main technical contribution of this article. With these notions and results at hand, we can proceed to define the weakest preconditions of probabilistic assertions straightforwardly.

Definition 15 (Weakest precondition of probabilistic assertion). *The weakest precondition of the probabilistic assertion is defined inductively as follows.*

1. $WP(C, r_1 \; rop \; r_2) = pt(C, r_1) \; rop \; pt(C, r_2)$
2. $WP(C, \neg\Phi) = \neg WP(C, \Phi)$
3. $WP(C, \Phi_1 \wedge \Phi_2) = WP(C, \Phi_1) \wedge WP(C, \Phi_2)$

The following theorem shows that $WP(C, \Phi)$ defined above is indeed a precondition of the probabilistic assertion Φ given command C.

Theorem 1. $\mu \models^I WP(C, \Phi)$ *iff* $[\![C]\!]\mu \models^I \Phi$.

Proof. We prove it by structural induction:

1. $\mu \models^I WP(C, r_1 \; rop \; r_2)$ iff $\mu \models^I pt(C, r_1) \; rop \; pt(C, r_2)$ iff $[\![pt(C, r_1)]\!]^I_\mu \; rop \; [\![pt(C, r_2)]\!]^I_\mu = \top$ iff $[\![r_1]\!]^I_{[\![C]\!]\mu} \; rop \; [\![r_2]\!]^I_{[\![C]\!]\mu} = \top$ iff $[\![C]\!]\mu \models^I r_1 \; rop \; r_2$.
2. $\mu \models^I WP(C, \neg\Phi)$ iff $\mu \models^I \neg WP(C, \Phi)$ iff $\mu \not\models^I WP(C, \Phi)$ iff $[\![C]\!]\mu \not\models^I \Phi$ iff $[\![C]\!]\mu \models^I \neg\Phi$.
3. $\mu \models^I WP(C, \Phi_1 \wedge \Phi_2)$ iff $\mu \models^I WP(C, \Phi_1) \wedge WP(C, \Phi_2)$ iff $\mu \models^I WP(C, \Phi_1)$ and $\mu \models^I WP(C, \Phi_2)$ iff $[\![C]\!]\mu \models^I \Phi_1$ and $[\![C]\!]\mu \models^I \Phi_2$ iff $[\![C]\!]\mu \models^I \Phi_1 \wedge \Phi_2$. □

The last theorem inspires us to define a relatively complete PHL proof system with probabilistic assertion in a uniform manner: $\vdash \{WP(C, \Phi)\}C\{\Phi\}$ for every command C and for every probabilistic assertion Φ. We will follow the idea with some minor adjustments.

Definition 16 (Proof system of probabilistic formulas PHL). *The proof system of PHL with probabilistic assertions consists of the following inference rules:*

$$SKIP: \quad \overline{\vdash \{\Phi\}\mathtt{skip}\{\Phi\}}$$

$$AS: \quad \overline{\vdash \{WP(X \leftarrow E,\ \Phi)\}X \leftarrow E\{\Phi\}}$$

$$PAS: \quad \overline{\vdash \{WP(X \xleftarrow{\$} \{a_1{:}r_1,...,a_n{:}r_n\},\ \Phi)\}X \xleftarrow{\$} \{a_1{:}r_1,...,a_n{:}r_n\}\{\Phi\}}$$

$$SEQ: \quad \frac{\vdash \{\Phi\}C_1\{\Phi_1\} \quad \vdash \{\Phi_1\}C_2\{\Phi_2\}}{\vdash \{\Phi\}C_1;C_2\{\Phi_2\}}$$

$$IF: \quad \overline{\vdash \{WP(\mathtt{if}\ B\ \mathtt{then}\ C_1\ \mathtt{else}\ C_2,\ \Phi)\}\mathtt{if}\ B\ \mathtt{then}\ C_1\ \mathtt{else}\ C_2\{\Phi\}}$$

$$WHILE: \quad \overline{\vdash \{WP(\mathtt{while}\ B\ \mathtt{do}\ C,\ \Phi)\}\mathtt{while}\ B\ \mathtt{do}\ C\{\Phi\}},$$

$$CONS: \quad \frac{\models \Phi' \to \Phi \quad \vdash \{\Phi\}C\{\Psi\} \quad \models \Psi \to \Psi'}{\vdash \{\Phi'\}C\{\Psi'\}}$$

With the rule (CONS) in our proof system, we can treat $WP(C,\Phi)$ as a semantic notion: if $\models WP(C,\Phi) \leftrightarrow \Psi$, then Ψ is conceived as the weakest precondition of Φ with the command C. Some examples of inferences about Hoare triples based on our PHL are given below.

• *Example 6.* $\vdash \{\top\}\mathtt{while}\ \top\ \mathtt{do}\ \mathtt{skip}\{\mathbb{P}(\top) = 0\}$ is derivable in our proof system. This is because $pt(\mathtt{while}\ \top\ \mathtt{do}\ \mathtt{skip}, \mathbb{P}(\top)) = 0$ and $pt(\mathtt{while}\ \top\ \mathtt{do}\ \mathtt{skip}, 0) = 0$, which implies that $WP(\mathtt{while}\ \top\ \mathtt{do}\ \mathtt{skip}, \mathbb{P}(\top) = 0) = (0 = 0)$. By the rule (while) we derive $\vdash \{0 = 0\}\mathtt{while}\ \top\ \mathtt{do}\ \mathtt{skip}\{\mathbb{P}(\top) = 0\}$, then by the rule (CONS), we derive $\vdash \{\top\}\mathtt{while}\ \top\ \mathtt{do}\ \mathtt{skip}\{\mathbb{P}(\top) = 0\}$.

• *Example 7.* Let C be the following program:

$$X \leftarrow \{\frac{1}{3} : 0, \frac{2}{3} : 1\}; \mathtt{if}\ X = 0\ \mathtt{then}\ (\mathtt{while}\ \top\ \mathtt{do}\ \mathtt{skip})\ \mathtt{else}\ \mathtt{skip}$$

$\vdash \{\top\}C\{\mathbb{P}(\top) \leq \frac{2}{3}\}$ is derivable in our proof system. This is because $pt(C, \mathbb{P}(\top)) = \frac{2}{3}\mathbb{P}(\top)$. Then we know $WP(C, \mathbb{P}(\top) \leq \frac{2}{3}) = \frac{2}{3}\mathbb{P}(\top) \leq \frac{2}{3}$. Then by applying rules in our proof system we get $\vdash \{\frac{2}{3}\mathbb{P}(\top) \leq \frac{2}{3}\}C\{\mathbb{P}(\top) \leq \frac{2}{3}\}$. Note that $\models \top \to (\frac{2}{3}\mathbb{P}(\top) \leq \frac{2}{3})$. We then use (CONS) to derive $\vdash \{\top\}C\{\mathbb{P}(\top) \leq \frac{2}{3}\}$.

Next, we show that $WP(C,\Phi)$ is the 'weakest' precondition.

Lemma 4. *If* $\models \{\Phi\}C\{\Psi\}$, *then* $\models \Phi \to WP(C,\Psi)$.

Proof. Let μ be a probabilistic state such that $\mu \models \Phi$. Then $[\![C]\!]\mu \models \Psi$. Then by Theorem 1 we know $\mu \models WP(C,\Psi)$. □

Lemma 5. *For an arbitrary command C and an arbitrary probabilistic formula Φ, we have* $\vdash \{WP(C,\Phi)\}C\{\Phi\}$.

Proof. We prove it by induction on the structure of the command C. When C is `skip`, assignment, random assignment, IF or WHILE command, we can directly derive them from the proof system PHL. The only case to show is the sequential command, which means that we need to prove $\vdash \{WP(C_1;C_2,\Phi)\}C_1;C_2\{\Phi\}$.

By Definition 15, we have $WP(C_1;C_2,\Phi) = WP(C_1,WP(C_2,\Phi))$. By the induction hypothesis, we have $\vdash \{WP(C_2,\Phi)\}C_2\{\Phi\}$ and $\vdash \{WP(C_1,WP(C_2,\Phi))\}C_1\{WP(C_2;\Phi)\}$. By the inference rule *SEQ*, we can conclude that $\vdash \{WP(C_1,WP(C_2,\Phi))\}C_1;C_2\{\Phi\}$ which implies that $\vdash \{WP(C_1;C_2,\Phi)\}C_1;C_2\{\Phi\}$. □

Putting everything together, we can finally show the soundness and relative completeness of our PHL as follows.

Theorem 2. *The proof system of PHL with probabilistic assertions is sound and complete.*

Proof. Soundness: the soundness of SKIP, AS, PAS, IF, WHILE follows from Theorem 1. The soundness of SEQ and CON can be proved by simple deduction.

Completeness: Assume $\models \{\Phi\}C\{\Psi\}$, then $\models \Phi \to WP(C,\Psi)$ by Lemma 4. Then by $\vdash \{WP(C,\Psi)\}C\{\Psi\}$ and CONS we know $\vdash \{\Phi\}C\{\Psi\}$. □

Now we have a relatively complete proof system of PHL. We have defined it in a minimalist fashion, which makes it seem not very useful in practice at first sight. However, since it is relatively complete, all valid derivation rules of Hoare triples are deducible in our proof system. For example, the following rule from Chadha *et al.* [6] is deducible and useful in practice.

$$IF_C : \frac{\{\Phi_1\}C_1\{\mathbb{P}(\phi) = r_1\} \qquad \{\Phi_2\}C_2\{\mathbb{P}(\phi) = r_2\}}{\{(\Phi_1/B) \wedge (\Phi_2/\neg B)\}\texttt{if } B \texttt{ then } C_1 \texttt{ else } C_2\{\mathbb{P}(\phi) = r_1 + r_2\}}$$

Here, Φ_1/B is a straightforward transfer of conditional terms to conditional formulas. The readers may also find some deducible rules about the loop in [4,22,23].

3.3 Extension and Discussion

Remove Infinite Addition. Some readers may feel unsatisfactory with the infinite addition in rational expressions. In this subsection, we show how to remove it in our syntax without loss of completeness. We will replace the infinite addition with a star operation and use it to define a proper rational expression. We call a rational expression without infinite addition the basic rational expression. Intuitively, we will define the proper rational expression as the least subset of rational expression which extends the basic rational expression and is closed under the weakest preterm calculus. We define proper rational expression by mutual structural induction with proper function. Compared to a rational expression with infinite additions, a proper rational expression defined in this means is easier to implement by a theorem prover like Coq.

Definition 17 (proper rational expression).

1. *Every basic rational expression is a proper rational expression.*
2. *If f is a proper function, then f^* is a proper rational expression.*

A proper function is a function that maps natural numbers to proper rational expressions. It includes two classes of primitive functions and is closed under the conditional operation and the weakest preterm calculus.

Definition 18 (proper function).

1. *For all boolean expressions B and command C, the function $S_{B,C}$ that maps a natural number i to*

   $$S_{B,C}(i) := \mathbb{P}(wp_B^C(i))(pt((IF)^i, \mathbb{P}(\phi))/(wp_B^C(i)))$$

   is a proper function. Here IF is short for if B then C else skip.
2. *For all boolean expressions B, command C and deterministic assertion ϕ, the function $T_{B,C,\phi}$ that maps a natural number i to*

   $$T_{B,C,\phi}(i) := f_{C,B}^i(S_{B,C}^*)\prod_{j=0}^{i-1} f_{C,B}^j(\mathbb{P}(wp_B^C(\infty)))$$

   is a proper function.
3. *If f is a proper function and B is a boolean expression, then f^B is a proper function, where $f^B(i) := f(i)/B$.*
4. *If f is a proper function and C is a command, then $pt(C, f)$ is a proper function, where $pt(C, f)(i) := pt(C, f(i))$.*

To extend conditional terms and the weakest preterm calculus to proper expressions, we let $f^*/B := (f^B)^*$ and $pt(C, f^*) := pt(C, f)^*$ for all proper functions f. Semantically, we let $[\![f^*]\!]_\mu^I := \sum_{i=0}^{\infty} [\![f(i)]\!]_\mu^I$ for all probabilistic states μ and interpretation I. This finishes the definition of proper rational expression.

Loop Invariant. Some readers may miss an axiom of the WHILE command in the form of loop invariant. This kind of axiom will be more useful than the current WHILE axiom in mechanized program verification. We now introduce two loop invariant axioms as follows:

$$\text{LOOP-1}\ \frac{\{\mathbb{P}(\phi)=1\}C\{\mathbb{P}(\neg\phi)=0\}}{\{\mathbb{P}(\phi)=1\}\texttt{while}\ B\ \texttt{do}\ C\{\mathbb{P}(\neg\phi)=0 \wedge \mathbb{P}(B)=0\}}$$

$$\text{LOOP-r}\ \frac{\{\mathbb{P}(\phi)=1\}C\{\mathbb{P}(\neg\phi)=0\} \qquad \{\mathbb{P}(\phi)=0\}C\{\mathbb{P}(\phi)=0\}}{\{\mathbb{P}(\phi)=r\}\texttt{while}\ B\ \texttt{do}\ C\{(\mathbb{P}(\neg\phi)\le 1-r) \wedge (\mathbb{P}(\phi)\le r) \wedge \mathbb{P}(B)=0\}}$$

LOOP-1 says that starting from a state that fully satisfies ϕ, if the execution of C does not decrease the probability of ϕ in all terminated executions, then the execution of $\texttt{while}\ B\ \texttt{do}\ C$ does not decrease the probability of ϕ in all terminated executions. LOOP-r says that in addition to the premise of LOOP-1, if starting from a state which completely does not satisfy ϕ implies that the execution of C does not increase the probability of ϕ, then the execution of $\texttt{while}\ B\ \texttt{do}\ C$ on a state which partially satisfies ϕ preserves the probability of ϕ. The proof of the soundness of these axioms is easy, and we leave it to the readers. What is not obvious is whether these axioms can replace the WHILE axiom without loss of completeness. We leave this issue for future work.

Comparison to Expectation-Based PHL. We have mentioned in Sect. 1 that the completeness of expectation-based PHL with the While loop is proved by Batz *et al.* [5]. It is worthwhile to clarify the difference between our work and theirs. Although the deterministic assertion ϕ is expressible in both work, there are significant differences in dealing with the probabilistic assertion. For example, suppose that we are interested in whether $\mathbb{P}(\phi) > \frac{1}{2}$ is true after the execution of C. The weakest pre-expectation calculus in Batz *et al.* [5] allows us to compute $wp[\![C]\!](\phi)$, which is the weakest pre-expectation of ϕ with C. By definition, $wp[\![C]\!](\phi)$ is a function that maps a deterministic state to a rational number. Applying $wp[\![C]\!](\phi)$ to a deterministic state S will give us the number which represents the probability of ϕ being true in the probabilistic state $[\![C]\!]S$. This means that we can use the weakest pre-expectation calculus to check whether $\mathbb{P}(\phi) > \frac{1}{2}$ is true for a given command and a *deterministic* state. If we are given a probabilistic assertion such as $\mathbb{P}(\phi) > \frac{1}{3}$, then the preexpectation calculus is unable to tell us whether $\mathbb{P}(\phi) > \frac{1}{2}$ is true after executing C from a probabilistic state that satisfies $\mathbb{P}(\phi) > \frac{1}{3}$. On the other hand, our weakest preterm calculus can solve the problem by checking the validity of $\{\mathbb{P}(\phi) > \frac{1}{3}\}C\{\mathbb{P}(\phi) > \frac{1}{2}\}$.

4 Conclusions and Future Work

Probabilistic Hoare logic is particularly useful in fields such as formal verification of probabilistic programs. The studies on PHL can be classified into two approaches: satisfaction-based and expectation-based. In this paper we establish the weakest precondition calculus for satisfaction-based PHL and use it to solve the relative completeness problem of satisfaction-based probabilistic Hoare logic, which has been open since 1979. In the future, we are interested in extending our PHL to a probabilistic dynamic logic.

A Appendix

Lemma 1. For all $i \geq 0$, if $S \models \neg wp(C^0, \neg B) \wedge \ldots \wedge \neg wp(C^{i-1}, \neg B) \wedge wp(C^i, \neg B)$, then $[\![\texttt{while}\ B\ \texttt{do}\ C]\!]\mu_S = [\![(\texttt{if}\ B\ \texttt{then}\ C\ \texttt{else}\ \texttt{skip})^i]\!]\mu_S$.

Proof. We first prove the case where $i = 0$. Here we have $S \models wp(C^0, \neg B)$. That is, $S \models \neg B$. It is then trivial to prove $[\![\texttt{while}\ B\ \texttt{do}\ C]\!]\mu_S = [\![(\texttt{if}\ B\ \texttt{then}\ C\ \texttt{else}\ \texttt{skip})^0]\!]\mu_S$.

We now prove the case with $i=2$, other cases are similar. Assume $i=2$. Then we have $S \models \neg wp(C^0,\neg B) \wedge \neg wp(C^1,\neg B) \wedge wp(C^2,\neg B)$. Then we know $S \not\models wp(\texttt{skip},\neg B)$, $S \not\models wp(C,\neg B)$ and $S \models wp(C;C,\neg B)$. Therefore, $S \models \neg B$, $[\![C]\!]S \not\models \neg B$ and $[\![C;C]\!]S \models \neg B$. We then know $\downarrow_B([\![C;C]\!]S) = \mathbf{0}$.

It's easy to see that $sp([\![C]\!]\circ \downarrow_B(\mu_S)) \subseteq sp([\![C]\!](\mu_S))$. Then from $S \models wp(C^2,\neg B)$ we deduce $[\![C^2]\!]S \models \neg B$, which implies $[\![(C\circ \downarrow_B)^2]\!]S \models \neg B$.

Then $\sum_{i=0}^{\infty} \downarrow_{\neg B}(([\![C]\!]\circ \downarrow_B)^i(\mu_S)) = \downarrow_{\neg B}(([\![C]\!]\circ \downarrow_B)^0(\mu_S)) + \downarrow_{\neg B}(([\![C]\!]\circ \downarrow_B)^1(\mu_S)) + \downarrow_{\neg B}(([\![C]\!]\circ \downarrow_B)^2(\mu_S)) + \downarrow_{\neg B}(([\![C]\!]\circ \downarrow_B)^3(\mu_S)) + \ldots = \downarrow_{\neg B}(([\![C]\!]\circ \downarrow_B)^0(\mu_S)) + \downarrow_{\neg B}(([\![C]\!]\circ \downarrow_B)^1(\mu_S)) + \downarrow_{\neg B}(([\![C]\!]\circ \downarrow_B)^2(\mu_S)) = \downarrow_{\neg B}(\mu_S) + \downarrow_{\neg B}\circ[\![C]\!]\circ \downarrow_B(\mu_S) + ([\![C]\!]\circ \downarrow_B)^2(\mu_S)$.

On the other hand, $[\![\texttt{if}\ B\ \texttt{then}\ C\ \texttt{else}\ \texttt{skip}]\!]^2\mu_S =$
$[\![\texttt{if}\ B\ \texttt{then}\ C\ \texttt{else}\ \texttt{skip}]\!]([\![C]\!]\circ \downarrow_B(\mu_S) + \downarrow_{\neg B}(\mu_S)) =$
$[\![C]\!]\circ \downarrow_B([\![C]\!]\circ \downarrow_B(\mu_S) + \downarrow_{\neg B}(\mu_S)) + \downarrow_{\neg B}([\![C]\!]\circ \downarrow_B(\mu_S) + \downarrow_{\neg B}(\mu_S)) =$
$[\![C]\!]\circ \downarrow_B [\![C]\!]\circ \downarrow_B(\mu_S) + [\![C]\!]\circ \downarrow_B(\downarrow_{\neg B}(\mu_S)) + \downarrow_{\neg B}[\![C]\!]\circ \downarrow_B(\mu_S) + \downarrow_{\neg B}(\downarrow_{\neg B}(\mu_S)) =$
$([\![C]\!]\circ \downarrow_B)^2(\mu_S) + \mathbf{0} + \downarrow_{\neg B}[\![C]\!]\circ \downarrow_B(\mu_S) + \downarrow_{\neg B}(\mu_S) =$
$\downarrow_{\neg B}(\mu_S) + \downarrow_{\neg B}[\![C]\!]\circ \downarrow_B(\mu_S) + ([\![C]\!]\circ \downarrow_B)^2(\mu_S)$. □

Lemma 3 [characterization lemma of preterms] Given a probabilistic state μ, an interpretation I, a command C and a rational expression r,

$$[\![pt(C,r)]\!]^I_\mu = [\![r]\!]^I_{[\![C]\!]\mu}$$

Proof. 1. $[\![pt(C,a)]\!]^I_\mu = [\![a]\!]^I_\mu = a = [\![a]\!]^I_{[\![C]\!]\mu}$.

2. $[\![pt(C,\mathfrak{r})]\!]^I_\mu = [\![\mathfrak{r}]\!]^I_\mu = I(\mathfrak{r}) = [\![\mathfrak{r}]\!]^I_{[\![C]\!]\mu}$.

3. $[\![pt(C,r_1\ aop\ r_2)]\!]^I_\mu [\![pt(C,r_1)\ aop\ pt(C,r_2)]\!]^I_\mu =$
$[\![pt(C,r_1)\ ]\!]^I_\mu aop [\![\ pt(C,r_2)]\!]^I_\mu = [\![r_1]\!]^I_{[\![C]\!]\mu} aop [\![r_2]\!]^I_{[\![C]\!]\mu}$.

4. $[\![pt(C,\sum_{i=0}^{\infty} r_i)]\!]^I_\mu = [\![\sum_{i=0}^{\infty} pt(C,r_i)]\!]^I_\mu = \sum_{i=0}^{\infty} [\![pt(C,r_i)]\!]^I_\mu$.

5. $[\![pt(\texttt{skip},\mathbb{P}(\phi))]\!]^I_\mu = [\![\mathbb{P}(\phi)]\!]^I_\mu = [\![\mathbb{P}(\phi)]\!]^I_{[\![\texttt{skip}]\!]\mu}$.

6. $[\![pt(X\leftarrow E,\mathbb{P}(\phi))]\!]^I_\mu = [\![\mathbb{P}(\phi[X/E])]\!]^I_\mu = [\![\mathbb{P}(wp(X\leftarrow E,\phi))]\!]^I_\mu$. Since $S \models \phi[X/E]$ iff $S \models wp(X\leftarrow E,\phi)$ iff $[\![X\leftarrow E]\!]S \models \phi$, we know $[\![\mathbb{P}(\phi[X/E])]\!]^I_\mu = [\![\mathbb{P}(\phi)]\!]^I_{[\![X\leftarrow E]\!]\mu}$.

7. PAS: For the sake of simplicity, we assume $n=2$. No generality is lost with this assumption. We have $[\![pt(X \xleftarrow{\$} \{a_1:k_1,a_2:k_2\},\mathbb{P}(\phi))]\!]^I_\mu = [\![a_1\mathbb{P}(\phi[X/k_1]\wedge \neg\phi[X/k_2]) + a_2\mathbb{P}(\neg\phi[X/k_1]\wedge\phi[X/k_2]) + (a_1+a_2)\mathbb{P}(\phi[X/k_1]\wedge\phi[X/k_2])]\!]^I_\mu$. Without loss generality, assume $sp(\mu) = \{S_1,S_2,S_3,S_4\}$, $S_1 \models^I \phi[X/k_1]\wedge\neg\phi[X/k_2]$, $S_2 \models^I \neg\phi[X/k_1]\wedge\phi[X/k_2]$, $S_3 \models^I \phi[X/k_1]\wedge\phi[X/k_2]$, $S_4 \models^I \neg\phi[X/k_1]\wedge\neg\phi[X/k_2]$, $\mu(S_i) = b_i$. Then we have $[a_1\mathbb{P}(\phi[X/k_1]\wedge\neg\phi[X/k_2]) + a_2\mathbb{P}(\neg\phi[X/k_1]\wedge\phi[X/k_2]) + (a_1+a_2)\mathbb{P}(\phi[X/k_1]\wedge\phi[X/k_2])]\!]^I_\mu = a_1b_1 + a_2b_2 + (a_1+a_2)b_3$.

We further have $[\![X \xleftarrow{\$} \{a_1:k_1,a_2:k_2\}]\!](S_1) \models^I \mathbb{P}(\phi) = a_1$, $[\![X \xleftarrow{\$} \{a_1:k_1,a_2:k_2\}]\!](S_2) \models^I \mathbb{P}(\phi) = a_2$, $[\![X \xleftarrow{\$} \{a_1:k_1,a_2:k_2\}]\!](S_3) \models^I \mathbb{P}(\phi) = a_1+a_2$ and $[\![X \xleftarrow{\$} \{a_1:k_1,a_2:k_2\}]\!](S_4) \models^I \mathbb{P}(\phi) = 0$. Then we know $[\![X \xleftarrow{\$} \{a_1:k_1,a_2:k_2\}]\!](\mu) \models^I \mathbb{P}(\phi) = a_1b_1 + a_2b_2 + (a_1+a_2)b_3$. This means that $[\![\mathbb{P}(\phi)]\!]^I_{[\![X \xleftarrow{\$} \{a_1:k_1,a_2:k_2\}]\!]\mu} =$
$a_1b_1 + a_2b_2 + (a_1+a_2)b_3 = [\![pt(X \xleftarrow{\$} \{a_1:k_1,a_2:k_2\},\mathbb{P}(\phi))]\!]^I_\mu$.

8. SEQ: $[\![pt(C_1;C_2,\mathbb{P}(\phi))]\!]^I_\mu = [\![pt(C_1,pt(C_2,\mathbb{P}(\phi)))]\!]^I_\mu =$
$[\![pt(C_2,\mathbb{P}(\phi))]\!]^I_{[\![C_1]\!]\mu} = [\![\mathbb{P}(\phi)]\!]^I_{[\![C_2]\!][\![C_1]\!]\mu} = [\![\mathbb{P}(\phi)]\!]^I_{[\![C_1;C_2]\!]\mu}$.
9. IF: $[\![pt(\texttt{if } B \texttt{ then } C_1 \texttt{ else } C_2,\mathbb{P}(\phi))]\!]^I_\mu =$
$[\![pt(C_1,\mathbb{P}(\phi))/B+pt(C_2,\mathbb{P}(\phi))/(\neg B)]\!]^I_\mu =$
$[\![pt(C_1,\mathbb{P}(\phi))/B]\!]^I_\mu + [\![pt(C_2,\mathbb{P}(\phi))/(\neg B)]\!]^I_\mu =$
$[\![pt(C_1,\mathbb{P}(\phi))]\!]^I_{\downarrow_B\mu} + [\![pt(C_2,\mathbb{P}(\phi))]\!]^I_{\downarrow_{\neg B}\mu} =$
$[\![\mathbb{P}(\phi)]\!]^I_{[\![C_1]\!]\downarrow_B\mu} + [\![\mathbb{P}(\phi)]\!]^I_{[\![C_2]\!]\downarrow_{\neg B}\mu} =$
$[\![\mathbb{P}(\phi)]\!]^I_{[\![C_1]\!]\downarrow_B\mu+[\![C_2]\!]\downarrow_{\neg B}\mu} = [\![\mathbb{P}(\phi)]\!]^I_{[\![\texttt{if } B \texttt{ then } C_1 \texttt{ else } C_2]\!]\mu}$
10. WHILE: Without loss of generality, let $sp(\mu) = \{S_{0,\infty}, S_{0,0}, S_{0,1},\ldots\}$, in which $S_{0,i} \models wp(i)$ and $S_{0,\infty} \models wp(\infty)$. Equivalently, we may let $\mu_{S_{0,i}} = \downarrow_{wp(i)} (\mu)$ and $\mu_{S_{0,\infty}} = \downarrow_{wp(\infty)} (\mu)$.
Then we know $\mu(S_{0,i}) = [\![\mathbb{P}(wp(i))]\!]_\mu$, $\mu(S_{0,\infty}) = [\![\mathbb{P}(wp(\infty))]\!]_\mu$.
That is, $\mu = [\![\mathbb{P}(wp(\infty))]\!]_\mu \mu_{S_{0,\infty}} + \sum_{i=0}^{\infty} [\![\mathbb{P}(wp(i))]\!]_\mu \mu_{S_{0,i}}$.
Then $[\![\mathbb{P}(\phi)]\!]_{[\![WL]\!]\mu}$

$$= [\![\mathbb{P}(\phi)]\!]_{[\![WL]\!](\mu(S_{0,\infty})\mu_{S_{0,\infty}} + \sum_{i=0}^{\infty} \mu(S_{0,i})\mu_{S_{0,i}})}$$

$$= [\![\mathbb{P}(\phi)]\!]_{[\![WL]\!](\mu(S_{0,\infty})\mu_{S_{0,\infty}})} + [\![\mathbb{P}(\phi)]\!]_{[\![WL]\!] \sum_{i=0}^{\infty} \mu(S_{0,i})\mu_{S_{0,i}}}$$

$$= \mu(S_{0,\infty})[\![\mathbb{P}(\phi)]\!]_{[\![WL]\!](\mu_{S_{0,\infty}})} + \sum_{i=0}^{\infty} \mu(S_{0,i})[\![\mathbb{P}(\phi)]\!]_{[\![WL]\!]\mu_{S_{0,i}}}$$

By Lemma 1 we know $[\![WL]\!]\mu_{S_{0,i}} = [\![(IF)^i]\!]\mu_{S_{0,i}}$.
Therefore, $[\![\mathbb{P}(\phi)]\!]_{[\![WL]\!]\mu_{S_{0,i}}} = [\![\mathbb{P}(\phi)]\!]_{[\![(IF)^i]\!]\mu_{S_{0,i}}}$
By induction hypothesis we know

$$[\![\mathbb{P}(\phi)]\!]_{[\![(IF)^i]\!]\mu_{S_{0,i}}} = [\![pt((IF)^i,\mathbb{P}(\phi))]\!]_{\mu_{S_{0,i}}}$$

Then we have

$$[\![\mathbb{P}(\phi)]\!]_{[\![WL]\!]\mu_{S_{0,i}}} = [\![pt((IF)^i,\mathbb{P}(\phi))]\!]_{\mu_{S_{0,i}}}$$

Moreover,

$$[\![pt((IF)^i,\mathbb{P}(\phi))]\!]_{\mu_{S_{0,i}}} = [\![pt((IF)^i,\mathbb{P}(\phi))]\!]_{\downarrow_{wp(i)}(\mu)}$$

$$= [\![pt((IF)^i,\mathbb{P}(\phi))/(wp(i))]\!]_\mu$$

At this stage we know $[\![\mathbb{P}(\phi)]\!]_{[\![WL]\!]\mu} =$

$$\mu(S_{0,\infty})[\![\mathbb{P}(\phi)]\!]_{[\![WL]\!](\mu_{S_{0,\infty}})} + \sum_{i=0}^{\infty} \mu(S_{0,i})[\![\mathbb{P}(\phi)]\!]_{[\![WL]\!]\mu_{S_{0,i}}}$$

in which $\sum_{i=0}^{\infty}\mu(S_{0,i})[\![\mathbb{P}(\phi)]\!]_{[\![WL]\!]\mu_{S_{0,i}}} =$
$[\![\sum_{i=0}^{\infty}(\mathbb{P}(wp(i))(pt((IF)^i,\mathbb{P}(\phi))/(wp(i))))]\!]_\mu$
Note that SUM is short for

$$\sum_{i=0}^{\infty}(\mathbb{P}(wp(i))(pt((IF)^i,\mathbb{P}(\phi))/(wp(i)))).$$

Then $[\![\mathbb{P}(\phi)]\!]_{[\![WL]\!]\mu} =$

$$\mu(S_{0,\infty})[\![\mathbb{P}(\phi)]\!]_{[\![WL]\!](\mu_{S_{0,\infty}})} + [\![SUM]\!]_\mu$$

$$= [\![\mathbb{P}(wp(\infty))]\!]_\mu[\![\mathbb{P}(\phi)]\!]_{[\![WL]\!](\mu_{S_{0,\infty}})} + [\![SUM]\!]_\mu.$$

It remains to study $[\![\mathbb{P}(\phi)]\!]_{[\![WL]\!]\mu_{S_{0,\infty}}}$.
Note that $[\![WL]\!] = [\![IF;WL]\!]$.
We then know $[\![\mathbb{P}(\phi)]\!]_{[\![WL]\!]\mu_{S_{0,\infty}}}$
$= [\![\mathbb{P}(\phi)]\!]_{[\![IF;WL]\!]\mu_{S_{0,\infty}}}$
$= [\![\mathbb{P}(\phi)]\!]_{[\![WL]\!][\![IF]\!]\mu_{S_{0,\infty}}}$
$= [\![\mathbb{P}(\phi)]\!]_{[\![WL]\!][\![\texttt{if } B \texttt{ then } C \texttt{ else skip}]\!]\mu_{S_{0,\infty}}}$
$= [\![\mathbb{P}(\phi)]\!]_{[\![WL]\!][\![C]\!]\mu_{S_{0,\infty}}}$.
Here we also note that $[\![C]\!]\mu_{S_{0,\infty}} = [\![C]\!]\downarrow_{wp(\infty)}(\mu)$.
Without loss of generality, let $sp([\![C]\!]\mu_{S_{0,\infty}}) = \{S_{1,\infty}, S_{1,0}, S_{1,1},\ldots,\}$, in which $S_{1,i} \models wp(i)$ and $S_{1,\infty} \models wp(\infty)$.
Then by repeating the reasoning on the cases of $S_{0,i}$, we know
$[\![\mathbb{P}(\phi)]\!]_{[\![WL]\!][\![C]\!]\mu_{S_{0,\infty}}} =$

$$[\![\mathbb{P}(wp(\infty))]\!]_{[\![C]\!]\mu_{S_{0,\infty}}}[\![\mathbb{P}(\phi)]\!]_{[\![WL]\!](\mu_{S_{1,\infty}})} + [\![SUM]\!]_{[\![C]\!]\mu_{S_{0,\infty}}} =$$

$$[\![\mathbb{P}(wp(\infty))]\!]_{[\![C]\!]\downarrow_{wp(\infty)}(\mu)}[\![\mathbb{P}(\phi)]\!]_{[\![WL]\!](\mu_{S_{1,\infty}})} + [\![SUM]\!]_{[\![C]\!]\downarrow_{wp(\infty)}(\mu)}$$

By induction hypothesis, we know

$$[\![\mathbb{P}(wp(\infty))]\!]_{[\![C]\!]\downarrow_{wp(\infty)}(\mu)} = [\![pt(C,\mathbb{P}(wp(\infty)))]\!]_{\downarrow_{wp(\infty)}(\mu)}$$

$$= [\![pt(C,\mathbb{P}(wp(\infty)))/wp(\infty)]\!]_\mu$$

and $[\![SUM]\!]_{[\![C]\!]\downarrow_{wp(\infty)}(\mu)} = [\![pt(C,SUM)/wp(\infty)]\!]_\mu$.
It then remains to study $[\![\mathbb{P}(\phi)]\!]_{[\![WL]\!]\mu_{S_{1,\infty}}}$.
By repeating the reasoning on the cases of $S_{0,i}$, we know
$[\![\mathbb{P}(\phi)]\!]_{[\![WL]\!][\![C]\!]\mu_{S_{1,\infty}}} =$

$$[\![\mathbb{P}(wp(\infty))]\!]_{[\![C]\!]\mu_{S_{1,\infty}}}[\![\mathbb{P}(\phi)]\!]_{[\![WL]\!](\mu_{S_{2,\infty}})} + [\![SUM]\!]_{[\![C]\!]\mu_{S_{1,\infty}}}$$

in which $\mu_{S_{2,\infty}} = \downarrow_{wp(\infty)} ([\![C]\!](\mu_{S_{1,\infty}}))$.
Let $\mu_{S_{i+1,\infty}} = \downarrow_{wp(\infty)} ([\![C]\!](\mu_{S_{i,\infty}}))$.
Repeat the above procedure to infinity we get $[\![pt(WL, \mathbb{P}(\phi))]\!]_{\mu} = [\![\mathbb{P}(\phi)]\!]_{[\![WL]\!](\mu)}$.

References

1. Apt, K.R.: Ten years of hoare's logic: a survey part II: nondeterminism. Theor. Comput. Sci. **28**, 83–109 (1984). https://doi.org/10.1016/0304-3975(83)90066-X
2. Apt, K.R., de Boer, F.S., Olderog, E.: Verification of Sequential and Concurrent Programs. Texts in Computer Science. Springer (2009). https://doi.org/10.1007/978-1-84882-745-5
3. Apt, K.R., Olderog, E.R.: Fifty years of hoare's logic. Formal Aspects Comput. **31**, 751–807 (2019)
4. Barthe, G., Espitau, T., Gaboardi, M., Grégoire, B., Hsu, J., Strub, P.-Y.: An assertion-based program logic for probabilistic programs. In: Ahmed, A. (ed.) ESOP 2018. LNCS, vol. 10801, pp. 117–144. Springer, Cham (2018). https://doi.org/10.1007/978-3-319-89884-1_5
5. Batz, K., Kaminski, B.L., Katoen, J., Matheja, C.: Relatively complete verification of probabilistic programs: an expressive language for expectation-based reasoning. Proc. ACM Program. Lang. **5**(POPL), 1–30 (2021). https://doi.org/10.1145/3434320
6. Chadha, R., Cruz-Filipe, L., Mateus, P., Sernadas, A.: Reasoning about probabilistic sequential programs. Theoret. Comput. Sci. **379**(1–2), 142–165 (2007)
7. Cook, S.A.: Soundness and completeness of an axiom system for program verification. SIAM J. Comput. **7**(1), 70–90 (1978)
8. Corin, R., den Hartog, J.: A probabilistic hoare-style logic for game-based cryptographic proofs (extended version). Cryptology ePrint Archive (2005)
9. Den Hartog, J., de Vink, E.P.: Verifying probabilistic programs using a hoare like logic. Int. J. Found. Comput. Sci. **13**(03), 315–340 (2002)
10. Deng, Y., Feng, Y.: Formal semantics of a classical-quantum language. Theor. Comput. Sci. **913**, 73–93 (2022). https://doi.org/10.1016/J.TCS.2022.02.017
11. Dijkstra, E.W.: Guarded commands, nondeterminacy and formal derivation of programs. Commun. ACM **18**(8), 453–457 (1975). https://doi.org/10.1145/360933.360975
12. Dijkstra, E.W.: A Discipline of Programming. Prentice-Hall (1976). https://www.worldcat.org/oclc/01958445
13. Foley, M., Hoare, C.A.R.: Proof of a recursive program: Quicksort. Comput. J. **14**(4), 391–395 (1971). https://doi.org/10.1093/COMJNL/14.4.391
14. Hoare, C.A.R.: Procedures and parameters: an axiomatic approach. In: Engeler, E. (ed.) Symposium on Semantics of Algorithmic Languages, Lecture Notes in Mathematics, vol. 188, pp. 102–116. Springer (1971). https://doi.org/10.1007/BFB0059696
15. Hoare, C.A.R.: An axiomatic basis for computer programming. Commun. ACM **12**(10), 576–580 (1969)
16. Hoare, C.A.R.: Procedures and parameters: an axiomatic approach. In: Proceedings of Symposium on the Semantics of Algorithmic Languages. Lecture Notes in Mathematics, vol. 188, pp. 102–116. Springer (1971)
17. Jones, C.: Probabilistic Nondeterminism. Ph.D. thesis, University of Edinburgh (1990)
18. Liu, J., et al.: Formal verification of quantum algorithms using quantum hoare logic. In: Dillig, I., Tasiran, S. (eds.) CAV 2019. LNCS, vol. 11562, pp. 187–207. Springer, Cham (2019). https://doi.org/10.1007/978-3-030-25543-5_12

19. Morgan, C., McIver, A.: pGCL: formal reasoning for random algorithms. S. Afr. Comput. J. (1999)
20. Morgan, C., McIver, A., Seidel, K.: Probabilistic predicate transformers. ACM Trans. Program. Lang. Syst. **18**(3), 325–353 (1996). https://doi.org/10.1145/229542.229547
21. Ramshaw, L.: Formalizing the analysis of algorithms. Ph.D. thesis, Stanford University (1979)
22. Rand, R., Zdancewic, S.: VPHL: a verified partial-correctness logic for probabilistic programs. Electron. Notes Theor. Comput. Sci. **319**, 351–367 (2015)
23. Smolka, S., Foster, N., Hsu, J., Kappé, T., Kozen, D., Silva, A.: Guarded kleene algebra with tests: verification of uninterpreted programs in nearly linear time. Proc. ACM Program. Lang. **4**(POPL), 61:1–61:28 (2020). https://doi.org/10.1145/3371129
24. Sun, X., Su, X., Bian, X., Cui, A.: On the relative completeness of satisfaction-based probabilistic hoare logic with while loop (2024). https://arxiv.org/abs/2406.16054
25. Unruh, D.: Quantum relational hoare logic. Proc. ACM Program. Lang. **3**(POPL), 33:1–33:31 (2019). https://doi.org/10.1145/3290346
26. Winskel, G.: The Formal Semantics of Programming Languages. The MIT Press (1993)
27. Ying, M.: Floyd-hoare logic for quantum programs. ACM Trans. Program. Lang. Syst. **33**(6), 19:1–19:49 (2011). https://doi.org/10.1145/2049706.2049708
28. Zhou, L., Yu, N., Ying, M.: An applied quantum hoare logic. In: McKinley, K.S., Fisher, K. (eds.) Proceedings of the 40th ACM SIGPLAN Conference on Programming Language Design and Implementation, PLDI 2019, Phoenix, AZ, USA, 22–26 June 2019, pp. 1149–1162. ACM (2019). https://doi.org/10.1145/3314221.3314584

Weakest Precondition and Relative Completeness of Some Fragments of Probabilistic Relational Hoare Logic

Xin Sun[1], Zhiguang Zhao[1(✉)], and Alina Mingalieva[2]

[1] School of Mathematics and Statistics, Taishan University, Taian, China
Zhaozhiguang23@gmail.com

[2] Institute of Economics and Finance, Catholic University of Lublin, Lublin, Poland
mingalieva.alina@wp.pl

Abstract. Probabilistic relational Hoare logic (PRHL) plays an important role in the formal verification of probabilistic algorithms. The relative completeness of PRHL remains unsolved since 2009. In this paper, we make progress on this open problem. Based on the investigation of weakest precondition we design four new proof systems of PRHL and prove that they are relatively complete for programs that are structurally similar.

Keywords: Hoare logic · probabilistic program · relative completeness

1 Introduction

Hoare logic provides a formalization with logical rules for reasoning about the correctness of programs. It was originally designed by C. A. R. Hoare in 1969 in his seminal paper [20] and extended by himself in [21]. The underpinning idea captures the precondition and postcondition of executing a certain program. The precondition describes the property on which the command is based. The postcondition describes the property to which the command must lead after each correct execution. Hoare logic has become one of the most influential tools in formal verification of programs in the past decades. It has been successfully applied in the analysis of deterministic [20,21], non-deterministic [16], recursive [18,19], probabilistic [13,26,28] and quantum programs [15,29,30]. The idea of Hoare logic also inspired the birth of dynamic logic [25], which in turn has found successful applications in reasoning about knowledge [3,4,17,23], games [12,24], and quantum information [5–8,27]. A comprehensive review of Hoare logic is referred to Apt and Olderog [1].

Probabilistic Relational Hoare Logic (PRHL) [2,9–11] is a variant of Hoare logic to reason about relational properties of probabilistic programs. In PRHL, we will consider a pair of commands and binary relations on states, so the Hoare triple will be written in the form $\{\phi\}C_1 \sim C_2\{\psi\}$, which intuitively means that if the two starting states (S_1, S_2) satisfy the relation ϕ, then after the execution of C_1 on S_1 and C_2 on S_2, the ending states satisfy the relation ψ. Since C_1 and C_2 are probabilistic programs, the ending states will be the probability distributions of the states. Therefore, the relation ψ is a relation

J. Wang et al. (Eds.): DaLí 2025, LNCS 16472, pp. 79–95, 2026.
https://doi.org/10.1007/978-3-032-22626-6_5

over distributions, which is obtained by a lifting from the relation over states to the relation over distributions. This lifting is defined formally using probabilistic coupling, a fundamental tool from probability theory [22].

PRHL has been successfully applied in the formal verification of cryptographic algorithms and differential privacy. However, all existing PRHL remain relatively incomplete. Relative completeness of Hoare logic means that if we had an oracle capable of checking the validity of assertions, then we could prove all valid Hoare triples. This notion was first introduced by Cook in [14] and is now the most widely accepted notion of the completeness of Hoare logic. The relatively completeness problem of PRHL seems quite difficult. To the best of our knowledge, the only relatively complete probabilistic relational program logic is the eRHL of [2], which reasons about relational *expectation* of pairs of probabilistic programs. In eRHL, a Hoare triple takes the form $Z : \{f\}C_1 \sim C_2\{g\}$ where Z is a set of auxiliary variables and where $f, g : Z \to \mathbb{S} \times \mathbb{S} \to [0, \infty]$, the pre- and post-expectation, which maps an auxiliary variable and two states to a non-negative extended real number. Although eRHL is claimed to be relatively complete and we can embed PRHL into eRHL, it does not imply that PRHL is relatively complete because to prove valid Hoare triples of PRHL in eRHL, we will have to use expectations which are not expressible in PRHL. Moreover, there is no well-defined syntax of expectation in eRHL. In eRHL, every function $f : Z \to \mathbb{S} \times \mathbb{S} \to [0, \infty]$ is an expectation. This treatment leaves the construction of the syntaxt of expectations as an open problem. In particular, it is desirable to have a constructive, countable and expressive syntax for expectations. The relatively completeness of eRHL will be convincing only if the proof system of eRHL uses well-defined expectations.

This article makes steps towards a relatively complete PRHL. We design four new PRHL proof systems and prove that they are relatively complete for programs that are structurally similar. The basic notions of PRHL are introduced in Sect. 2. The new proof systems are presented, and the relative completeness is proved in Sect. 3.

2 Syntax and Semantics of PRHL

This section formally introduces the syntax and semantics of assertions and commands, as well as the validity of Hoare triples in PRHL.

2.1 Arithmetic and Boolean Expressions

Let $\mathbb{PV} = \{X, Y, Z, \ldots\}$ be a finite set of program variables. Let $\mathbb{Z}$ be the set of integers. The set of arithmetic operators (aop) is $\{+, -, \times, \ldots\} \subseteq \mathbb{Z} \times \mathbb{Z} \to \mathbb{Z}$.

Definition 1 (Arithmetic expressions). *Given a set of program variables $\mathbb{PV}$, we define the arithmetic expression E as follows:*

$$E := n \mid X \mid (E \; aop \; E).$$

This syntax allows an arithmetic expression to be an integer n, a program variable X, or a composition of two arithmetic expressions built by an arithmetic operation.

The set of Boolean constants is $\mathbb{B} = \{\top, \bot\}$. We use relational operators (rop) $\{>, <,$
$\geq, =, \leq, ...\} \subseteq \mathbb{Z} \times \mathbb{Z} \to \mathbb{B}$ and logical operators (lop) $\wedge, \vee, \to$ to define Boolean expressions.

Definition 2 (Boolean expressions). *The Boolean expression is defined as follows:*

$$B := \top \mid \bot \mid (E\ rop\ E) \mid \neg B \mid (B\ lop\ B).$$

The expression $E\ rop\ E$ represents the truth value determined by the binary relation *rop* on two arithmetic expressions. $B\ lop\ B$ is the expression obtained by applying a logical operator to two Boolean expressions.

The semantics of arithmetic and Boolean expressions is defined on deterministic states S, which are maps $S : \mathbb{PV} \to \mathbb{Z}$. Let $\mathbb{S}$ be the set of all deterministic states. Each state $S \in \mathbb{S}$ is a description of the value of each program variable. Consequently, the semantics of the arithmetic expressions is $[\![E]\!] : \mathbb{S} \to \mathbb{Z}$, which maps each deterministic state to an integer. Analogously, the semantics of Boolean expressions is $[\![B]\!] : \mathbb{S} \to \mathbb{B}$ that maps each state to a Boolean value.

Definition 3 (Semantics of deterministic expressions). *The semantics of arithmetic and Boolean expressions are defined inductively as follows:*

$$\begin{aligned}
[\![X]\!]S &= S(X) \\
[\![n]\!]S &= n \\
[\![E_1\ aop\ E_2]\!]S &= [\![E_1]\!]S\ aop\ [\![E_2]\!]S \\
[\![\top]\!]S &= \top \\
[\![\bot]\!]S &= \bot \\
[\![E_1\ rop\ E_2]\!]S &= [\![E_1]\!]S\ rop\ [\![E_2]\!]S \\
[\![\neg B]\!]S &= \neg[\![B]\!]S \\
[\![B_1\ lop\ B_2]\!]S &= [\![B_1]\!]S\ lop\ [\![B_2]\!]S
\end{aligned}$$

As mentioned above, the interpretation of an arithmetic expression is an integer. A program variable X in a deterministic state is interpreted as its value in the state. A constant is always itself over any state. An arithmetic expression $E_1\ aop\ E_2$ is mapped to the integer calculated by the operator *aop* applied to the interpretation of E_1 and the interpretation of E_2 in the state. The Boolean expressions can be understood in a similar way. For example, let S be a state such that $S(X) = 1$ and $S(Y) = 2$. Then $[\![X + 1]\!]S = 2$ and $[\![(X + 2 \leq 3) \wedge (X + Y = 3)]\!]S = \top$.

2.2 Commands

Commands are actions that we perform on program states. They change a deterministic state to a probability distribution of deterministic states. In this article, we only deal with loop-free commands.

Definition 4 (command). *The commands are defined inductively as follows:*

$$C := \texttt{skip} \mid X \leftarrow E \mid X \overset{\$}{\leftarrow} R \mid C_1;C_2 \mid \texttt{if}\, B\; \texttt{then}\, C_1\; \texttt{else}\, C_2$$

where $R = \{a_1 : k_1, \cdots, a_n : k_n\}$ *in which* $\{k_1, \cdots, k_n\}$ *is a set of integers and* $a_1, \ldots, a_n$ *are real numbers such that* $0 < a_i < 1$ *and* $a_1 + \ldots + a_n = 1$.

The command skip represents a null command that does not do anything. $X \leftarrow E$ is the deterministic assignment. $X \overset{\$}{\leftarrow} R$ can be read as a value k_i is chosen with probability a_i and is assigned to X. $C_1;C_2$ is the sequential composition of C_1 and C_2 as usual. The last expression is the conditional choice.

The semantics of commands are defined on probabilistic states. It shows how different commands *update* probabilistic states. A probabilistic state, denoted by μ, is a probability sub-distribution on deterministic states. Let $D(\mathbb{S})$ refer to the set of probability sub-distributions on $\mathbb{S}$. Thus, each $\mu : \mathbb{S} \to [0,1]$ requires that $\sum_{S \in \mathbb{S}} \mu(S) \leq 1$. For a deterministic state $S \in \mathbb{S}$, μ_S is a special probabilistic state that assigns the value of 1 to S and the value of 0 to any other state. A deterministic state S is a support for μ if $\mu(S) > 0$. The set of all supports of μ is denoted by $sp(\mu)$. Addition and scalar multiplication of sub-distributions means that they are added or scalar multiplied point-wise.

Definition 5 (semantics of commands). *The semantics of commands is a function* $[\![C]\!] : D(\mathbb{S}) \longrightarrow D(\mathbb{S})$. *It is defined inductively as follows:*

- $[\![\texttt{skip}]\!](\mu) = \mu$
- $[\![X \leftarrow E]\!](\mu) = \sum_{S \in \mathbb{S}} \mu(S) \cdot \mu_{S[X \mapsto [\![E]\!]S]}$
- $[\![X \overset{\$}{\leftarrow} \{a_1 : k_1, ..., a_n : k_n\}]\!](\mu) = \sum_{i=1}^{n} a_i [\![X \leftarrow k_i]\!](\mu)$
- $[\![C_1;\ C_2]\!](\mu) = [\![C_2]\!]([\![C_1]\!](\mu))$
- $[\![\texttt{if}\, B\; \texttt{then}\, C_1\; \texttt{else}\, C_2]\!](\mu) = [\![C_1]\!](\downarrow_B(\mu)) + [\![C_2]\!](\downarrow_{\neg B}(\mu))$

The command skip changes nothing. We write $S[X \mapsto [\![E]\!]S]$ to denote the state which assigns variables the same values as S except that the variable X is assigned the value $[\![E]\!]S$. $\downarrow_B(\mu)$ denotes the distribution μ restricted to those states where B is true. Formally, $\downarrow_B(\mu) = \nu$ with $\nu(S) = \mu(S)$ if $[\![B]\!]S = \top$ and $\nu(S) = 0$ otherwise. We can write $[\![C]\!]S$ to denote $[\![C]\!]\mu_S$ if the initial state is deterministic. In general, if $\mu = [\![C]\!]S$, $S' \in sp(\mu)$ and $\mu(S') = a$, then it means that executing command C from state S will terminate on state S' with probability a.

• *Example 1* Let $R = \{\frac{1}{2} : 0, \frac{1}{2} : 1\}$ and let S be a deterministic state such that $S(X) = 1$. If we run the command $X \overset{\$}{\leftarrow} R$ on S, then distribution $[\![X \overset{\$}{\leftarrow} \{\frac{1}{2} : 0, \frac{1}{2} : 1\}]\!]\mu_S = \frac{1}{2}(\mu_{S[X \mapsto 0]}) + \frac{1}{2}(\mu_{S[X \mapsto 1]})$ is obtained.

2.3 Relational Arithmetic Expression and Assertion

Definition 6 (relational arithmetic expression). *Given a set of program variables* $\mathbb{PV}$*, we define the relational arithmetic expression e as follows:*

$$e := n \mid X\langle 1\rangle \mid X\langle 2\rangle \mid e \ aop \ e$$

where $n \in \mathbb{Z}$ *and* $X\langle 1\rangle, X\langle 2\rangle$ *are called tagged variables for all* $X \in \mathbb{PV}$.

Definition 7 (semantics of relational arithmetic expression). *Let* (S_1, S_2) *be a pair of deterministic states. The semantics of relational arithmetic expressions is defined as follows,*

$$\begin{aligned} [\![n]\!](S_1,S_2) &= n \\ [\![X\langle 1\rangle]\!](S_1,S_2) &= S_1(X) \\ [\![X\langle 2\rangle]\!](S_1,S_2) &= S_2(X) \\ [\![e_1 \ aop \ e_2]\!](S_1,S_2) &= [\![e_1]\!](S_1,S_2) \ aop \ [\![e_2]\!](S_1,S_2) \end{aligned}$$

The precondition and postcondition in Hoare triples are assertions. In the context of PRHL, an assertion describes a relation between a pair of deterministic states.

Definition 8 (assertion). *Assertions are defined as follows*

$$\phi := e \ rop \ e \mid \neg\phi \mid \phi \ lop \ \phi$$

The basic assertions here are quantifier-free Boolean combinations of arithmetic expressions. First-order quantification is allowed in assertions. This is because in Hoare logic, first-order quantification applies to logical variables but not program variables.

Definition 9 (semantics of assertion). *Let* S_1, S_2 *be any pair of deterministic states,*

- $(S_1,S_2) \models e_1 \ rop \ e_2$ *if* $[\![e_1]\!](S_1,S_2) \ rop \ [\![e_2]\!](S_1,S_2) = \top$;
- $(S_1,S_2) \models \neg\phi$, *if not* $(s_1,s_2) \models \phi$;
- $(S_1,S_2) \models \phi_1 \wedge \phi_2$ *if* $(s_1,s_2) \models \phi_1$ *and* $(s_1,s_2) \models \phi_2$;

2.4 Coupling and Lifting

The validity of Hoare triples in PRHL is defined on the concept of coupling, which is originally developed in probability theory for analyzing Markov chains.

Definition 10 (projection). *The project functions* π_1 *and* π_2 *are defined as follows: For any* μ *be a sub-distribution over* $A \times B$,

$$\pi_1(\mu)(a) = \sum_{b \in B} \mu(a,b)$$

and

$$\pi_2(\mu)(b) = \sum_{a \in A} \mu(a,b)$$

We call $\pi_1(\mu)$ and $\pi_2(\mu)$ the projection of μ to A and B respectively.

Definition 11 (coupling). *Let μ_1, μ_2 be sub-distributions over A, B respectively. μ over $A \times B$ is a coupling of μ_1 and μ_2, if $\mu_1 = \pi_1(\mu)$ and $\mu_2 = \pi_2(\mu)$. In this case, we write $\mu \triangleright \mu_1 \overline{\wedge} \mu_2$.*

Definition 12 (lifting). *The lifting of a relation $R \subseteq A \times B$ is a relation $R^\# \subseteq D(A) \times D(B)$ such that $(\mu_1, \mu_2) \in R^\#$ whenever there exists $\mu \in D(A \times B)$ satisfying that*

- $\mu \triangleright \mu_1 \overline{\wedge} \mu_2$;
- $sp(\mu) \subseteq R$.

Here we also say that μ is a *witnesses* of $\mu_1 R^\# \mu_2$. Valid Hoare triples in PRHL relate two output distributions by lifting the post-condition.

Definition 13 (PRHL judgment). *A Hoare triple of PRHL $\{\phi\}C_1 \sim C_2\{\psi\}$ is valid, written as $\models \{\phi\}C_1 \sim C_2\{\psi\}$, if for all state pairs (S_1, S_2),*

$$\text{if } (S_1, S_2) \models \phi \text{ then } ([\![C_1]\!]S_1, [\![C_2]\!]S_2) \in \psi^\#.$$

3 Proof System and Weakest Precondition

Intuitively, we say that two programs are coordinated if their flow charts are isomorphic. The precise definition of coordinated programs is inductively defined as follows.

Definition 14 (coordinated programs). *We use the following rules to define coordinated programs $C_1 \approx C_2$*

(SKIP): `skip` $\approx$ `skip`.
(ASSN): $X \leftarrow E \approx Y \leftarrow F$.
(RND): $X \xleftarrow{\$} D_1 \approx Y \xleftarrow{\$} D_2$.
(SEQ): *If $C_1 \approx C_3$ and $C_2 \approx C_4$, then $C_1; C_2 \approx C_3; C_4$.*
(IF): *If $C_1 \approx C_3$, $C_1 \approx C_4$, $C_2 \approx C_3$ and $C_2 \approx C_4$, then* `if` B_1 `then` C_1 `else` $C_2 \approx$ `if` B_2 `then` C_3 `else` C_4.

Coordinated straight programs ($\approx_{cs}$) is the subset of coordinated programs which is generated by the above rules without IF. Coordinated conditional programs ($\approx_{cc}$) is the subset of coordinated programs that is generated by the above rules without SEQ. Coordinated deterministic programs ($\approx_{cd}$) is the subset of coordinated programs that are generated by the above rules without RND. Coordinated random-conditional independent programs ($\approx_c^{rci}$) is the subset of coordinated programs that satisfies the random-conditional independent property (RCI): for all programs of the form $X \xleftarrow{\$} R$; `if` B `then` C_1 `else` C_2, it is the case that X does not appear in B.

3.1 Proof System

Now we introduce the proof system $PRHL_c$ for coordinated programs. It consists of the following rules.

Skip :

$$\frac{}{\vdash \{\phi\}\texttt{skip} \sim \texttt{skip}\{\phi\}}$$

Assn :

$$\frac{}{\vdash \{\phi[E\langle 1\rangle, F\langle 2\rangle / X\langle 1\rangle, Y\langle 2\rangle]\} X \leftarrow E \sim Y \leftarrow F\{\phi\}}$$

Here $E\langle 1\rangle$ is the tagged variant of E, which is obtained by replacing all the program variables in E by its corresponding $\langle 1\rangle$-tagged variable. $F\langle 2\rangle$ is defined similarly.

Rnd :

$$\frac{}{\vdash \{\bigvee_{\overline{f}\in[cp(D_1,D_2)]}(\bigwedge_{i\in sp(f)} \phi[fst(i), snd(i)/X\langle 1\rangle, Y\langle 2\rangle])\} X \xleftarrow{\$} D_1 \sim Y \xleftarrow{\$} D_2\{\phi\}}$$

where $D_1 = \{r_1 : c_1, \ldots, r_m : c_m\}, D_2 = \{t_1 : d_1, \ldots, t_n : d_n\}$. Here $[cp(D_1, D_2)]$ is the equivalence class of coupling of D_1 and D_2, where two couplings are equivalent if they have the same support. $\overline{f}$ is the equivalence of f. $fst(i)$ and $snd(i)$ mean the first and second components of i, respectively.

Seq :

$$\frac{\vdash \{\phi\}C_1 \sim C_2\{\psi\} \quad \vdash \{\psi\}C_3 \sim C_4\{\theta\}}{\vdash \{\phi\}C_1; C_3 \sim C_2; C_4\{\theta\}}$$

IF :

$$\frac{\begin{array}{ll}\vdash \{\phi \wedge B\langle 1\rangle \wedge B'\langle 2\rangle\}C_1 \sim C_3\{\psi\} & \vdash \{\phi \wedge B\langle 1\rangle \wedge \neg B'\langle 2\rangle\}C_1 \sim C_4\{\psi\} \\ \vdash \{\phi \wedge \neg B\langle 1\rangle \wedge B'\langle 2\rangle\}C_2 \sim C_3\{\psi\} & \vdash \{\phi \wedge \neg B\langle 1\rangle \wedge \neg B'\langle 2\rangle\}C_2 \sim C_4\{\psi\}\end{array}}{\vdash \{\phi\}\texttt{if } B \texttt{ then } C_1 \texttt{ else } C_2 \sim \texttt{if } B' \texttt{ then } C_3 \texttt{ else } C_4\{\psi\}}$$

Cons :

$$\frac{\models \phi \to \phi' \quad \vdash \{\phi'\}C_1 \sim C_2\{\psi'\} \quad \models \psi' \to \psi}{\vdash \{\phi\}C_1 \sim C_2\{\psi\}}$$

PRHL$_c$ contains one rule for each constructor of coordinated programs and one logical rule (Cons) dealing with the entailment of assertions. Here (Assn), (Rnd), (Seq) and (Cons) denote assignment, random assignment, sequential composition and consequence, respectively. The rules (Skip), (Assn), (Seq), and (Cons) are already used in other existing PRHL proof systems [10, 11]. The rule (Skip) simply says that the `skip` command preserves the pre-condition. The rule (Assn) is the usual Hoare-style rule: If ϕ is initially true with $E\langle 1\rangle$ and $F\langle 2\rangle$ replaced by $X\langle 1\rangle$ and $Y\langle 2\rangle$, then ϕ is true after the respective assignment command. The rule (Cons) is the usual consequence rule in Hoare logic, allowing us to strengthen the pre-condition and weaken the post-condition.

The rule (Seq) resembles the rule for sequential composition in Hoare logic. However, there are significant differences between them when we deeply investigate their semantics. The intermediate assertion ψ is interpreted differently in the two premises: in the first it is a post-condition and interpreted as a relation between distributions of states, while in the second it is a pre-condition and interpreted as a relation between states. This difference makes the proof of completeness of the (Seq) rule more difficult than that of other rules.

The rules (Rnd) and (IF) are quite different from their counterparts in other existing PRHL proof systems [10,11]. They seem complicated at first sight, but they are all sound and indispensable in proving relative completeness. In fact, it will be shown in Theorems 2 and 3 that they are the rules that capture the concept of weakest precondition.

We further define some subsets/variants of PRHL$_c$. PRHL$_{cs}$ is the proof system for the coordinated straight program. It is defined by removing the (IF) rule from PRHL$_c$. PRHL$_{cc}$ is the proof system for the coordinated conditional program. It is defined by removing the (Seq) rule from PRHL$_c$. RHL$_{cd}$ is the proof system for the coordinated deterministic program. It is defined by removing the (Rnd) rule from PRHL$_c$.

In the following, we will show that PRHL$_c$ is sound, which implies that all PRHL$_{cs}$, PRHL$_{cc}$ and PRHL$_{cd}$ are sound. In this paper, the relative completeness of PRHL$_{cs}$, PRHL$_{cc}$ and PRHL$_{cd}$ is also proved. We don't know whether PRHL$_c$ is relatively complete at the current stage. On the other hand, we will prove that PRHL$_c$ is relatively complete for coordinated programs which satisfy the RCI property.

Theorem 1. *PRHL$_c$ is sound. That is, for all coordinated programs $C_1 \approx C_2$, if $\vdash \{\phi\}C_1 \sim C_2\{\psi\}$, then $\models \{\phi\}C_1 \sim C_2\{\psi\}$.*

Proof. We prove this by structural induction.

(Skip) Let S_1, S_2 be arbitrary states such that $(S_1, S_2) \models \phi$. In this case we need to prove that $([\![skip]\!]S_1, [\![skip]\!]S_2) \in \phi^\sharp$. That is, we need to prove $(\mu_{S_1}, \mu_{S_2}) \in \phi^\sharp$, which is true because $\mu_{(S_1,S_2)} \triangleright \mu_{S_1} \overline{\wedge} \mu_{S_2}$ and all the support of $\mu_{(S_1,S_2)}$ satisfies ϕ.

(Assn) Let S_1, S_2 be arbitrary states such that $(S_1, S_2) \models \phi[E\langle 1\rangle, F\langle 2\rangle / X\langle 1\rangle, Y\langle 2\rangle]$. Let $\mu_{S'_1} = [\![X \leftarrow E]\!]S_1$ and $\mu_{S'_2} = [\![Y \leftarrow F]\!]S_2$. Then $(S'_1, S'_2) \models \phi$. Then we know $\mu_{S'_1,S'_2}$ is a witness of $(\mu_{S'_1}, \mu_{S'_2}) \in \phi^\sharp$.

(Rnd) Let (S_1, S_2) be an arbitrary pair of states on which

$$(S_1, S_2) \models \bigvee_{\overline{f} \in [cp(D_1,D_2)]} (\bigwedge_{i \in sp(f)} \phi[fst(i), snd(i)/X\langle 1\rangle, Y\langle 2\rangle]).$$

Then we know there is a coupling f of (D_1, D_2) such that

$$(S_1, S_2) \models \bigwedge_{i \in sp(f)} \phi[fst(i), snd(i)/X\langle 1\rangle, Y\langle 2\rangle].$$

Let $\mu^f_{S_1,S_2}$ be the distribution induced by (S_1, S_2) and f. That is,

$$\mu^f_{S_1,S_2}([\![X \leftarrow c_i]\!]S_1, [\![Y \leftarrow d_j]\!]S_2) = f(c_i, d_j).$$

Let $\mu_1 = [\![X \xleftarrow{\$} D_1]\!]S_1$, $\mu_2 = [\![Y \xleftarrow{\$} D_2]\!]S_2$. We then know that $\mu^f_{S_1,S_2}$ is a coupling of (μ_1, μ_2) because μ_i is isomorphic to D_i.
Moreover, from $(S_1,S_2) \models \bigwedge_{i \in sp(f)} \phi[fst(i), snd(i)/X\langle 1\rangle, Y\langle 2\rangle]$ we know that for all the support (c_m, d_n) of f, it holds that $(S_1,S_2) \models \phi[c_m, d_n/X\langle 1\rangle, Y\langle 2\rangle]$. Therefore, $([\![X \leftarrow c_m]\!]S_1, [\![Y \leftarrow d_n]\!]S_2) \models \phi$. Note that the support of f and the support of $\mu^f_{S_1,S_2}$ is in one-to-one correspondence, i.e. (c_m, d_n) is in correspondence with $([\![X \leftarrow c_m]\!]S_1, [\![Y \leftarrow d_n]\!]S_2)$. We then know all the support of $\mu^f_{S_1,S_2}$ satisfies ϕ, which implies that $\mu^f_{S_1,S_2}$ is a witness of $(\mu_1, \mu_2) \in \phi^\sharp$.

(Seq) By I.H. we have $\models \{\phi\}C_1 \sim C_2\{\psi\}$ and $\models \{\psi\}C_3 \sim C_4\{\theta\}$.
Let (S_1,S_2) be an arbitrary pair of states such that $(S_1,S_2) \models \phi$. From $\models \{\phi\}C_1 \sim C_2\{\psi\}$ we know $([\![C_1]\!]S_1, [\![C_2]\!]S_2) \in \psi^\sharp$. Let μ' be a witness of $([\![C_1]\!]S_1, [\![C_2]\!]S_2) \in \psi^\sharp$. Let (S_{1i}, S_{2i}) be an arbitrary support of μ', then we know $(S_{1i}, S_{2i}) \models \psi$. Now from $\models \{\psi\}C_3 \sim C_4\{\theta\}$ we deduce $([\![C_3]\!]S_{1i}, [\![C_4]\!]S_{2i}) \in \theta^\sharp$. Let μ_i be a witness of $([\![C_3]\!]S_{1i}, [\![C_4]\!]S_{2i}) \in \theta^\sharp$. Now we define

$$\mu = \mu'(S_{1i}, S_{2i}) \cdot \mu_i$$

Then μ is a witness of $([\![C_1]\!][\![C_3]\!]S_1, [\![C_2]\!][\![C_4]\!]S_2) \in \theta^\sharp$. This proves $\models \{\phi\}C_1; C_3 \sim C_2; C_4\{\theta\}$.

(IF) Let (S_1,S_2) be an arbitrary pair of states which satisfies ϕ.

1. If $S_1 \models B$ and $S_2 \models B'$, then $(S_1,S_2) \models B\langle 1\rangle \wedge B'\langle 2\rangle$. Then from $\models \{\phi \wedge B\langle 1\rangle \wedge B'\langle 2\rangle\}C_1 \sim C_3\{\psi\}$ we deduce $([\![C_1]\!]S_1, [\![C_3]\!]S_2) \in \psi^\sharp$, which further implies that $([\![\texttt{if } B \texttt{ then } C_1 \texttt{ else } C_2]\!]S_1, [\![\texttt{if } B' \texttt{ then } C_3 \texttt{ else } C_4]\!]S_2) \in \psi^\sharp$ because $([\![\texttt{if } B \texttt{ then } C_1 \texttt{ else } C_2]\!]S_1, [\![\texttt{if } B' \texttt{ then } C_3 \texttt{ else } C_4]\!]S_2) = ([\![C_1]\!]S_1, [\![C_3]\!]S_2)$.
2. Other cases are similar.

(Cons) Trivial. □

3.2 Relative Completeness

In order to prove relative completeness, we make use of the notion of weakest precondition. Semantically, given an assertion ψ and a pair of programs (C_1, C_2), the weakest precondition of ψ with (C_1, C_2) is an assertion θ such that $\models \{\theta\}C_1 \sim C_2\{\psi\}$ and for all assertion ϕ, if $\models \{\phi\}C_1 \sim C_2\{\psi\}$ then $\models \phi \rightarrow \theta$. The following is the syntactic notion of the weakest precondition. We will prove that semantically it is indeed a precondition and it is the weakest.

Definition 15. *The weakest precondition operator wp for coordinated programs is defined inductively on the structure of coordinated programs as follows:*

1. $wp(skip \sim skip, \phi) = \phi$.
2. $wp(X \leftarrow E \sim Y \leftarrow F, \phi) = \phi[E\langle 1\rangle, F\langle 2\rangle / X\langle 1\rangle, Y\langle 2\rangle]$.
3. $wp(X \xleftarrow{\$} D_1 \sim Y \xleftarrow{\$} D_2, \phi) = \bigvee_{\overline{f} \in [cp(D_1,D_2)]}(\bigwedge_{i \in sp(f)} \phi[fst(i), snd(i)/X\langle 1\rangle, Y\langle 2\rangle])$.

4. $wp(C_1;C_3 \sim C_2;C_4,\phi) = wp(C_1 \sim C_2, wp(C_3 \sim C_4,\phi))$.
5. $wp(\textit{if } B \textit{ then } C_1 \textit{ else } C_2 \sim \textit{if } B' \textit{ then } C_3 \textit{ else } C_4,\phi) =$
$(B\langle 1\rangle \wedge B'\langle 2\rangle \wedge wp(C_1 \sim C_3,\phi)) \vee (B\langle 1\rangle \wedge \neg B'\langle 2\rangle \wedge wp(C_1 \sim C_4,\phi)) \vee$
$(\neg B\langle 1\rangle \wedge B'\langle 2\rangle \wedge wp(C_2 \sim C_3,\phi)) \vee (\neg B\langle 1\rangle \wedge \neg B'\langle 2\rangle \wedge wp(C_2 \sim C_4,\phi))$.

The following theorem states that the weakest precondition is a precondition.

Theorem 2. *It holds that* $\vdash \{wp(C_1 \sim C_2,\phi)\}C_1 \sim C_2\{\phi\}$.

Proof. 1. The cases for (Skip), (Assn), and (Rnd) are trivial.

2. (Seq) By I.H. we have $\vdash \{wp(C_1 \sim C_2, wp(C_3 \sim C_4,\phi))\}C_1 \sim C_2\{wp(C_3 \sim C_4,\phi))\}$ and $\vdash \{wp(C_3 \sim C_4,\phi))\}C_3 \sim C_4\{\phi\}$. Then by (Seq) we know $\vdash \{wp(C_1 \sim C_2, wp(C_3 \sim C_4,\phi))\}C_1;C_3 \sim C_2;C_4\{\phi\}$.
3. (IF) Note that $\vDash (((B\langle 1\rangle \wedge B'\langle 2\rangle \wedge wp(C_1 \sim C_3,\phi)) \vee (B\langle 1\rangle \wedge \neg B'\langle 2\rangle \wedge wp(C_1 \sim C_4,\phi)) \vee (\neg B\langle 1\rangle \wedge B'\langle 2\rangle \wedge wp(C_2 \sim C_3,\phi)) \vee (\neg B\langle 1\rangle \wedge \neg B'\langle 2\rangle \wedge wp(C_2 \sim C_4,\phi))) \wedge (B\langle 1\rangle \wedge B'\langle 2\rangle)) \leftrightarrow (B\langle 1\rangle \wedge B'\langle 2\rangle \wedge wp(C_1 \sim C_3,\phi))$. By I.H. we have $\vdash \{wp(C_1 \sim C_3,\phi)\}C_1 \sim C_3\{\phi\}$. Then by (Cons) we know

$$\vdash \{B\langle 1\rangle \wedge B'\langle 2\rangle \wedge wp(C_1 \sim C_3,\phi)\}C_1 \sim C_3\{\phi\}.$$

Again by (Cons) we know $\vdash \{((B\langle 1\rangle \wedge B'\langle 2\rangle \wedge wp(C_1 \sim C_3,\phi)) \vee (B\langle 1\rangle \wedge \neg B'\langle 2\rangle \wedge wp(C_1 \sim C_4,\phi)) \vee (\neg B\langle 1\rangle \wedge B'\langle 2\rangle \wedge wp(C_2 \sim C_3,\phi)) \vee (\neg B\langle 1\rangle \wedge \neg B'\langle 2\rangle \wedge wp(C_2 \sim C_4,\phi))) \wedge (B\langle 1\rangle \wedge B'\langle 2\rangle)\}C_1 \sim C_3\{\phi\}$.
Similarly, we know $\vdash \{((B\langle 1\rangle \wedge B'\langle 2\rangle \wedge wp(C_1 \sim C_3,\phi)) \vee (B\langle 1\rangle \wedge \neg B'\langle 2\rangle \wedge wp(C_1 \sim C_4,\phi)) \vee (\neg B\langle 1\rangle \wedge B'\langle 2\rangle \wedge wp(C_2 \sim C_3,\phi)) \vee (\neg B\langle 1\rangle \wedge \neg B'\langle 2\rangle \wedge wp(C_2 \sim C_4,\phi))) \wedge (B\langle 1\rangle \wedge \neg B'\langle 2\rangle)\}C_1 \sim C_4\{\phi\}$,
$\vdash \{((B\langle 1\rangle \wedge B'\langle 2\rangle \wedge wp(C_1 \sim C_3,\phi)) \vee (B\langle 1\rangle \wedge \neg B'\langle 2\rangle \wedge wp(C_1 \sim C_4,\phi)) \vee (\neg B\langle 1\rangle \wedge B'\langle 2\rangle \wedge wp(C_2 \sim C_3,\phi)) \vee (\neg B\langle 1\rangle \wedge \neg B'\langle 2\rangle \wedge wp(C_2 \sim C_4,\phi))) \wedge (\neg B\langle 1\rangle \wedge B'\langle 2\rangle)\}C_2 \sim C_3\{\phi\}$ and
$\vdash \{((B\langle 1\rangle \wedge B'\langle 2\rangle \wedge wp(C_1 \sim C_3,\phi)) \vee (B\langle 1\rangle \wedge \neg B'\langle 2\rangle \wedge wp(C_1 \sim C_4,\phi)) \vee (\neg B\langle 1\rangle \wedge B'\langle 2\rangle \wedge wp(C_2 \sim C_3,\phi)) \vee (\neg B\langle 1\rangle \wedge \neg B'\langle 2\rangle \wedge wp(C_2 \sim C_4,\phi))) \wedge (\neg B\langle 1\rangle \wedge \neg B'\langle 2\rangle)\}C_2 \sim C_4\{\phi\}$.
Then by (IF) we know $\vdash \{((B\langle 1\rangle \wedge B'\langle 2\rangle \wedge wp(C_1 \sim C_3,\phi)) \vee (B\langle 1\rangle \wedge \neg B'\langle 2\rangle \wedge wp(C_1 \sim C_4,\phi)) \vee (\neg B\langle 1\rangle \wedge B'\langle 2\rangle \wedge wp(C_2 \sim C_3,\phi)) \vee (\neg B\langle 1\rangle \wedge \neg B'\langle 2\rangle \wedge wp(C_2 \sim C_4,\phi)))\}\text{if } B \text{ then } C_1 \text{ else } C_2 \sim \text{if } B' \text{ then } C_3 \text{ else } C_4\{\phi\}$. □

The next theorem (Theorem 3) states that the weakest precondition is the weakest for coordinated straight programs. Some parts of the proof are difficult. We will use structural induction to simultaneously prove Theorem 3 and Lemma 1. The induction is on the structure of $C_1 \approx_{cs} C_2$ for coordinated straight programs.

Lemma 1 (intermediate lemma). *Assume* $C_1 \approx_{cs} C_2$*, if* $C_3 \approx_{cs} C_4$ *and* $([\![C_1]\!][\![C_3]\!]S_1, [\![C_2]\!][\![C_4]\!]S_2) \in \phi^\sharp$*, then there exists* μ *which is a coupling of* $([\![C_3]\!]S_1, [\![C_4]\!]S_2)$ *such that for all* $(S_3,S_4) \in sp(\mu)$*, it holds that that* $([\![C_1]\!]S_3, [\![C_2]\!]S_4) \in \phi^\sharp$.

Proof. We prove by induction on the structure of (C_1,C_2).

1. If (C_1, C_2) is $(\texttt{skip}, \texttt{skip})$, then $([\![C_1]\!][\![C_3]\!]S_1, [\![C_2]\!][\![C_4]\!]S_2) \in \phi^\sharp$ implies $([\![C_3]\!]S_1, [\![C_4]\!]S_2) \in \phi^\sharp$, which means that there exists μ which is a coupling of $([\![C_3]\!]S_1, [\![C_4]\!]S_2)$ such that for all $(S_3, S_4) \in sp(\mu)$, it holds that that $(S_3, S_4) \models \phi$. It is easy to see that $([\![\texttt{skip}]\!]S_3, [\![\texttt{skip}]\!]S_4) \in \phi^\sharp$.
2. Assume C_1 is $X \leftarrow E$ and C_2 is $Y \leftarrow F$. Let μ' be the coupling of $([\![X \leftarrow E]\!][\![C_3]\!]S_1, [\![Y \leftarrow F]\!][\![C_4]\!]S_2)$ whose supports satisfy ϕ. We define the desired coupling μ of $([\![C_3]\!]S_1, [\![C_4]\!]S_2)$ via the following procedure.

   Let (S_x, S_y) be an arbitrary support of μ'. Let $\{S_{x1}, \ldots, S_{xm}\}$ be the set of all supports of $[\![C_3]\!]S_1$ such that $[\![X \leftarrow E]\!]S_{xj} = S_x$. Let $\{S_{y1}, \ldots, S_{yn}\}$ be the set of all supports of $[\![C_4]\!]S_2$ such that $[\![Y \leftarrow F]\!]S_{yk} = S_y$. Then we define

   $$\mu(S_{xj}, S_{yk}) = [\![C_3]\!]S_1(S_{xj}) \cdot [\![C_4]\!]S_2(S_{yk}).$$

   It is easy to see that all the support of μ satisfies $([\![X \leftarrow E]\!]S_{xj}, [\![Y \leftarrow F]\!]S_{yk}) = (S_x, S_y) \models \phi$.
   Now we prove that μ is a coupling of $([\![C_3]\!]S_1, [\![C_4]\!]S_2)$. We need to prove $\pi_1(\mu) = [\![C_3]\!]S_1$ and $\pi_2(\mu) = [\![C_4]\!]S_2$. We have

   $$\pi_1(\mu)(S_{xj}) = \sum_{y,k} [\![C_3]\!]S_1(S_{xj}) \cdot [\![C_4]\!]S_2(S_{yk}) = [\![C_3]\!]S_1(S_{xj}) \cdot \sum_{y,k} [\![C_4]\!]S_2(S_{yk})$$

   $$= [\![C_3]\!]S_1(S_{xj}) \cdot \sum_{y} [\![Y \leftarrow F]\!][\![C_4]\!]S_2(S_y) = [\![C_3]\!]S_1(S_{xj})$$

   This prove $\pi_1(\mu) = [\![C_3]\!]S_1$. The case for $\pi_2(\mu) = [\![C_4]\!]S_2$ is similar.
3. Assume C_1 is $X \xleftarrow{\$} \{a_1 : k_1, \cdots, a_m : k_m\}$ and C_2 is $Y \xleftarrow{\$} \{b_1 : l_1, \cdots, b_n : l_n\}$. Let μ' be the coupling of $([\![C_1]\!][\![C_3]\!]S_1, [\![C_2]\!][\![C_4]\!]S_2)$ whose supports satisfy ϕ. We define the desired coupling μ of $([\![C_3]\!]S_1, [\![C_4]\!]S_2)$ via the following procedure.
   Let (S_{ui}, S_{vj}) be an arbitrary support of μ'. Without loss of generality, we assume $S_{ui}(X) = k_u, S_{vj}(Y) = l_v$.
   Let S_{uip} range over the supports of $[\![C_3]\!]S_1$ such that $[\![X \leftarrow k_u]\!]S_{uip} = S_{ui}$. Let S_{vjq} range over the supports of $[\![C_4]\!]S_2$ such that $[\![Y \leftarrow l_v]\!]S_{vjq} = S_{vj}$.
   Then we define

   $$\mu(S_{uip}, S_{vjq}) = [\![C_3]\!]S_1(S_{uip}) \cdot [\![C_4]\!]S_2(S_{vjq}).$$

   Now we prove that all the support of μ satisfies $([\![C_1]\!]S_{uip}, [\![C_2]\!]S_{vjq}) \in \phi^\sharp$. Note that $[\![C_1]\!]S_{uiq}$ is a distribution with supports $\{S_{1x_1}, \ldots, S_{mx_m}\}$ and $[\![C_1]\!]S_{uiq}(S_{kx_k}) = a_k$, $[\![C_2]\!]S_{vjq}$ is a distribution with supports $\{S_{1y_1}, \ldots, S_{ny_n}\}$ and $[\![C_2]\!]S_{vjq}(S_{ky_k}) = b_k$. Let μ_1 be a distribution where $\mu_1(S_{ix_k}, S_{jy_h}) = a_i b_j$. Then μ_1 is a witness of $([\![C_1]\!]S_{uip}, [\![C_2]\!]S_{vjq}) \in \phi^\sharp$.

   Finally, we prove that μ is a coupling of $([\![C_3]\!]S_1, [\![C_4]\!]S_2)$. We need to prove $\pi_1(\mu) = [\![C_3]\!]S_1$ and $\pi_2(\mu) = [\![C_4]\!]S_2$. We have

   $$\pi_1(\mu)(S_{uip}) = \sum_{v,j,q} [\![C_3]\!]S_1(S_{uip}) \cdot [\![C_4]\!]S_2(S_{vjq}) =$$

$$[\![C_3]\!]S_1(S_{uip}) \cdot \sum_{v,j,q} [\![C_4]\!]S_2(S_{vjq}).$$

Since S_{vjq} ranges over all the supports of $[\![C_4]\!]S_2$, we know

$$\sum_{v,j,q} [\![C_4]\!]S_2(S_{vjq}) = 1.$$

This prove $\pi_1(\mu) = [\![C_3]\!]S_1$. The case for $\pi_2(\mu) = [\![C_4]\!]S_2$ is similar.

4. Assume C_1 is $C_{12};C_{11}$, C_2 is $C_{22};C_{21}$. By I.H. there exists μ_1 which is a coupling of $([\![C_{12}]\!][\![C_3]\!]S_1, [\![C_{22}]\!][\![C_4]\!]S_2)$ such that for all $(S_3,S_4) \in sp(\mu_1)$, it holds that $([\![C_{11}]\!]S_3, [\![C_{21}]\!]S_4) \in \phi^\sharp$. By the I.H. of Theorem 3 we then know $(S_3,S_4) \models wp(C_{11} \sim C_{21}, \phi)$. Therefore, $([\![C_{12}]\!][\![C_3]\!]S_1, [\![C_{22}]\!][\![C_4]\!]S_2) \in wp(C_{11} \sim C_{21}, \phi)^\sharp$. Now by I.H. we know there exists μ_2 which is a coupling of $([\![C_3]\!]S_1, [\![C_4]\!]S_2)$ such that for all $(S_5,S_6) \in sp(\mu_2)$, it is true that $([\![C_{12}]\!]S_5, [\![C_{22}]\!]S_6) \in wp(C_{11} \sim C_{21}, \phi)^\sharp$. Let μ_3 be a coupling of $([\![C_{12}]\!]S_5, [\![C_{22}]\!]S_6)$ such that all the supports of μ_3 satisfies $wp(C_{11} \sim C_{21}, \phi)$. Let (S_i,S_j) be an arbitrary support of μ_3, then $(S_i,S_j) \models wp(C_{11} \sim C_{21}, \phi)$. Then by Theorem 2 and soundness we know $\models \{wp(C_{11} \sim C_{21}, \phi)\}C_{11} \sim C_{21}\{\phi\}$. Now from $(S_i,S_j) \models wp(C_{11} \sim C_{21}, \phi)$ we deduce $([\![C_{11}]\!]S_i, [\![C_{21}]\!]S_j) \in \phi^\sharp$, which means that there exists a coupling μ_{ij} of $([\![C_{11}]\!]S_i, [\![C_{21}]\!]S_j)$ whose support satisfies ϕ. Now we define a coupling μ of $([\![C_{11}]\!][\![C_{12}]\!]S_5, [\![C_{21}]\!][\![C_{22}]\!]S_6)$ as follows:

$$\mu = \sum_{i,j} \mu_3(S_i,S_j) \cdot \mu_{ij}$$

It can be verified that all the support of μ satisfies ϕ. We then know

$$([\![C_{11}]\!][\![C_{12}]\!]S_5, [\![C_{21}]\!][\![C_{22}]\!]S_6) \in \phi^\sharp.$$

□

Theorem 3. *Assume $C_1 \approx_{cs} C_2$, if $\models \{\psi\}C_1 \sim C_2\{\phi\}$, then $\models \psi \rightarrow wp(C_1 \sim C_2, \phi)$.*

Proof. 1. (Skip) Assume $\models \{\psi\}skip \sim skip\{\phi\}$, we need to prove that $\models \psi \rightarrow \phi$. Let (S_1,S_2) be arbitrary states such that $(S_1,S_2) \models \psi$. By $\models \{\psi\}skip \sim skip\{\phi\}$ we know $([\![skip]\!]S_1, [\![skip]\!]S_2) \in \phi^\sharp$. Since μ_{S_1,S_2} is the only coupling of μ_{S_1} and μ_{S_2}, we know that $\mu_{S_1,S_2} \models \phi$, i.e. $(S_1,S_2) \models \phi$. This proves that $\models \psi \rightarrow \phi$.

2. (Assn) Let (S_1,S_2) be arbitrary states such that $(S_1,S_2) \models \psi$. Let $\mu_{S'_1} = [\![X \leftarrow E]\!]S_1$ and $\mu_{S'_2} = [\![Y \leftarrow F]\!]S_2$. Then by $\models \{\psi\}X \leftarrow E \sim Y \leftarrow F\{\phi\}$ we know $(\mu_{S'_1}, \mu_{S'_2}) \in \phi^\sharp$. Since $\mu_{S'_1,S'_2}$ is the only coupling of $(\mu_{S'_1}, \mu_{S'_2})$, we know $(\mu_{S'_1}, \mu_{S'_2}) \models \phi$. That is, $([\![X \leftarrow E]\!]S_1, [\![Y \leftarrow F]\!]S_2) \models \phi$, which implies that $(S_1,S_2) \models \phi[E\langle 1\rangle, F\langle 2\rangle / X\langle 1\rangle, Y\langle 2\rangle]$.

3. (Rnd) Let (S_1,S_2) be an arbitrary pair of states such that $(S_1,S_2) \models \psi$. Let $\mu_1 = [\![X \xleftarrow{\$} D_1]\!]S_1$, $\mu_2 = [\![Y \xleftarrow{\$} D_2]\!]S_2$. From $\models \{\psi\}X \xleftarrow{\$} D_1 \sim Y \xleftarrow{\$} D_2\{\phi\}$ we know $(\mu_1,\mu_2) \in \phi^\sharp$, which means that there is a coupling μ of (μ_1,μ_2) such that $(S'_1,S'_2) \models \phi$ for all $(S'_1,S'_2) \in supp(\mu)$. Note that there is a one-to-one correspondence between the coupling of (μ_1,μ_2) and the coupling of (D_1,D_2). Let f_μ be

the coupling of (D_1, D_2) which corresponds to μ. Let $(c_m, d_n) \in supp(f_\mu)$. Then $(S_1, S_2) \vdash \phi[c_m, d_n/X\langle 1\rangle, Y\langle 2\rangle]$. We further know that

$$(S_1, S_2) \vdash \bigvee_{f \in [cp(D_1, D_2)]} (\bigwedge_{i \in sp(f)} \phi[fst(i), snd(i)/X\langle 1\rangle, Y\langle 2\rangle]).$$

4. (SEQ) Assume $\models \{\phi\}C_1; C_2 \sim C_3; C_4\{\psi\}$ and $(S_1, S_2) \models \phi$. Then $([\![C_1; C_2]\!]S_1, [\![C_3; C_4]\!]S_2) \in \psi^\sharp$, which means that $([\![C_2]\!][\![C_1]\!]S_1, [\![C_4]\!][\![C_3]\!]S_2) \in \psi^\sharp$. Then by Lemma 1 we know there exists μ which is a coupling of $([\![C_1]\!]S_1, [\![C_3]\!]S_2)$ such that for all $(S_3, S_4) \in supp(\mu)$, it holds that that $([\![C_2]\!]S_3, [\![C_4]\!]S_4) \in \psi^\sharp$. Let ϕ' be the formula which characterizes (S_3, S_4). We then know $\models \{\phi'\}C_2 \sim C_4\{\psi\}$. By I.H. we know $\models \phi' \to wp(C_2 \sim C_4, \psi)$. Hence $(S_3, S_4) \models wp(C_2 \sim C_4, \psi)$. Since (S_3, S_4) is an arbitrary support of μ, we then know $([\![C_1]\!]S_1, [\![C_3]\!]S_2) \in wp(C_2 \sim C_4, \psi)^\sharp$. Let ϕ'' be the formula which characterizes (S_1, S_2). Then we know $\models \{\phi''\}C_1 \sim C_3\{wp(C_2 \sim C_4, \psi)\}$. Then by I.H. we know $(S_1, S_2) \models wp(C_1 \sim C_3, wp(C_2 \sim C_4, \psi))$. □

Just like in Hoare logic, as long as we have a syntactic notion of the weakest precondition, we can easily prove that our proof system is relatively complete.

Theorem 4. *PRHL$_{cs}$ is relatively complete. That is, for all coordinated straight programs $C_1 \approx_{cs} C_2$, if $\models \{\phi\}C_1 \sim C_2\{\psi\}$, then $\vdash \{\phi\}C_1 \sim C_2\{\psi\}$.*

Proof. Assume $\models \{\phi\}C_1 \sim C_2\{\psi\}$, then by Theorem 3 we know $\models \phi \to wp(C_1 \sim C_2, \psi)$. By Theorem 2 we know $\vdash \{wp(C_1 \sim C_2, \psi)\}C_1 \sim C_2\{\psi\}$. Now by (Cons) we derive $\vdash \{\phi\}C_1 \sim C_2\{\psi\}$.

The following corollary shows the characterizing feature of the weakest precondition. It is also used implicitly several times in later proofs.

Corollary 1. *For all states S_1, S_2 and coordinated straight programs $C_1 \approx_{cs} C_2$, $(S_1, S_2) \models wp(C_1 \sim C_2, \phi)$ iff $([\![C_1]\!]S_1, [\![C_2]\!]S_2) \in \phi^\sharp$.*

Proof. $(\Rightarrow)$ By Theorem 2 and soundness we know $\models \{wp(C_1 \sim C_2, \phi)\}C_1 \sim C_2\{\phi\}$. Now from $(S_1, S_2) \models wp(C_1 \sim C_2, \phi)$ we deduce $([\![C_1]\!]S_1, [\![C_2]\!]S_2) \in \phi^\sharp$.

$(\Leftarrow)$ Assume $([\![C_1]\!]S_1, [\![C_2]\!]S_2) \in \phi^\sharp$. Let ϕ' be the formula such that $[\![\phi']\!] = \{(S_1, S_2)\}$. Note that such formula always exist and is unique up to equivalence because we assume the number of program variables to be finite. Then $(S_1, S_2) \models \phi'$. Then we know $\models \{\phi'\}C_1 \sim C_2\{\phi\}$. Hence we have $\models \phi' \to wp(C_1 \sim C_2, \phi)$. We then know $(S_1, S_2) \models wp(C_1 \sim C_2, \phi)$. □

Using a similar strategy, we can prove that both PRHL$_{cc}$ and PRHL$_{cd}$ are relatively complete in the following 4 theorems.

Theorem 5. *Assume $C_1 \approx_{cc} C_2$, if $\models \{\psi\}C_1 \sim C_2\{\phi\}$, then $\models \psi \to wp(C_1 \sim C_2, \phi)$.*

Proof. We prove by induction on the structure of $C_1 \approx_{cc} C_2$. The case for Skip, Assn, Rnd are the same as in the proof of Theorem 3. Now we prove the case for IF.

1. (IF) Assume $\models \{\psi\}\texttt{if}\ B\ \texttt{then}\ C_{11}\ \texttt{else}\ C_{12} \sim \texttt{if}\ B'\ \texttt{then}\ C_{21}\ \texttt{else}\ C_{22}\{\phi\}$. Let (S_1,S_2) be an arbitrary pair of states such that $(S_1,S_2) \models \psi$. Then $([\![\texttt{if}\ B\ \texttt{then}\ C_{11}\ \texttt{else}\ C_{12}]\!]S_1, [\![\texttt{if}\ B'\ \texttt{then}\ C_{21}\ \texttt{else}\ C_{22}]\!]S_2) \in \phi^\sharp$.
   (a) If $S_1 \models B$ and $S_2 \models B'$, then

$$([\![\texttt{if}\ B\ \texttt{then}\ C_{11}\ \texttt{else}\ C_{12}]\!]S_1, [\![\texttt{if}\ B'\ \texttt{then}\ C_{21}\ \texttt{else}\ C_{22}]\!]S_2)$$

$$= ([\![C_{11}]\!]S_1, [\![C_{21}]\!]S_2).$$

   This means that $([\![C_{11}]\!]S_1, [\![C_{21}]\!]S_2) \in \phi^\sharp$. Let ϕ' be the formula such that $[\![\phi']\!] = \{(S_1,S_2)\}$. Note that such a formula always exist and is unique up to equivalence because we assume the number of program variables to be finite. Then $(S_1,S_2) \models \phi'$. Then we know $\models \{\phi'\}C_{11} \sim C_{12}\{\phi\}$. Hence we have $\models \phi' \rightarrow wp(C_{11} \sim C_{12}, \phi)$. We then know $(S_1,S_2) \models wp(C_{11} \sim C_{12}, \phi)$. It then follows that $(S_1,S_2) \models B\langle 1\rangle \wedge B'\langle 2\rangle \wedge wp(C_{11} \sim C_{21}, \phi)$.
   (b) Other cases are similar. □

Theorem 6. *$PRHL_{cc}$ is relatively complete. That is, for all coordinated conditional programs $C_1 \approx_{cc} C_2$, if $\models \{\phi\}C_1 \sim C_2\{\psi\}$, then $\vdash \{\phi\}C_1 \sim C_2\{\psi\}$.*

Proof. Assume $\models \{\phi\}C_1 \sim C_2\{\psi\}$, then by Theorem 5 we know $\models \phi \rightarrow wp(C_1 \sim C_2, \psi)$. By Theorem 2 we know $\vdash \{wp(C_1 \sim C_2, \psi)\}C_1 \sim C_2\{\psi\}$. Now by (Cons) we derive $\vdash \{\phi\}C_1 \sim C_2\{\psi\}$. □

Theorem 7. *Assume $C_1 \approx_{cd} C_2$, if $\models \{\psi\}C_1 \sim C_2\{\phi\}$, then $\models \psi \rightarrow wp(C_1 \sim C_2, \phi)$.*

Proof. We prove by induction on the structure of $C_1 \approx_{cd} C_2$. The case for Skip and Assn are the same as in the proof of Theorem 3. The case for IF is the same as in the proof of Theorem 5. Now we prove the case for Seq.

Assume $\models \{\phi\}C_1;C_2 \sim C_3;C_4\{\psi\}$ and $(S_1,S_2) \models \phi$. Then $([\![C_1;C_2]\!]S_1, [\![C_3;C_4]\!]S_2) \in \psi^\sharp$, which means that $([\![C_2]\!][\![C_1]\!]S_1, [\![C_4]\!][\![C_3]\!]S_2) \in \psi^\sharp$. Let $[\![C_1]\!]S_1 = S_3, [\![C_3]\!]S_2 = S_4$. Let ϕ' be the formula which characterizes (S_3,S_4). We then know $\models \{\phi'\}C_2 \sim C_4\{\psi\}$. By I.H. we know $\models \phi' \rightarrow wp(C_2 \sim C_4, \psi)$. Hence $(S_3,S_4) \models wp(C_2 \sim C_4, \psi)$.

We then know $([\![C_1]\!]S_1, [\![C_3]\!]S_2) \in wp(C_2 \sim C_4, \psi)^\sharp$. Let ϕ'' be the formula which characterizes (S_1,S_2). Then we know $\models \{\phi''\}C_1 \sim C_3\{wp(C_2 \sim C_4, \psi)\}$. Then by I.H. we know $(S_1,S_2) \models wp(C_1 \sim C_3, wp(C_2 \sim C_4, \psi))$. □

Theorem 8. *$PRHL_{cd}$ is relatively complete. That is, for all coordinated conditional programs $C_1 \approx_{cd} C_2$, if $\models \{\phi\}C_1 \sim C_2\{\psi\}$, then $\vdash \{\phi\}C_1 \sim C_2\{\psi\}$.*

Proof. Assume $\models \{\phi\}C_1 \sim C_2\{\psi\}$, then by Theorem 7 we know $\models \phi \rightarrow wp(C_1 \sim C_2, \psi)$. By Theorem 2 we know $\vdash \{wp(C_1 \sim C_2, \psi)\}C_1 \sim C_2\{\psi\}$. Now by (Cons) we derive $\vdash \{\phi\}C_1 \sim C_2\{\psi\}$. □

So far we have proved that 3 subsets of $PRHL_c$ are relatively complete, but we cannot prove that $PRHL_c$ is relatively complete at the current stage. The difficulty lies in proof of the Intermediate Lemma when dealing with conditional choice. To avoid the difficulty, we make use of the random-conditional independent (RCI) property.

Lemma 2 (RCI-intermediate lemma). *Assume* $C_1 \approx_c^{rci} C_2$*, if* $C_3 \approx_c^{rci} C_4$ *and* $([\![C_1]\!][\![C_3]\!]S_1,\ [\![C_2]\!][\![C_4]\!]S_2) \in \phi^\sharp$*, then there exists* μ *which is a coupling of* $([\![C_3]\!]S_1, [\![C_4]\!]S_2)$ *such that for all* $(S_3,S_4) \in sp(\mu)$*, it holds that that* $([\![C_1]\!]S_3, [\![C_2]\!]S_4) \in \phi^\sharp$.

Proof. We prove by induction of the structure of $C_1 \approx_c^{rci} C_2$. Other cases are the same as in the proof of intermediate lemma. Now we deal with the case in which C_1 is $\texttt{if}\, B\, \texttt{then}\, C_{11}\, \texttt{else}\, C_{12}$, C_2 is $\texttt{if}\, B'\, \texttt{then}\, C_{21}\, \texttt{else}\, C_{22}$.

Assume $([\![\texttt{if}\, B\, \texttt{then}\, C_{11}\, \texttt{else}\, C_{12}]\!][\![C_3]\!]S_1, [\![\texttt{if}\, B'\, \texttt{then}\, C_{21}\, \texttt{else}\, C_{22}]\!][\![C_4]\!]S_2) \in \phi^\sharp$.

Then there exist μ_1 which is a coupling of $([\![\texttt{if}\, B\, \texttt{then}\, C_{11}\, \texttt{else}\, C_{12}]\!][\![C_3]\!]S_1, [\![\texttt{if}\, B'\, \texttt{then}\, C_{21}\, \texttt{else}\, C_{22}]\!][\![C_4]\!]S_2)$ such that for all $(S_3,S_4) \in sp(\mu_1)$ it is true that $(S_3,S_4) \vDash \phi$.

Now we proceed our proof by induction of the structure of $C_3 \approx_c^{rci} C_4$.

1. If (C_3,C_4) is $(\texttt{skip},\texttt{skip})$, then $([\![C_1]\!][\![\texttt{skip}]\!]S_1, [\![C_2]\!][\![\texttt{skip}]\!]S_2) \in \phi^\sharp$ implies $([\![C_1]\!]S_1,\ [\![C_2]\!]S_2) \in \phi^\sharp$. Let $\mu = \mu_{(S_1,S_2)}$. Then μ is a coupling of $([\![\texttt{skip}]\!]S_1, [\![\texttt{skip}]\!]S_2)$ and the only support of μ is (S_1,S_2). We already know that $([\![C_1]\!]S_1, [\![C_2]\!]S_2) \in \phi^\sharp$.
2. Assume C_3 is $X \leftarrow E$ and C_4 is $Y \leftarrow F$. Then $([\![C_1]\!][\![X \leftarrow E]\!]S_1, [\![C_2]\!][\![Y \leftarrow F]\!]S_2) \in \phi^\sharp$. Let $S_3 = [\![X \leftarrow E]\!]S_1$, $S_4 = [\![Y \leftarrow F]\!]S_2$. Then $\mu_{(S_3,S_4)}$ is a coupling of $([\![X \leftarrow E]\!]S_1, [\![Y \leftarrow F]\!]S_2)$ and its only support is (S_3,S_4). It's easy to see that $([\![C_1]\!]S_3, [\![C_2]\!]S_4) = [\![C_1]\!][\![X \leftarrow E]\!]S_1, [\![C_2]\!][\![Y \leftarrow F]\!]S_2) \in \phi^\sharp$
3. Assume C_3 is $X \xleftarrow{\$} \{a_1 : k_1, \cdots, a_m : k_m\}$ and C_4 is $Y \xleftarrow{\$} \{b_1 : l_1, \cdots, b_n : l_n\}$. By the RCI property we know that X does not appear in B and Y does not appear in B'.
   (a) Assume $S_1 \vDash B$ and $S_2 \vDash B'$. Then $([\![C_1]\!][\![C_3]\!]S_1, [\![C_2]\!][\![C_4]\!]S_2) = ([\![C_{11}]\!][\![C_3]\!]S_1, [\![C_{21}]\!][\![C_4]\!]S_2)$, which implies that $([\![C_{11}]\!][\![C_3]\!]S_1, [\![C_{21}]\!][\![C_4]\!]S_2) \in \phi^\sharp$. Then by I.H. we know ther exists exists μ which is a coupling of $([\![C_3]\!]S_1, [\![C_4]\!]S_2)$ such that for all $(S_3,S_4) \in sp(\mu)$, it holds that that $([\![C_{11}]\!]S_3, [\![C_{21}]\!]S_4) \in \phi^\sharp$. Note that by RCI we know $S_3 \vDash B$ and $S_4 \vDash B'$. Therefore $([\![C_1]\!]S_3, [\![C_2]\!]S_4) = ([\![C_{11}]\!]S_3, [\![C_{21}]\!]S_4) \in \phi^\sharp$.
   (b) Other cases are similar.
4. Assume C_3 is $C_{31};C_{32}$ and C_4 is $C_{41};C_{42}$. The proof is similar to the proof in Lemma 1 where C_1 is $C_{12};C_{11}$ and C_2 is $C_{22};C_{21}$.
5. Assume C_3 is $\texttt{if}\, B_3\, \texttt{then}\, C_{31}\, \texttt{else}\, C_{32}$, C_4 is $\texttt{if}\, B_4\, \texttt{then}\, C_{41}\, \texttt{else}\, C_{42}$.
   (a) If $S_1 \vDash B_3$ and $S_2 \vDash B_4$, then $([\![C_3]\!]S_1, [\![C_4]\!]S_2) = ([\![C_{31}]\!]S_1, [\![C_{41}]\!]S_2)$ and $([\![C_1]\!][\![C_3]\!]S_1,\ [\![C_2]\!][\![C_4]\!]S_2) = ([\![C_1]\!][\![C_{31}]\!]S_1, [\![C_2]\!][\![C_{41}]\!]S_2)$. By I.H. we know there exists μ which is a coupling of $([\![C_{31}]\!]S_1, [\![C_{41}]\!]S_2)$ such that for all $(S_3,S_4) \in sp(\mu)$, it holds that that $([\![C_1]\!]S_3, [\![C_2]\!]S_4) \in \phi^\sharp$. Note that μ is also a coupling of $([\![C_3]\!]S_1, [\![C_4]\!]S_2)$.
   (b) Other cases are similar.

Theorem 9. *Assume* $C_1 \approx_c^{rci} C_2$*, if* $\vDash \{\psi\}C_1 \sim C_2\{\phi\}$*, then* $\vDash \psi \rightarrow wp(C_1 \sim C_2, \phi)$.

Proof. The case for Skip, Assn, Rnd, Seq are the same as in the proof of Theorem 3. The case for IF is the same as in the proof of Theorem 5. □

Theorem 10. *$PRHL_c$ is relatively complete for coordinated programs which satisfies the random-conditional independent property.*

Proof. Assume $\models \{\phi\}C_1 \sim C_2\{\psi\}$, then by Theorem 9 we know $\models \phi \rightarrow wp(C_1 \sim C_2, \psi)$. By Theorem 2 we know $\vdash \{wp(C_1 \sim C_2, \psi)\}C_1 \sim C_2\{\psi\}$. Now by (Cons) we derive $\vdash \{\phi\}C_1 \sim C_2\{\psi\}$. □

4 Conclusion and Future Research

The relative completeness of PRHL has not been established since its birth in 2009. This paper proves for the first time that some fragments of PRHL are relatively complete. In the future, we will study whether we can expand those fragments without losing relative completeness.

Acknowledgement. The research of the author is supported by Shandong Provincial Natural Science Foundation, China (project number: ZR2023QF021).

References

1. Apt, K.R., Olderog, E.R.: Fifty years of hoare's logic. Formal Aspects Comput. **31**, 751–807 (2019)
2. Avanzini, M., Barthe, G., Davoli, D., Grégoire, B.: A quantitative probabilistic relational hoare logic. Proc. ACM Program. Lang. **9**(POPL), 1167–1195 (2025)
3. Baltag, A., Christoff, Z., Rendsvig, R.K., Smets, S.: Dynamic epistemic logics of diffusion and prediction in social networks. Stud. Logica **107**(3), 489–531 (2019)
4. Baltag, A., Gierasimczuk, N., Özgün, A., Sandoval, A.L.V., Smets, S.: A dynamic logic for learning theory. J. Log. Algebraic Methods Program. **109** (2019)
5. Baltag, A., Smets, S.: LQP: the dynamic logic of quantum information. Math. Struct. Comput. Sci. **16**(3), 491–525 (2006)
6. Baltag, A., Smets, S.: A dynamic-logical perspective on quantum behavior. Stud. Logica **89**(2), 187–211 (2008)
7. Baltag, A., Smets, S.: Quantum logic as a dynamic logic. Synthese **179**(2), 285–306 (2011)
8. Baltag, A., Smets, S.: The dynamic turn in quantum logic. Synthese **186**(3), 753–773 (2012)
9. Barthe, G., Espitau, T., Grégoire, B., Hsu, J., Stefanesco, L., Strub, P.Y.: Relational reasoning via probabilistic coupling. In: Davis, M., Fehnker, A., McIver, A., Voronkov, A. (eds.) Logic for Programming, Artificial Intelligence, and Reasoning, pp. 387–401. Springer, Heidelberg (2015)
10. Barthe, G., Grégoire, B., Béguelin, S.Z.: Formal certification of code-based cryptographic proofs. In: Shao, Z., Pierce, B.C. (eds.) Proceedings of the 36th ACM SIGPLAN-SIGACT Symposium on Principles of Programming Languages, POPL 2009, Savannah, GA, USA, 21–23 January 2009, pp. 90–101. ACM (2009)
11. Barthe, G., Hsu, J.: Probabilistic Couplings from Program Logics, pp. 145–184. Cambridge University Press (2020)
12. van Benthem, J., Ghosh, S., Liu, F.: Modelling simultaneous games in dynamic logic. Synthese **165**(2), 247–268 (2008)
13. Chadha, R., Cruz-Filipe, L., Mateus, P., Sernadas, A.: Reasoning about probabilistic sequential programs. Theoret. Comput. Sci. **379**(1–2), 142–165 (2007)

14. Cook, S.A.: Soundness and completeness of an axiom system for program verification. SIAM J. Comput. **7**(1), 70–90 (1978)
15. Deng, Y., Feng, Y.: Formal semantics of a classical-quantum language. Theor. Comput. Sci. **913**, 73–93 (2022)
16. Dijkstra, E.W.: Guarded commands, nondeterminacy and formal derivation of programs. Commun. ACM **18**(8), 453–457 (1975)
17. Ditmarsch, H., Hoek, W., Kooi, B.: Dynamic Epistemic Logic. Springer (2007)
18. Foley, M., Hoare, C.A.R.: Proof of a recursive program: quicksort. Comput. J. **14**(4), 391–395 (1971)
19. Hoare, C.A.R.: Procedures and parameters: an axiomatic approach. In: Engeler, E. (ed.) Symposium on Semantics of Algorithmic Languages. Lecture Notes in Mathematics, vol. 188, pp. 102–116. Springer (1971)
20. Hoare, C.A.R.: An axiomatic basis for computer programming. Commun. ACM **12**(10), 576–580 (1969)
21. Hoare, C.A.R.: Procedures and parameters: an axiomatic approach. In: Proceedings of Symposium on the Semantics of Algorithmic Languages. Lecture Notes in Mathematics, vol. 188, pp. 102–116. Springer (1971)
22. Lindvall, T.: Lectures on the Coupling Method. Courier Corporation (2002)
23. Özgün, A., Berto, F.: Dynamic hyperintensional belief revision. Rev. Symb. Log. **14**(3), 766–811 (2021)
24. Pacuit, E.: Dynamic models of rational deliberation in games. In: van Benthem, J., Ghosh, S., Verbrugge, R. (eds.) Models of Strategic Reasoning - Logics, Games, and Communities. Lecture Notes in Computer Science, vol. 8972, pp. 3–33. Springer (2015)
25. Pratt, V.R.: Semantical considerations on Floyd-Hoare logic. In: 17th Annual Symposium on Foundations of Computer Science, Houston, Texas, USA, 25–27 October 1976, pp. 109–121. IEEE Computer Society (1976)
26. Rand, R., Zdancewic, S.: VPHL: a verified partial-correctness logic for probabilistic programs. Electron. Notes Theor. Comput. Sci. **319**, 351–367 (2015)
27. Sun, X., He, F.: A first step to the categorical logic of quantum programs. Entropy **22**(2), 144 (2020)
28. Sun, X., Su, X., Bian, X., Cui, A.: On the relative completeness of satisfaction-based probabilistic hoare logic with while loop (2024). https://arxiv.org/abs/2406.16054
29. Sun, X., Su, X., Bian, X., Wu, H.: On the relative completeness of satisfaction-based quantum hoare logic (2024). https://arxiv.org/abs/2405.01940
30. Ying, M.: Floyd-Hoare logic for quantum programs. ACM Trans. Program. Lang. Syst. **33**(6), 19:1–19:49 (2011)

Virtual Group Knowledge and Group Belief in Topological Evidence Models

Alexandru Baltag[1], Malvin Gattinger[1], and Djanira Gomes[2(✉)]

[1] ILLC, University of Amsterdam, Amsterdam, The Netherlands
[2] Institute of Computer Science, University of Bern, Bern, Switzerland
djaniradsgomes@gmail.com

Abstract. We study notions of (virtual) group knowledge and group belief within multi-agent evidence models, obtained by extending the topological semantics of evidence-based belief and fallible knowledge from individuals to groups. We completely axiomatize and show the decidability of the logic of ("hard" and "soft") group evidence, and do the same for an especially interesting fragment of it: the logic of group knowledge and group belief. We also extend these languages with dynamic evidence-sharing operators, and completely axiomatize the corresponding logics, showing that they are co-expressive with their static bases.

1 Introduction

A natural framework for reasoning about knowledge in distributed systems is Epistemic Logic: an umbrella term for modal logics that formalize notions of knowledge and belief for rational agents. Traditionally, these logics are interpreted on *relational (Kripke) models*, according to Hintikka's semantics [20]. It is also useful to have notions of knowledge and belief associated with *groups* [18]. The best-known are *distributed* and *common knowledge*. The first is inherently linked to communication: it describes what a group of agents *could come to know* after sharing their individual information with the group [19]. This "virtual" or "potential" aspect is made explicit in Dynamic Epistemic Logic [4,10,14], with dynamic operators for information sharing [1,3,12,16].

Recently, *topological models* for epistemic logics have gained popularity, see e.g. [5,9,11,13,15,22,23,25]. An advantage of topological semantics is that it comes with a natural, semantical notion of *evidence*, making the evidential basis of knowledge and belief apparent.

In this paper we use multi-agent topological evidence models, or *topo-e-models*, which explicitly represent the topology of evidence [7,8]. One way to interpret knowledge and belief in topo-e-models is to apply the *interior semantics* of McKinsey and Tarski [21] to (a basis for) the so-called *dense-open topology*. This restricts the evidential topology to *dense open sets*, which represent *"uncontroversial" evidential justifications*: pieces of evidence consistent with all other evidence. Belief amounts to having such a justification, and (fallible, defeasible) knowledge is interpreted as *correctly justified belief* [7].

J. Wang et al. (Eds.): DaLí 2025, LNCS 16472, pp. 96–114, 2026.
https://doi.org/10.1007/978-3-032-22626-6_6

A natural continuation of this research is to extend the framework to the multi-agent case and to incorporate a notion of group knowledge. It has long been noticed [5,15,22,25] that the most straightforward such extension is obtained by applying the same definitions (as for individual knowledge and belief) to the *join topology*, obtained by pooling together all the individual evidence. One objection [15,22] raised against this notion is that it loses the main characteristic property of classical distributed knowledge, namely *Group Monotonicity* (saying that *a group potentially knows everything known by any subgroup*): in topo-e-models, group knowledge is *not* monotonic with respect to group inclusion. In fact, a group may sometimes know even *less* than *any* of its members [7].[1]

Nevertheless, in this paper we argue that Monotonicity will have to fail for *any* realistic notion of group knowledge. Since fallible knowledge is not fully introspective, agents cannot separate it from other beliefs in order to share it; so the best they can do is to share all their *evidence*. And it turns out that the topological notion of group knowledge matches the knowledge that can be obtained after evidence-sharing. In this sense, topological group knowledge accurately captures the *group's epistemic potential*: its true "virtual" knowledge. In a nutshell: the failure of Group Monotonicity is a "feature", not a "bug".

Given this fact, it becomes imperative to study *the laws governing this natural notion of group knowledge, and the corresponding concept of group belief*. In this paper, we provide *complete and decidable axiomatizations* of these notions, as well as of the related concepts of *group evidence*. While our axioms of group evidence are the expected ones (similar to any other distributed attitude in Epistemic Logic), it turns out that virtual group knowledge obeys new interesting laws, that can be seen as subtle forms of weakening Group Monotonicity. The completeness proof for the logic of group knowledge is also more intricate, relying on a new representation result. In order to make explicit the sense in which our notion captures a group's epistemic potential, we add *dynamic evidence-sharing modalities*, and we completely axiomatize the resulting dynamic logics.

The paper is structured as follows. Section 2 presents topo-e-models and defines our key epistemic notions. Section 3 gives the syntax, semantics and axiomatizations of our logics, and states our completeness/decidability results. Section 4 contains some conclusions and an open question for future work.

Extended Version. This paper is based on the Master thesis of the third author [17]. The original proofs in [17] use somewhat different notations and definitions than the ones adopted here. We provide an extended version of the paper at https://arxiv.org/abs/2509.00184 with full proof details fitting the current version.

2 Topological Knowledge and Evidence-Sharing

In this section we introduce *multi-agent topological evidence models*, and define the notions of *hard and soft evidence, knowledge and belief*, and their *natural*

[1] In order to resolve this, two alternatives of this semantics have been proposed [15,22], both ensuring the validity of the Group Monotonicity property.

extensions to groups. We discuss the crucial differences between virtual group knowledge and the standard concept of distributed knowledge, and we explain and defend the first from a communication-based perspective, formalized in terms of an *evidence-sharing update.* The presentation is purely semantical-mathematical: we postpone the introduction of our formal languages to Sect. 3.

Topological Prerequisites. We first recall some basic topological notions. Given a set X, a *topology* τ is a family of subsets of X, called *open sets.* The *closed* sets are given by their complements: $\bar{\tau} = \{X \setminus U \mid U \in \tau\}$. The topology τ by definition contains $\emptyset$ and X as elements, and is closed under finite intersections and arbitrary unions. A set X equipped with a topology τ is called a *topological space*, denoted (X, τ).

Given a space X, every set $A \subseteq X$ has an *interior* and a *closure*, which are computed by the *interior* and *closure* operators $Int_\tau, Cl_\tau : \mathcal{P}(X) \to \mathcal{P}(X)$, respectively. The *interior* $Int_\tau(A)$ of $A \subseteq X$ is the union of all open subsets of A; the *closure* $Cl_\tau(A)$ is its dual:

$$Int_\tau(A) = \bigcup\{U \in \tau \mid U \subseteq A\}$$

$$Cl_\tau(A) = \bigcap\{C \in \bar{\tau} \mid A \subseteq C\}.$$

While the interior of A is the largest open set contained in A, its closure is the least closed set containing A.

A family $\mathcal{B} \subseteq \tau$ is a *topological basis* for a topological space (X, τ) if every non-empty open subset of X can be written as a union of elements of $\mathcal{B}$. A *subbasis* for (X, τ) is a family $\mathcal{B} \subseteq \tau$, whose closure under finite intersections forms a basis for (X, τ). Given any family of subsets $\mathcal{E} \subseteq \mathcal{P}(X)$, we obtain the *generated topology* by closing $\mathcal{E}$ under finite intersections and, subsequently, under arbitrary unions. The topology generated by $\mathcal{E}$ is the smallest topology τ on X s.t. $\mathcal{E} \subseteq \tau$. The *join* $\bigvee_{i \in I} \tau_i$ of a family $\{\tau_i\}_{i \in I}$ of topologies on the same set X is defined as the topology generated by the union $\bigcup_{i \in I} \tau_i$.[2]

Topology-Partition Pairs, Local Density and the Dense-Open Topology. We shall consider *topology-partition pairs* (τ, Π), consisting of a topology τ on a set X and a partition $\Pi = \{\Pi(x) \mid x \in X\}$ of X (where each x belongs to a unique partition cell $\Pi(x)$), s.t. all partition cells are open (i.e., $\Pi \subseteq \tau$). For every open set $U \in \tau$, we denote by $\Pi(U) := \bigcup\{\Pi(x) \mid x \in U\}$ the union of cells of all points in U. For a point $x \in X$, we say that U is *locally dense in* $\Pi(x)$ (or "locally dense at x") if $Cl_\tau(U) \supseteq \Pi(x)$. We say that U is *locally dense in* Π (or just "locally dense", when Π is understood) if U is locally dense at all its points, i.e., $Cl_\tau(U) \supseteq \Pi(U)$. It is easy to see that the family

$$\tau^{dense(\Pi)} := \{U \in \tau \mid Cl_\tau(U) \supseteq \Pi(U)\} \cup \{\emptyset\}$$

(consisting of all locally dense open sets and $\emptyset$) is itself a topology, called *the dense-open topology* for (τ, Π). Once again, when Π is understood from context, we skip it and just write τ^{dense} instead.

[2] This is the same as the *supremum* of the family $\{\tau_i\}_{i \in I}$ in the lattice of all topologies on X with inclusion.

2.1 Knowledge and Belief in Multi-Agent Topo-Evidence Models

Topological evidence models [7,8] are a variant of the *evidence models* defined by [13], in which the role of the topology is stressed and the definition of belief is streamlined (to ensure its consistency). While [7] studied these notions within a single-agent setting, this has been generalized to multi-agent models in [5,15,22].

Vocabulary: Atoms, Agents and Groups. Throughout this paper, we fix a vocabulary, consisting of: a finite or countable set Prop of *atomic formulas* $p, q, \ldots$, intuitively denoting "ontic facts": non-epistemic features of the world; and a finite set $A = \{1, 2 \ldots, n\}$ of *agents*, labeled by numbers, and denoted by meta-variables $i, j, k, \ldots$. A *group* is a non-empty set of agents (i.e., any $I \subseteq A$ with $I \neq \emptyset$). We use capital letters $I, J, K, \ldots$ as meta-variables for groups.

Definition 1 (Topo-E-Models). *A* multi-agent topological evidence model *(or "topo-e-model", for short) is a tuple* $\mathfrak{M} = (X, \Pi_1, \ldots, \Pi_n, \tau_1, \ldots, \tau_n, [\![\cdot]\!])$ *(or* $(X, \Pi_i, \tau_i, [\![\cdot]\!])_{i \in A}$ *for short), where:*

- *X is a set of* states *(or "possible worlds");*
- *For each $i \in A$, the family $\Pi_i \subseteq \mathcal{P}(X)$ is a partition of X, called* agent i's information partition, *and consisting of mutually disjoint partition cells. Every state $x \in X$ belongs to a* unique cell $\Pi_i(x) \in \Pi_i$, *representing the* private information*—the "hard evidence"—possessed by agent i in state x. The states $x' \in \Pi_i(x)$ are said to be* indistinguishable *from x by agent i;*
- *For each $i \in A$, $\tau_i \subseteq \mathcal{P}(X)$ is a topology on X, called* agent i's evidential topology, *and subject to the constraint that $\Pi_i \subseteq \tau_i$ ("hard evidence is evidence"). The non-empty open sets ($U \in \tau_i$ s.t. $U \neq \emptyset$) represent* agent i's ("soft") evidence. *For any state $x \in X$, $\tau_i^*(\Pi_i(x)) := \{U \in \tau_i \mid \emptyset \neq U \subseteq \Pi_i(x)\}$ is the collection of all* soft evidence possessed by agent i at state x; *while $\tau_i^*(x) := \{U \in \tau_i \mid x \in U \subseteq \Pi_i(x)\}$ is the collection of* agent i's factive ("true") evidence at state x.[3] *We denote by Cl_i and Int_i the* closure Cl_{τ_i} and interior Int_{τ_i} operators *with respect to agent i's evidential topology τ_i.*
- $[\![\cdot]\!] : X \to \mathcal{P}(\mathsf{Prop})$ *is a* valuation *function, mapping each atomic formula $p \in \mathsf{Prop}$ to the set $[\![p]\!] \subseteq X$ of states "satisfying" p.*

The intuition is that in state x, each agent $i \in A$ has some "hard" evidence $\Pi_i(x)$, as well as some pieces of "soft" evidence $U \in \tau_i^*(\Pi_i(x))$. Since $x \in \Pi_i(x)$, the hard evidence is *infallibly true* (i.e., true with absolute certainty),[4] while soft evidence can be false (when $x \notin U$); moreover, two pieces of soft evidence $U, V \in \tau_i^*(\Pi_i(x))$ may be mutually inconsistent (when $U \cap V = \emptyset$).

[3] For the consistency of our notation, note that $\tau_i^*(\Pi_i(x)) = \bigcup\{\tau_i^*(y) : y \in \Pi_i(x)\}$.

[4] This is the reason we assigned *only one piece of (private) hard evidence* $\Pi_i(x)$ to each agent i at each state x. In principle, one can of course have many pieces of hard evidence; but, since they are mutually consistent (being all true in the actual world), the agent can just combine all of them by taking their intersection.

Subbasis Presentation. The evidential topology is sometimes specified using a *designated subbasis* $\mathcal{E}_i^0 \subseteq \mathcal{P}(X)$, with $\emptyset \notin \mathcal{E}_i^0$. Intuitively, the sets $U \in \mathcal{E}_i^0$ represent the "basic" or *"primary" evidence*: the pieces of evidence that are *directly observable*. The agent then forms the family $\mathcal{E}_i$ of *conjunctive evidence* by taking the closure of $\mathcal{E}_i^0$ under finite intersections.[5] Finally, she forms the topology τ_i, as the family of *disjunctive evidence* (also known as "arguments"), by closing $\mathcal{E}_i$ under unions.[6] While the subbasis presentation is computationally less demanding, the distinction between primary evidence and indirect (conjunctive or disjunctive) evidence does not play any role in the semantics.

Propositions and Operators. A *proposition* in model $\mathfrak{M} = (X, \Pi_i, \tau_i, [\![\cdot]\!])_{i \in A}$ is a set of states $P \in \mathcal{P}(X)$. An example are *atomic propositions*: those of the form $[\![p]\!]$, for $p \in \mathsf{Prop}$. Note that the family $\mathcal{P}(X)$ forms a Boolean algebra, with the operations of set-complementation, intersection and union. Next, we define a number of (unary) *propositional operators* $\Gamma : \mathcal{P}(X) \to \mathcal{P}(X)$.

Hard Evidence Gives Infallible Knowledge. Given a proposition $P \subseteq X$, we say that *an agent i has hard evidence for P (or "infallibly knows" P) at state x* if P is true at all states that are indistinguishable for i from x, i.e., if $\Pi_i(x) \subseteq P$. Formally, the proposition "agent i infallibly knows P" is denoted by

$$[\forall]_i(P) := \{x \in X \mid \Pi_i(x) \subseteq P\}.$$

This is an absolutely certain, "infallible" type of knowledge, hence it is *factive*, i.e., we have $[\forall]_i(P) \subseteq P$, and *fully (=positively and negatively) introspective*, i.e., we have $[\forall]_i(P) = [\forall]_i([\forall]_i(P))$ and $X - [\forall]_i(P) = [\forall]_i(X - [\forall]_i(P))$.

Interior as "Soft Evidence" Operator. We say that *agent i has factive evidence for P at state x* if there is some $U \in \tau_i^*(x)$ with $U \subseteq P$;[7] equivalently, if $x \in Int_i(P)$. The proposition "i has factive evidence for P" is denoted by:

$$\Box_i(P) := \{x \in X \mid \exists U \in \tau_i : x \in U \subseteq P\} = Int_i(P).$$

This attitude is again *factive*, i.e., $\Box_i(P) = Int_i(P) \subseteq P$, and *positively (but not negatively) introspective*, i.e., $\Box_i(P) = \Box_i(\Box_i(P))$. The dual of $\Box_i$ is denoted by $\Diamond_i(P)$ and matches *topological closure*: $\Diamond_i(P) := X - \Box_i(X - P) = Cl_i(P)$.

Justified Belief. According to the dense-interior semantics [7,8], *rational agents base their beliefs only on "uncontroversial" evidence*: those pieces of evidence that are not contradicted by any other evidence available to them.[8] *Agent i believes P at state x* if i has such "uncontroversial" evidence for P: some $U \in \tau_i^*(\Pi_i(x))$ s.t. $U \subseteq P$ and $U \cap V \neq \emptyset$ for all $V \in \tau_i^*(\Pi_i(x))$. It is easy to see that an open

[5] Note that $X = \bigcap \emptyset \in \mathcal{E}_i$.

[6] Note that τ_i equals the topology *generated* by $\mathcal{E}_i^0$.

[7] Requiring $U \in \tau_i^*(x)$ with $U \subseteq P$ is in fact equivalent to requiring $U \in \tau_i$ with $x \in U \subseteq P$, as $U \in \tau_i^*(x)$ implies that $x \in U$ and that $U \in \tau_i$ and, conversely, the existence of an $U \in \tau_i$ with $x \in U \subseteq P$ implies the existence of an $U' \in \tau_i^*(x)$ with $U' \subseteq P$. Hence, throughout the paper, we use the two specifications interchangeably.

[8] Note that "uncontroversial" does not mean "factive": such evidence can be false.

subset $U \subseteq P$ is an uncontroversial piece of evidence for P at x for agent i iff U is *locally dense at* x with respect to (τ_i, Π_i), i.e., iff $Cl_i(U) \supseteq \Pi_i(x)$. In this case, U can be thought of as a *justification* for (believing) P: one that "coheres" with all the available evidence. Equivalently, *P is believed at x iff its interior is locally dense at x.* The operator for *agent i's belief* is denoted by

$$B_i(P) \;:=\; \{x \in X \mid \Pi_i(x) \subseteq Cl_i(Int_i(P))\},$$

while its dual $\langle B_i \rangle(P) := X - B_i(X - P)$ captures "doxastic possibility".

Fallible Knowledge.[9] We say that *an agent i "knows" P at state x* if she has a *factive justification* (= true uncontroversial evidence) for P: there is some $U \in \tau_i^*(x)$, with $U \subseteq P$, and $Cl_i(U) \supseteq \Pi_i(x)$. Equivalently, iff x is in the locally dense interior of P for i: $x \in Int_i(P)$ and $Cl_i(Int_i(P)) \supseteq \Pi_i(x)$. We denote by $K_i(P)$ the proposition "i knows P":

$$K_i(P) \;:=\; \{x \in X \mid \exists U \in \tau_i : x \in U \subseteq P \text{ and } Cl_i(U) \supseteq \Pi_i(x)\}.$$

In words: *knowledge is correctly justified belief.*[10] In contrast to $[\forall]_i(P)$, this type of knowledge is "defeasible": it can be defeated by "misleading" evidence [7]. Its dual $\langle K_i \rangle(P) \;:=\; X - K_i(X - P)$ captures a notion of "soft epistemic possibility".

Knowledge as Interior in the Dense-Open Topology. We characterized knowledge $K_i(P)$ of a proposition P as the locally dense interior of P (for i). Equivalently, we can characterize knowledge as the *interior in the dense-open topology* τ_i^{dense}:

$$K_i(P) = Int_{\tau_i^{dense}}(P).$$

That is, under our characterization, $K_i(P)$ coincides with the interior of P in agent i's topology of locally dense open sets.

Connections Between Operators. For $P \subseteq X$ and $i \in A$, we have:

$$[\forall]_i(P) \subseteq \Box_i(P), \qquad K_i(P) \subseteq B_i(P).$$

In words: *hard evidence is also soft evidence*, and *agents believe the things they know.* More interestingly, we have the following equations, which will allow us to define belief and knowledge as abbreviations in one of our formal languages:

$$B_i(P) = [\forall]_i \Diamond_i \Box_i(P), \qquad K_i(P) = \Box_i(P) \cap B_i(P), \qquad B_i(P) = \langle K_i \rangle K_i(P).$$

The first equation follows directly from the characterizations of $[\forall]_i$, $\Diamond_i$, and $\Box_i$. The second states that having a correct justification of P amounts to having a justification for P, as well as a piece of factive evidence U for P.[11] Finally,

[9] Notions of knowledge that do not imply absolute certainty are called *fallible*. In our setting, only the "hard" evidence $\Pi_i(x)$ provides "infallible" knowledge.

[10] Note the difference between *correctly* justified belief and *true* justified belief [25].

[11] The left-to-right inclusion of this equation is immediate; for the converse inclusion, recall that agent i has a justification for P at x iff $Int_i(P)$ is locally dense at x. By definition, $Int_i(P)$ contains U, which contains x, hence, the justification is correct.

the last equation says that belief is also definable in terms of fallible knowledge: *belief is the "soft possibility" of knowledge.*[12]

2.2 Group Evidence, Group Belief and Group Knowledge

The most natural way to generalize the above notions from individual agents $i \in A$ to *groups* $I \subseteq A$ is to pool together all the hard and soft evidence possessed by agents in I into a group partition Π_I and a group evidential topology τ_I.

Group Evidence: Join Partition and Join Topology. Given a group $I \subseteq A$ and a topo-e-model $\mathfrak{M} = (X, \Pi_i, \tau_i, [\![\cdot]\!])_{i\in A}$, *group I's hard evidence at a state* $x \in X$ is the intersection (conjunction) of all individual group members' hard evidence at state x. The group's hard-evidence sets form again a partition Π_I, called *group I's partition*, which coincides with the join (supremum) $\bigvee \Pi_i$ of all individual partitions (in the lattice of partitions on X with inclusion):[13]

$$\Pi_I : = \bigvee_{i\in I} \Pi_i = \{\Pi_I(x) \mid x \in X\}, \quad \text{where} \quad \Pi_I(x) := \bigcap_{i\in I} \Pi_i(x).$$

Similarly, *group I's evidential topology* τ_I is just the *join topology*

$$\tau_I \;:=\; \bigvee_{i\in I} \tau_i \;(= \text{the topology generated by the union } \bigcup_{i\in I} \tau_i).$$

To motivate this, note that τ_I is also generated by *the group's "joint evidence"*, i.e. by the family of all non-empty intersections $\bigcap_{i\in I} U_i \neq \emptyset$ of individual pieces of soft evidence U_i possessed by any of the group's members $i \in I$. As before, we use Int_I and Cl_I for the interior and closure operators w.r.t. τ_I.

Group Operators. A *group operator* on a set X is a group-indexed family $\Gamma = \{\Gamma_I\}_{I\subseteq A, I\neq\emptyset}$ of propositional operators $\Gamma_I : \mathcal{P}(X) \to \mathcal{P}(X)$. As usual, when $I = \{i\}$ is a singleton consisting of a single agent, we write Γ_i instead of $\Gamma_{\{i\}}$.

Examples: Group Evidence, Group Belief, Group Knowledge. As important examples, we define *group analogues* of all the individual attitudes, by simply *applying the same definitions to the group partition and the group's soft evidence*:

$$\begin{aligned}
[\forall]_I(P) &:= \{x \in X \mid \Pi_I(x) \subseteq P\},\\
\Box_I(P) &:= \{x \in X \mid \exists U \in \tau_I : x \in U \subseteq P\} = Int_I(P),\\
B_I(P) &:= \{x \in X \mid \exists U \in \tau_I : U \subseteq P \text{ and } Cl_I(U) \supseteq \Pi_I(x)\}\\
&= \{x \in X \mid \Pi_I(x) \subseteq Cl_I(Int_I(P))\},\\
K_I(P) &:= \{x \in X \mid \exists U \in \tau_I : x \in U \subseteq P \text{ and } Cl_I(U) \supseteq \Pi_I(x)\}\\
&= \{x \in X \mid x \in Int_I(P) \text{ and } \Pi_I(x) \subseteq Cl_I(Int_I(P))\}.
\end{aligned}$$

The *Diamond (possibility) operators* $\Diamond_I(P)$, $\langle B_I\rangle(P)$ and $\langle K_I\rangle(P)$ are defined in the same way (as De Morgan duals) for groups $I \subseteq A$ as for individuals $i \in A$.

---

[12] This observation was taken by Stalnaker as the basis of a version of knowledge-first epistemology, which differs from the more well-known Williamsonian knowledge-first conception, by the fact that it is positively introspective.

[13] This is the smallest partition Π_I that includes that union $\bigcup_{i\in I} \Pi_i$.

Group operators are connected in the same way as the individual ones: we have $[\forall]_I(P) \subseteq \Box_I(P)$, $K_I(P) \subseteq B_I(P)$, $B_I(P) = [\forall]_I \Diamond_I \Box_I(P)$, $K_I(P) = \Box_I(P) \cap B_I(P)$ and $B_I(P) = \langle K_I \rangle K_I(P)$. As a consequence, we will *define group belief and group knowledge as abbreviations* in one of our formal languages.

Group Knowledge is Interior in the Dense-Open Join Topology. Similar to the alternative characterization of individual knowledge K_i as interior w.r.t. the individual dense-open topology τ_i^{dense}, we can equivalently characterize group knowledge K_I as the *interior operator w.r.t. the dense-open topology* $\tau_I^{dense} = \tau_I^{dense(\Pi_I)}$ *associated to the join topology* τ_I: $K_I(P) = Int_{\tau_I^{dense}}(P)$.

Interpreting the Group Operators. The above definitions seem natural from a mathematical point of view. But what is the *interpretation* of these group operators? Are they just formal analogues of the individual ones, with no intrinsic meaning or practical application, or do they capture some useful group attitudes? To give a partial answer, we need the following generalized notions:

Monotonicity and Distributedness. A group operator Γ is *monotonic* if it satisfies the *Group Monotonicity* condition: $I \subseteq J$ implies $\Gamma_I(P) \subseteq \Gamma_J(P)$. The operator Γ is *distributed* if it satisfies the *Group Distributedness* condition:

$$x \in \Gamma_I(P) \text{ iff } x \in \bigcap_{i \in I} \Gamma_i(P_i) \text{ for some } (P_i)_{i \in I} \text{ s.t. } \bigcap_{i \in I} P_i \subseteq P.$$

Distributedness implies that *Γ's behavior on sets can be recovered from its behavior on singletons.*[14] Moreover, it is easy to see that *every distributed operator is monotonic.*

Example: Distributed Knowledge in Relational Structures. The standard example of a distributed operator is the classical relational concept of *distributed knowledge* D_I in a multi-agent epistemic Kripke model, defined as the *Kripke modality for the intersection of all agents' accessibility relations.* This notion satisfies Group Distributedness (and hence also Group Monotonicity).[15] This fits the intended meaning of D_I: a group's distributed knowledge is simply the result of "adding" or "aggregating" all the knowledge possessed by the individuals.

Group Evidence is Distributed Evidence. It is easy to see that our group evidences operators $[\forall]_I$ and $\Box_I$ are distributed (and thus also monotonic). This provides the promised interpretation: *a group's evidence is the result of "adding" or "aggregating" all the evidence possessed by the individuals.*

2.3 The "Problem" of Non-Monotonicity

Unfortunately, we cannot use the "distributed knowledge" interpretation for our topological group knowledge and belief operators: *neither K_I nor B_I are distributed group operators*, and *they do not even satisfy the weaker Group*

[14] In philosophical jargon, the distributed group operators are *summative* attitudes.

[15] Indeed, Group Monotonicity is the main axiom for D_I in standard Epistemic Logic.

Monotonicity property! Moreover, a group may even *fail to (know or even just) believe* facts that are *known by all* its members: in general, we have $\bigcap_{i \in I} K_i(P) \nsubseteq B_I(P)$, as shown by the following counterexample.

Example 2. Let $\mathfrak{M} = (X, \Pi_i, \tau_i, [\![\cdot]\!])_{i \in A}$ be given by: $\mathsf{Prop} = \{p\}$; $A = \{a, b\}$; $X = \{w_1, w_2, w_3, w_4\}$; $[\![p]\!] = \{w_1, w_2, w_4\}$; partitions $\Pi_a = \Pi_b = \{\{X\}\}$; and topologies τ_a and τ_b are generated respectively by subases $\mathcal{E}^0_a = \{\{w_2, w_4\}, \{w_3, w_4\}\}$ and $\mathcal{E}^0_b = \{\{w_1, w_2\}, \{w_1, w_3\}\}$, representing each agent's primary or "direct" evidence. We can then calculate the topologies τ_a, τ_b and $\tau_{\{a,b\}} = \tau_a \vee \tau_b = \mathcal{P}(X)$. Note that $\tau_{\{a,b\}}$ is the discrete topology, generated by $\mathcal{E}^0_A = \mathcal{E}^0_a \cup \mathcal{E}^0_b$ (Fig. 1).

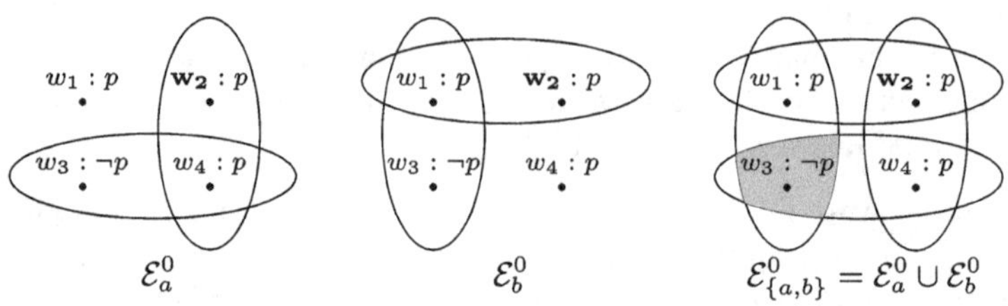

Fig. 1. The model from Example 2. For each topology, we draw only the primary evidence (the subbases $\mathcal{E}^0_a$, $\mathcal{E}^0_b$ and $\mathcal{E}^0_A = \mathcal{E}^0_a \cup \mathcal{E}^0_b$), and omit the single-cell partitions. Take $P = [\![p]\!] = \{w_1, w_2, w_4\}$. At w_2, a has τ_a-dense factive evidence $U_a = \{w_2, w_4\}$ for P, and b has τ_b-dense factive evidence $U_b = \{w_1, w_2\}$ for P, hence $w_2 \in K_a(P) \cap K_b(P)$. But $\{w_3\} = \{w_3, w_4\} \cap \{w_1, w_3\} \in \mathcal{E}_A$ is disjoint from P, hence $w_2 \notin B_{\{a,b\}}(P)$.

The failure of Group Monotonicity was taken as an *objection* against the topological definition of group knowledge [15,22,25]. Consequently, Ramírez [22] and Fernández [5,15] proposed alternative notions of group knowledge in topo-evidence models, designed to "save" Group Monotonicity. Here we only present Fernández' solution, because of its relevance for our discussion and our axioms.

Fernández' Approach: Topological Distributed Knowledge. In his Master thesis [15], Fernández proposes a different topological definition of group knowledge, later developed and investigated by Baltag et al. [5]. As we saw, individual knowledge K_i for agent i coincides with interior in agent i's dense-open topology τ_i^{dense}; while the virtual group knowledge operator K_I coincides with interior in the *group's* dense-open topology τ_I^{dense} (which is the dense-open topology for the pair (τ_I, Π_I), obtained by taking the joins of all the individual partitions and respectively all individual topologies). Fernández' proposal is to use instead the natural topological analogue of distributed knowledge D_I, as the *interior operator w.r.t. the join* $\bigvee_{i \in I} \tau_i^{dense}$ *of all individuals' dense-open topologies*:

$$D_I(P) := Int_{\bigvee_{i \in I} \tau_i^{dense}}(P).$$

This topological notion generalizes the relational definition of distributed knowledge in $S4$ (or $S5$) Kripke models,[16] and moreover *the topological D_I is indeed "distributed"* (in the above sense), and it thus also satisfies Group Monotonicity.

2.4 Dynamics: A Communication-Based View on Group Knowledge

In contrast to the mentioned authors, we will argue that the non-monotonic notion K_I fits better than D_I with a *communication-based interpretation of group knowledge*. In the context of distributed systems (see e.g. [18]), the concepts of knowledge and communication are intertwined. A realistic notion of "virtual" group knowledge should be "realizable" (as individual knowledge) through in-group communication. As we will see, K_I fulfills this desideratum (while D_I does not), so K_I is in fact more realistic and useful than D_I. To show this, we look at the group dynamics induced by evidence-exchange.

Evidence-Sharing Dynamics. For each group $I \subseteq A$, one can define an operator $\mathsf{share}(I)$ on topo-e-models, that represents the action of *sharing all evidence (soft and hard) within group I*. This is a "semi-public" action in the sense of [12]: intuitively, the outsiders $j \notin I$ know that this sharing is happening within group I, but they do not necessarily have access to the evidence that is being shared; in fact, it is common knowledge among *all* agents that this information-sharing event $\mathsf{share}(I)$ is happening; while the insiders $i \in I$ have more information: they gain common knowledge of which evidence is being shared among them. This is an "evidential" version of other group-sharing operators in the literature: the "deliberation" action in [16], the "share" action in [3], the "resolution" action in [1], or the semi-public sharing actions considered in [12].

Definition 3. *Given a topo-e-model $\mathfrak{M} = (X, \Pi_i, \tau_i, [\![\cdot]\!])_{i \in A}$ and a group $I \subseteq A$,* the updated model $\mathfrak{M}(\mathsf{share}_I) := (X, \Pi(\mathsf{share}_I), \tau(\mathsf{share}_I), [\![\cdot]\!])$ *has the same set of states and valuation, while the new partitions and topologies are given by:*

$$\begin{array}{lll} \tau_i(\mathsf{share}_I) = \tau_I, & \Pi_i(\mathsf{share}_I) \ = \Pi_I & \text{(for "insiders" } i \in I\text{)}, \\ \tau_j(\mathsf{share}_I) = \tau_j & \Pi_j(\mathsf{share}_I), = \Pi_j & \text{(for "outsiders" } j \notin I\text{)}, \end{array}$$

where τ_I is the group's topology, and Π_I is the group's partition.

Since the set of states X and the valuation $[\![\cdot]\!]$ stay the same when moving from $\mathfrak{M}$ to the updated model $\mathfrak{M}(\mathsf{share}_I)$, we can talk about the same semantic propositions $P \subseteq X$ in both models. However, the meaning of our operators $[\forall]_i, \Box_i, K_i, B_i$ differs in the two models! So we use $[\forall]_i^{\mathfrak{M}}, \Box_i^{\mathfrak{M}}, K_i^{\mathfrak{M}}, B_i^{\mathfrak{M}}$ to denote the operators in the model $\mathfrak{M}$, and $[\forall]_i^{\mathfrak{M}(\mathsf{share}_I)}$, $\Box_i^{\mathfrak{M}(\mathsf{share}_I)}$, $K_i^{\mathfrak{M}(\mathsf{share}_I)}$, $B_i^{\mathfrak{M}(\mathsf{share}_I)}$ to denote the operators in the updated model $\mathfrak{M}(\mathsf{share}_I)$.

With these notations, we can now make the following key observation:

[16] $S4$-frames are a special case of topological spaces (the Alexandroff spaces): the standard Kripke modality coincides with the interior operator in this case, and the relational definition of D_I coincides with Fernández' topological definition.

Proposition 4. *Let $\mathfrak{M} = (X, \Pi_i, \tau_i, [\![\cdot]\!])_{i \in A}$ be a topo-e-model. Then for every proposition $P \subseteq X$, every group $I \subseteq A$ and every group member $i \in I$, we have:*

$$[\forall]_i^{\mathfrak{M}(\mathsf{share}_I)}(P) = [\forall]_I^{\mathfrak{M}}(P), \qquad \Box_i^{\mathfrak{M}(\mathsf{share}_I)}(P) = \Box_I^{\mathfrak{M}}(P),$$

$$B_i^{\mathfrak{M}(\mathsf{share}_I)}(P) = B_I^{\mathfrak{M}}(P), \qquad K_i^{\mathfrak{M}(\mathsf{share}_I)}(P) = K_I^{\mathfrak{M}}(P).$$

In words: the individual *group members' hard information* $[\forall]_i$*, soft evidence* $\Box_i$*, knowledge* K_i*, and belief* B_i after evidence-sharing *match the corresponding* group *attitudes* $[\forall]_I$*,* $\Box_I$*,* K_I*,* B_I before *the evidence-sharing.*

Interpretation. This result provides a uniform interpretation of all our group operators: *they simply "pre-encode" the individual members' attitudes after in-group evidence-sharing*! This justifies our name of "virtual group knowledge", and vindicates our topological definition of K_I and B_I, from a communication-based perspective. *Topological group knowledge/belief is simply the knowledge/belief that the individual members could acquire by sharing all their evidence.*[17]

Why Not Directly Share Knowledge? At first sight, it might seem that Fernández' topological distributed knowledge D_I could be similarly given a communication-based interpretation, in terms of the epistemic situation after agents share all their *knowledge* (rather than evidence). Such knowledge-sharing actions were considered in [1,3,12,16], but all these proposals assumed an $S5$ setting, in which knowledge is absolutely certain and fully introspective: agents can infallibly distinguish what they know from what they don't know, so they can share exactly only what they know. In our non-$S5$ context, this is not realistic: when interested in fallible knowledge, we cannot assume such infallible powers of discrimination. Agents cannot be sure which of their beliefs (or pieces of evidence) are true and which are not, and they cannot select only those that constitute "knowledge". The best the agents can do is to either share all their *beliefs*, or else share all their *evidence* (as in share_I), and then use this to build new consistent and justified beliefs (and thus obtain new knowledge).

Fallible Knowledge must Violate Group Monotonicity. From a communication-based perspective on group knowledge, it would be questionable to impose Group Monotonicity on a *fallible* notion of knowledge. As widely recognized in the field of Belief Revision Theory [2], the dynamics of belief (and so also the dynamics of fallible knowledge) must be *non-monotonic*: if an agent fallibly knows a proposition, then further evidence might defeat that knowledge again. The failure of Group Monotonicity is then simply an inescapable consequence of this non-monotonic dynamics: after receiving new "soft" evidence from other members of the group, agents may radically revise their beliefs, and thus may lose some of their prior "knowledge". This is a feature, not a bug: any realistic notion of (fallible) group knowledge will invalidate Monotonicity.

[17] As we will see, at the syntactic level, the above equalities have to be replaced by more complex Reduction laws, because the same sentence φ may denote different sets of worlds in $\mathfrak{M}$ and in $\mathfrak{M}(\mathsf{share}_I)$.

A Concrete Scenario for Example 2. To illustrate this point more concretely, consider the following scenario, underlying the model in Example 2. Daisy was brutally murdered. Detective Bob is leading the case, and Alice is the jury foreperson in the murder trial. The accused is Daisy's husband: the lawyer Charles. The evidence at hand concerns whether Charles got caught in the act ($C := \{w_1, w_2\}$ in Example 2), as well as his intent to kill ($I := \{w_2, w_4\}$). Both killing and intent to kill are a crime. Therefore, both C and I individually imply that Charles is *guilty* (proposition p in Example 2). In the actual world (w_2), both C and I are factive. Charles is innocent only in world w_3.

But now suppose Bob's evidence was deemed inadmissible, hence Alice does not have access to it. Conversely, Bob does not have access to Alice's evidence, as it is obtained through witness testimony in court. Moreover, suppose Alice has the following evidence:

- $I = \{w_2, w_4\}$: Testimonial evidence from Charles' colleague reveals that Charles was inquiring at work about the legalities of collecting life insurance after sudden death. Moreover, he did this only a week after having taken out a life insurance policy for Daisy, and days before her death. This evidence of intent is factive.
- $\neg C = \{w_3, w_4\}$: Testimonial evidence from Charles' friend Ed provides an alibi for Charles at the time of the crime: they were watching tv at home. Ed lied under oath, and therefore this evidence is not factive.

Bob has the following evidence:

- $C = \{w_1, w_2\}$: An alcoholic, who was drunk when he witnessed the murdering of Daisy, identified Charles as the killer in a statement to the police. His statement was not confirmed by any third party: the evidence, although factive, is deemed inadmissible on grounds of being unreliable.
- $\neg I = \{w_1, w_3\}$: Charles handed over to the police his periodical handwritten love letters to Daisy, dating back more than ten years, as evidence against intent. The letters, which appeared to be (and were, in fact) fabricated over the past week, were deemed inadmissible.

Bob and Alice both individually know (fallibly) that Charles is guilty. However, sharing their evidence would result in reasonable doubt, since their (factive) individual evidence is defeated by some of the other's (non-factive) evidence.

3 Logics and Axiomatizations

In this section we introduce our logics for evidence, knowledge, belief and sharing of evidence, and present our main results on completeness and decidability.

3.1 The Logic of Group Evidence

Our language of group-evidence $\mathcal{L}_{\Box[\forall]_I}$ will have modalities for soft and hard (group) evidence. We also study a fragment $\mathcal{L}_{\Box[\forall]_{i,A}}$, obtained by restricting these modalities to individuals and the full group A.

Notational Convention. For concision, we use the symbol $\alpha \in \{A\} \cup A$ to denote either singletons $\{i\} \subseteq A$ or A itself, when considering notions of group evidence, knowledge, and belief restricted to individuals or the full group A.

Definition 5 (Syntax and Semantics of $\mathcal{L}_{\Box[\forall]_I}$ and $\mathcal{L}_{\Box[\forall]_{i,A}}$). *The language $\mathcal{L}_{\Box[\forall]_I}$ of group evidence is defined recursively as*

$$\varphi ::= p \mid \neg\varphi \mid \varphi \wedge \varphi \mid \Box_I \varphi \mid [\forall]_I \varphi$$

where $p \in \mathsf{Prop}$ and I is any group. The fragment $\mathcal{L}_{\Box[\forall]_{i,A}}$ of $\mathcal{L}_{\Box[\forall]_I}$ is obtained by restricting the modalities to $\Box_\alpha$ and $[\forall]_\alpha$, with $\alpha \in \{A\} \cup A$. For simplicity, we will write $[\forall]_i$ and $\Box_i$ instead of $[\forall]_{\{i\}}$ and $\Box_{\{i\}}$.

Given a topo-e-model $\mathfrak{M} = (X, \Pi_i, \tau_i, [\![\cdot]\!])_{i\in A}$, we define an interpretation *function $[\![\cdot]\!]^{\mathfrak{M}}$, mapping every formula φ of $\mathcal{L}_{\Box[\forall]_I}$ to a proposition $[\![\varphi]\!]^{\mathfrak{M}} \subseteq X$. The interpretation extends the valuation, so whenever the model is understood we can skip the superscript without ambiguity, writing $[\![\varphi]\!]$. The definition is by recursion on formulas: for atoms, $[\![p]\!]$ is just the valuation, and we let*

$$[\![\neg\varphi]\!] := X \setminus [\![\neg\varphi]\!], \qquad [\![\varphi \wedge \psi]\!] := [\![\varphi]\!] \cap [\![\psi]\!],$$
$$[\![\Box_I\varphi]\!] := \Box_I([\![\varphi]\!]) = Int_I([\![\varphi]\!]), \; [\![\forall_I]\varphi]\!] := [\forall_I]([\![\varphi]\!]) := \{x \in X : \Pi_I(x) \subseteq [\![\varphi]\!]\}.$$

The interpretation for $\mathcal{L}_{\Box[\forall]_{i,A}}$ is simply the restriction of $[\![\cdot]\!]$ to this language. As usual, we sometimes write $x \models \varphi$ for $x \in [\![\varphi]\!]$.

Abbreviations. The Boolean connectives $\vee$, $\rightarrow$, $\leftrightarrow$, and the modality $\Diamond_I$ (dual to $\Box_I$) are defined as abbreviations as usual. *Knowledge and belief are also abbreviations*: $B_I\varphi := [\forall]_I \Diamond_I \Box_I \varphi$ and $K_I\varphi := \Box_I\varphi \wedge B_I\varphi$. It is easy to see that we have $[\![B_I\varphi]\!] = B_I([\![\varphi]\!])$, $[\![K_I\varphi]\!] = K_I([\![\varphi]\!])$.

Theorem 6. *The proof system $\Box[\forall]_I$ from Table 1 is sound and complete for $\mathcal{L}_{\Box[\forall]_I}$ w.r.t. multi-agent topo-e-models, and the logic $\mathcal{L}_{\Box[\forall]_I}$ is decidable. All these properties are inherited by the proof system $\Box[\forall]_{i,A}$ and the logic $\mathcal{L}_{\Box[\forall]_{i,A}}$.*

Table 1. The proof system $\Box[\forall]_I$, where $I, J \subseteq A$ are groups. The proof system $\Box[\forall]_{i,A}$ for the fragment $\mathcal{L}_{\Box[\forall]_{i,A}}$ is obtained by restricting all axioms to $\mathcal{L}_{\Box[\forall]_{i,A}}$.

$(\mathsf{S4}_\Box)$	$\mathsf{S4}$ axioms and rules for $\Box_I$		
$(\mathsf{S5}_{[\forall]})$	$\mathsf{S5}$ axioms and rules for $[\forall]_I$		
Monotonicity	$\Box_J\varphi \rightarrow \Box_I\varphi$,	$[\forall]_J\varphi \rightarrow [\forall]_I\varphi$	(for $J \subseteq I$)
Inclusion	$[\forall]_I\varphi \rightarrow \Box_I\varphi$		

3.2 The Logic of Group Knowledge and Group Belief

To reason about knowledge and belief without explicitly mentioning notions of evidence, we introduce languages in which K_I and B_I are primitive operators.[18]

Definition 7 (Syntax and Semantics of $\mathcal{L}_{KB_I}$ and $\mathcal{L}_{KB_{i,A}}$). *The language $\mathcal{L}_{KB_I}$ of group knowledge and belief is defined recursively as*

$$\varphi ::= p \mid \neg\varphi \mid \varphi \wedge \varphi \mid B_I\varphi \mid K_I\varphi$$

where $p \in \mathsf{Prop}$ and I is any group. As before, the fragment $\mathcal{L}_{KB_{i,A}}$ of $\mathcal{L}_{KB_I}$ is obtained by restricting the modalities to B_α and K_α, for all $\alpha \in \{A\} \cup A$.

Given a topo-e-model $\mathfrak{M} = (X, \Pi_i, \tau_i, [\![\cdot]\!])_{i \in A}$, the interpretation *map $[\![\cdot]\!]^{\mathfrak{M}}$ is as before for atoms and Boolean connectives, while for B_I and K_I we use the corresponding semantic operators (with B_α and K_α as special cases):*

$$[\![B_I\varphi]\!] := B_I([\![\varphi]\!]) \qquad\qquad [\![K_I\varphi]\!] := K_I([\![\varphi]\!])$$

Theorem 8. *The proof system $\boldsymbol{KB_{i,A}}$ listed in Table 2 is sound and complete for $\mathcal{L}_{KB_{i,A}}$ w.r.t. multi-agent topo-e-models. Moreover, this logic is decidable.*

Table 2. The proof system $\boldsymbol{KB_{i,A}}$, where A is the group of all agents, $i \in A$ ranges over agents, and $\alpha \in \{A\} \cup A$ denotes either individual agents or the full group A.

(KB)	**Axioms & rules of normal modal logic for K & B**
Stalnaker's Epistemic-Doxastic Axioms:	
Truthfulness of knowledge (T)	$K_\alpha\varphi \to \varphi$
Pos. Intro. of knowledge (KK)	$K_\alpha\varphi \to K_\alpha K_\alpha\varphi$
Consistency of Beliefs (CB)	$B_\alpha\varphi \to \neg B_\alpha\neg\varphi$
Strong Pos. Intro. of beliefs (SPI)	$B_\alpha\varphi \to K_\alpha B_\alpha\varphi$
Strong Neg. Intro. of beliefs (SNI)	$\neg B_\alpha\varphi \to K_\alpha\neg B_\alpha\varphi$
Knowledge implies Belief (KB)	$K_\alpha\varphi \to B_\alpha\varphi$
Full Belief (FB)	$B_\alpha\varphi \to B_\alpha K_\alpha\varphi$
Group Knowledge Axioms:	
Super-Introspection (SI)	$B_i\varphi \to K_A B_i\varphi$
Weak Monotonicity (WM)	$(K_i\varphi \wedge B_A\varphi) \to K_A\varphi$
Consistency of group Belief with Distributed knowledge (CBD)	$(\bigwedge_{i\in A} K_i\varphi_i) \to \langle B_A\rangle(\bigwedge_{i\in A}\varphi_i)$ (where $\{\varphi_i \mid i \in A\}$ are arb. formulas)

[18] As already noted, B_I is definable in terms of K_I, so the belief operator is redundant. But our axioms are clearer when stated in terms of both modalities.

We briefly discuss the axioms. The first two groups contain generalizations (to multiple agents and groups) of Stalnaker's axioms and rules for (individual) knowledge and belief [24]. These axioms were shown in [6] to be complete for the topological interpretation, and their completeness for multiple agents was shown in [5,15]. All these axioms and rules are standard in epistemic-doxastic logic, except for the Full Belief axiom (FB), which is specific to Stalnaker's conception of *belief as the "subjective feeling" of knowledge.* Stalnaker calls this "strong belief", but we follow the terminology in [6], referring to it as "full belief". The intuition is that an agent "fully believes" φ when she *believes that she knows it*: from a *first-person* perspective, full belief and fallible knowledge are indistinguishable.

Moving on to the Group Knowledge axioms, Super-Introspection (SI) is a strengthening of ordinary (strong) introspection of beliefs, stating that *a group virtually knows the beliefs of its members.* Weak Monotonicity (WM) is a (valid) weakening of the (invalid) Group Monotonicity: individual knowledge of φ does imply virtual group knowledge of φ *provided that the group virtually believes* φ.

Finally, Consistency of group Belief with Distributed knowledge (CBD) says that *a group's virtual belief is consistent with its distributed knowledge.* In terms of Fernández' D-operator [15], this could be stated as $D_A\varphi \rightarrow \langle B_A\rangle\varphi$. Our language does not include a distributed knowledge modality, but (CBD) gives an equivalent statement in terms of conjunctions of individual pieces of knowledge.

Translation into the Languages of Evidence. Every formula φ of $\mathcal{L}_{KB_I}$ and $\mathcal{L}_{KB_{i,A}}$ can be translated into a formula $tr(\varphi)$ of the corresponding evidence languages $\mathcal{L}_{\Box[\forall]_I}$ and $\mathcal{L}_{\Box[\forall]_{i,A}}$: $tr(p) = p$, $tr(\neg\varphi) = \neg tr(\varphi)$, $tr(\varphi \wedge \psi) = tr(\varphi) \wedge tr(\psi)$, $tr(B_I\varphi) = \forall_I \Diamond_I \Box_I tr(\varphi)$, $tr(K_I\varphi) = \Box_I tr(\varphi) \wedge \forall_I \Diamond_I \Box_I tr(\varphi)$ (with B_i, K_i, B_A, K_A as special cases). This translation is *faithful*, i.e., $[\![tr(\varphi)]\!] = [\![\varphi]\!]$.

3.3 The Dynamic Logics of Evidence-Sharing

We now extend our languages with *dynamic modalities* $[\mathsf{share}_I]$ for evidence-sharing. Given the above completeness results, we only axiomatize two such logics: the extension of $\mathcal{L}_{\Box[\forall]_I}$ with $[\mathsf{share}_I]$ for arbitrary groups $I \subseteq A$; and the extension of $\mathcal{L}_{KB_{i,A}}$ with $[\mathsf{share}_A]$ for the full group A of all agents.

Definition 9 (Syntax and Semantics with $[\mathsf{share}_I]$). *The dynamic language $\mathcal{L}_{\Box[\forall]_I[\mathsf{share}_I]}$ is defined recursively as*

$$\varphi ::= p \mid \neg\varphi \mid \varphi \wedge \varphi \mid \Box_I\varphi \mid [\forall]_I\varphi \mid [\mathsf{share}_I]\varphi$$

(where $p \in \mathsf{Prop}$ and $I \subseteq A$ is any group); while $\mathcal{L}_{KB_{i,A}[\mathsf{share}_A]}$ is given by

$$\varphi ::= p \mid \neg\varphi \mid \varphi \wedge \varphi \mid K_i\varphi \mid K_A\varphi \mid [\mathsf{share}_A]\varphi$$

(where $p \in \mathsf{Prop}$, and $i \in A$ is any agent).

Given a topo-e-model $\mathfrak{M}$, the interpretation *map $[\![\cdot]\!]^{\mathfrak{M}}$ uses the clauses from Definition 5 for the static connectives, while for the dynamic modalities, we put*

$$[\![[\mathsf{share}_I]\varphi]\!]^{\mathfrak{M}} = [\![\varphi]\!]^{\mathfrak{M}(\mathsf{share}_I)}$$

and apply the special case $I = A$ of this clause to interpret $[\mathsf{share}_A]\varphi$.

Theorem 10. *The proof systems listed in Table 3 and Table 4 are sound and complete for the corresponding logics* $\mathcal{L}_{\Box[\forall]_I[\mathsf{share}_I]}$ *and* $\mathcal{L}_{KB_{i,A}[\mathsf{share}_A]}$ *w.r.t. multi-agent topo-e-models. Moreover, these logics are provably co-expressive with their static bases* $\mathcal{L}_{\Box[\forall]_I}$ *and respectively* $\mathcal{L}_{KB_{i,A}}$, *and thus they are decidable.*

Table 3. The proof system $\Box[\forall]_I[\mathsf{share}_I]$, where $I, J \subseteq A$ are groups, and we use the notation $J/{+}\,I := J \cup I$ when $I \cap J \neq \emptyset$, and $J/{+}\,I := J$ when $I \cap J = \emptyset$.

$(\Box[\forall]_I)$	**Axioms and rules of** $\Box[\forall]_I$	
$([\mathsf{share}_I])$	**Axioms and rules of normal modal logic for** $[\mathsf{share}_I]$	
	Reduction Axioms for $[\mathsf{share}_I]$:	
(Atomic Reduction)	$[\mathsf{share}_I]p \leftrightarrow p$	(for atomic propositions p)
(Negation Reduction)	$[\mathsf{share}_I]\neg\varphi \leftrightarrow \neg[\mathsf{share}_I]\varphi$	
($\Box$-Reduction)	$[\mathsf{share}_I]\Box_J\varphi \leftrightarrow \Box_{J/+I}[\mathsf{share}_I]\varphi$	
($\forall$-Reduction)	$[\mathsf{share}_I][\forall]_J\varphi \leftrightarrow [\forall]_{J/+I}[\mathsf{share}_I]\varphi$	

Table 4. System $\boldsymbol{KB}_{i,A}\,[\mathsf{share}_A]$, where $\alpha \in A \cup \{A\}$ is an individual or the full group.

$(\Box[\forall]_I)$	**Axioms and rules of** $\boldsymbol{KB}_{i,A}$	
$([\mathsf{share}_A])$	**Axioms and rules of normal modal logic for** $[\mathsf{share}_A]$	
	Reduction Axioms for $[\mathsf{share}_A]$:	
(Atomic Reduction)	$[\mathsf{share}_A]p \leftrightarrow p$	(for atomic propositions p)
(Negation Reduction)	$[\mathsf{share}_A]\neg\varphi \leftrightarrow \neg[\mathsf{share}_A]\varphi$	
(K-Reduction)	$[\mathsf{share}_A]K_\alpha\varphi \leftrightarrow K_A[\mathsf{share}_A]\varphi$	
(B-Reduction)	$[\mathsf{share}_A]B_\alpha\varphi \leftrightarrow B_A[\mathsf{share}_A]\varphi$	

As usual in DEL, there is also a Conjunction Reduction: $[\mathsf{share}_I](\varphi \wedge \psi) \leftrightarrow ([\mathsf{share}_I]\varphi \wedge [\mathsf{share}_I]\psi)$. But this is provable from the axioms and rules of normal modal logic for $[\mathsf{share}_I]$ together with the Negation Reduction axiom for $[share_I]$. Its special case $[\mathsf{share}_A](\varphi \wedge \psi) \leftrightarrow ([\mathsf{share}_A]\varphi \wedge [\mathsf{share}_A]\psi)$ is similarly provable from the normality of $[\mathsf{share}_A]$ and the Negation Reduction axiom for $[share_A]$.

3.4 Proofs of Completeness and Decidability

The full proofs are relegated to the extended online version of this paper.[19] Here we sketch a brief summary of the proof plan and the main ideas of the proofs.

[19] See https://arxiv.org/abs/2509.00184.

For each of the proof systems $\Box[\forall]_I$, $\Box[\forall]_{i,A}$, and $\boldsymbol{KB_{i,A}}$, we first show completeness w.r.t. *non-standard relational structures* (pseudo-models), which are tailored to the respective languages. For $\Box[\forall]_I$ and $\Box[\forall]_{i,A}$, this is done using appropriate versions of the standard modal technique of *filtration*, which gives us *finite pseudo-models* for $\mathcal{L}_{\Box[\forall]_I}$ and $\mathcal{L}_{\Box[\forall]_{i,A}}$. In the case of $\boldsymbol{KB_{i,A}}$, we use the classical method of *canonical structures*, obtaining an *(infinite) canonical pseudo-model for $\mathcal{L}_{KB_{i,A}}$*, having an additional special property (*max-density*).

The next step is to go back to the (intended) topo-e-models. For $\Box[\forall]_I$ and $\Box[\forall]_{i,A}$, we use a version of the well-known technique of *unraveling*, showing that every pseudo-model for these logics is modally equivalent to its unraveled "associated model": this is a tree-like relational model, which is itself equivalent to a multi-agent topo-e-model. This finishes the proof of Theorem 6.

For $\boldsymbol{KB_{i,A}}$, things are more complex: we have to first prove a representation theorem, showing that every pseudo-model for $\mathcal{L}_{KB_{i,A}}$ having the additional max-density property can be represented as a (p-morphic image of) a pseudo-model for $\mathcal{L}_{\Box[\forall]_{i,A}}$, in a way that preserves the truth of all formulas in $\mathcal{L}_{KB_{i,A}}$. This representation theorem is the key step, and its proof is non-trivial and uses an innovative technique.[20] Given this and the above unraveling result, we obtain completeness of $\boldsymbol{KB_{i,A}}$ w.r.t. topo-e-models. This concludes Theorem 8.

Finally, the completeness proof for the dynamic extensions (Theorem 10) follows a standard approach in Dynamic Epistemic Logic: we use the reduction axioms to show that these extensions are provably co-expressive with their static bases. Putting this together with Theorems 6 and 8, we obtain Theorem 10.

4 Conclusion

The key theoretical contribution of this paper is the *complete axiomatization* of non-monotonic, evidence-based notions of *(virtual) group knowledge and group belief*, in the shape of the logic $\boldsymbol{KB_{i,A}}$. Compared to previous attempts at topological accounts of group knowledge (corresponding to a *traditional* interpretation in terms of *distributed* knowledge), *the notion studied here is better suited to match the epistemic dynamics of knowledge induced by evidence-sharing.* This is a small step towards applying topological semantics to realistic, practical settings, such as distributed computing and the epistemology of social networks.

As an auxiliary tool, we also studied *the logic of group evidence* over the larger language $\mathcal{L}_{\Box[\forall]_I}$, and showed that it is sound and complete, as well as decidable. In its turn, this result was an important step in showing the completeness and decidability of the above-mentioned logic $\boldsymbol{KB_{i,A}}$.

Unfortunately, we do not have a completeness result for the full logic $\boldsymbol{KB_I}$ of group knowledge and belief for *arbitrary subgroups* $I \subseteq A$. All the above axioms have sound analogues for the general operators K_I and B_I. E.g., the following generalizations of Super-Introspection and Weak Monotonicity hold:

$$B_J\varphi \to K_I B_J\varphi, \qquad (K_J\varphi \wedge B_I\varphi) \to K_I\varphi, \qquad \text{(for groups } J \subseteq I \subseteq \text{ A).}$$

[20] Moreover, in contrast to the unraveling technique used for $\Box[\forall]_I$, it is not clear how to generalize this step to arbitrary subgroups, i.e., to the logic $\boldsymbol{KB_I}$.

Similarly, axiom (CBD) can be generalized to $(\bigwedge_{J \subseteq I} K_J \varphi_J) \to \langle B_I \rangle (\bigwedge_{J \subseteq I} \varphi_J)$.

But it is not at all clear that the resulting axiomatization is complete! Our proof methods do not seem to work for this extension. On the other hand, we know that *the logic* $\boldsymbol{KB_I}$ *is decidable* (since it can be translated into a fragment of the decidable logic $\Box[\forall]_{i,A}$), so there must exist a recursive axiomatization!

This leads to our outstanding unsolved problem:

Open Question. Find a complete proof system for the logic $\boldsymbol{KB_I}$.

The investigation of this intriguing question is left for future work.

Acknowledgements. We thank the anonymous referees for their helpful comments. D. Gomes is supported by the Swiss National Science Foundation (SNSF) under grant No. 10000440 (Epistemic Group Attitudes).

References

1. Ågotnes, T., Wáng, Y.N.: Resolving distributed knowledge. Artif. Intell. **252**, 1–21 (2017). https://doi.org/10.1016/j.artint.2017.07.002
2. Alchourrón, C.E., Gärdenfors, P., Makinson, D.: On the logic of theory change: partial meet contraction and revision functions. J. Symb. Log. **50**(2), 510–530 (1985). https://doi.org/10.2307/2274239
3. Baltag, A., Boddy, R., Smets, S.: Group knowledge in interrogative epistemology. In: van Ditmarsch, H., Sandu, G. (eds.) Jaakko Hintikka on Knowledge and Game-Theoretical Semantics. OCL, vol. 12, pp. 131–164. Springer, Cham (2018). https://doi.org/10.1007/978-3-319-62864-6_5
4. Baltag, A., van Ditmarsch, H.P., Moss, L.S.: Epistemic logic and information update. In: Adriaans, P., van Benthem, J. (eds.) Philosophy of Information. MIT Press (2008)
5. Baltag, A., Bezhanishvili, N., Fernández González, S.: Topological evidence logics: multi-agent setting. In: Özgün, A., Zinova, Y. (eds.) TbiLLC 2019. LNCS, vol. 13206, pp. 237–257. Springer, Cham (2022). https://doi.org/10.1007/978-3-030-98479-3_12
6. Baltag, A., Bezhanishvili, N., Özgün, A., Smets, S.: The topology of belief, belief revision and defeasible knowledge. In: Grossi, D., Roy, O., Huang, H. (eds.) LORI 2013. LNCS, vol. 8196, pp. 27–40. Springer, Heidelberg (2013). https://doi.org/10.1007/978-3-642-40948-6_3
7. Baltag, A., Bezhanishvili, N., Özgün, A., Smets, S.: Justified belief and the topology of evidence. In: Väänänen, J., Hirvonen, Å., de Queiroz, R. (eds.) WoLLIC 2016. LNCS, vol. 9803, pp. 83–103. Springer, Heidelberg (2016). https://doi.org/10.1007/978-3-662-52921-8_6
8. Baltag, A., Bezhanishvili, N., Özgün, A., Smets, S.: Justified belief, knowledge, and the topology of evidence. Synthese **200** (2022). https://doi.org/10.1007/s11229-022-03967-6
9. Baltag, A., Fiutek, V., Smets, S.: Beliefs and evidence in justification models. In: Beklemishev, L., Demri, S., Máté, A. (eds.) Advances in Modal Logic, pp. 156–176 (2016). http://www.aiml.net/volumes/volume11/Baltag-Fiutek-Smets.pdf

10. Baltag, A., Renne, B.: Dynamic epistemic logic. In: Zalta, E.N. (ed.) The Stanford Encyclopedia of Philosophy. Metaphysics Research Lab, Stanford University, Winter 2016 edn. (2016). https://plato.stanford.edu/archives/win2016/entries/dynamic-epistemic
11. Baltag, A., Renne, B., Smets, S.: The logic of justified belief change, soft evidence and defeasible knowledge. In: Ong, L., de Queiroz, R. (eds.) WoLLIC 2012. LNCS, vol. 7456, pp. 168–190. Springer, Heidelberg (2012). https://doi.org/10.1007/978-3-642-32621-9_13
12. Baltag, A., Smets, S.: Learning what others know. In: Kovacs, L., Albert, E. (eds.) LPAR23 Proceedings of the International Conference on Logic for Programming AI and Reasoning, vol. 73, pp. 90–110 (2020). https://doi.org/10.48550/arXiv.2109.07255
13. van Benthem, J., Pacuit, E.: Dynamic logics of evidence-based beliefs. Stud. Logica. **99**(1–3), 61–92 (2011). https://doi.org/10.1007/s11225-011-9347-x
14. van Ditmarsch, H., van der Hoek, W., Kooi, B.: Dynamic Epistemic Logic. Springer (2007)
15. Fernández González, S.: Generic Models for Topological Evidence Logics. Master's thesis, University of Amsterdam (2018). https://eprints.illc.uva.nl/id/eprint/1641/
16. Goldbach, R.: Modelling Democratic Deliberation. Master thesis, University of Amsterdam (2015). https://eprints.illc.uva.nl/id/eprint/946/
17. dos Santos Gomes, D.: Virtual group knowledge on topological evidence models. Master thesis, University of Amsterdam (2025). https://eprints.illc.uva.nl/id/eprint/2356/
18. Halpern, J.Y., Moses, Y.: Knowledge and common knowledge in a distributed environment. J. ACM **37**(3), 549–587 (1990). https://doi.org/10.1145/79147.79161
19. Halpern, J.Y., Moses, Y.: A guide to completeness and complexity for modal logics of knowledge and belief. Artif. Intell. **54**(3), 319–379 (1992). https://doi.org/10.1016/0004-3702(92)90049-4
20. Hintikka, J.: Knowledge and Belief. Cornell University Press, Ithaca (1962)
21. McKinsey, J.C.C., Tarski, A.: The algebra of topology. Ann. Math. **45**(1), 141–191 (1944). https://doi.org/10.2307/1969080
22. Ramírez Abarca, A.I.: Topological Models for Group Knowledge and Belief. Master's thesis, University of Amsterdam (2015). https://eprints.illc.uva.nl/id/eprint/2250/
23. Shi, C., Smets, S., Velázquez-Quesada, F.R.: Argument-based belief in topological structures. Electron. Proc. Theor. Comput. Sci. **251**, 489–503 (2017). https://doi.org/10.4204/EPTCS.251.36
24. Stalnaker, R.: On logics of knowledge and belief. Philos. Stud. Int. J. Philos. Anal. Tradit. **128**(1), 169–199 (2006). https://doi.org/10.1007/s11098-005-4062-y
25. Özgün, A.: Evidence in Epistemic Logic: a Topological Perspective. Ph.D. thesis, University of Amsterdam (2017). https://eprints.illc.uva.nl/id/eprint/2147/

On Fuzzy Topological Semantics

Muhammad Afaq Khan and Manuel A. Martins(✉)

CIDMA and Department of Mathematics, University of Aveiro, Aveiro, Portugal
{mafaqkhan,martins}@ua.pt

Abstract. Classical topological semantics for modal logic interprets necessity and possibility through the open sets in the underlying topology. However, the crisp truth values and sharp spatial boundaries of topological spaces limit this approach to dealing with uncertainty and vagueness appearing in real-time phenomena. To address this limitation, we propose a fuzzy extension of topological semantics that integrates fuzzy set theory into the interpretation of modal logic.

In this framework, each formula is associated with a fuzzy set that assigns a degree of truth to every point in space, reflecting graded rather than bivalent truth. Modal operators are redefined using fuzzy topological constructs, such as fuzzy interior and fuzzy closure, while logical connectives are interpreted through fuzzy operations. This generalization not only preserves the spatial intuition of classical approaches, but also enhances their capacity to model systems dealing with imprecise data.

1 Introduction

The study of many-valued propositional logics with modal operators can be traced back to the seminal work [1] by Melvin Fitting's, and was further advanced by several other logicians with different motivations (see for example, [2–9]). In some of these systems, due to the absence of an involutive negation (one that returns the original value when applied twice), the modal operators $\Box$ and $\Diamond$ are often treated as independent, with some variants opting to use only one. Another common syntactic feature is the use of truth constants to represent values in the underlying algebra, aiding in completeness proofs, as seen in Fitting's early work. Modal fuzzy logics typically employ relational semantics extending classical Kripke semantics [10,11]. In these generalizations, both the truth values at worlds and the accessibility relations may vary in degrees. While intuitively, this introduces challenges, such as the difficulty of axiomatizing Kripke semantics over certain algebras. For instance, modal product logic lacks a simple axiomatization, necessitating truth constants, projections, and even infinitary rules [7]. Similarly, modal logics based on standard finitary Łukasiewicz logic require infinitary rules for axiomatization [12].

Scott and Montague introduced neighborhood semantics [13,14] as a more flexible alternative, replacing accessibility relations with neighborhood collections at each world. This framework effectively captures non-normal modal logics. Recently, researchers have adapted this approach to modal fuzzy logics.

J. Wang et al. (Eds.): DaLí 2025, LNCS 16472, pp. 115–133, 2026.
https://doi.org/10.1007/978-3-032-22626-6_7

Cintula et al. [5] developed a neighborhood semantics tailored for fuzzy systems, and several works [15,16] have proposed general frameworks, particularly for fuzzy logics extending MTL.

Parallel to logic, topology emerged as a foundational branch of mathematics, dealing with structures and spatial properties [17]. Topological semantics for modal logic was introduced by McKinsey and Tarski [18], providing completeness results for systems like $S4$ by interpreting modal operators through topological notions such as interior and closure [19]. However, the intuitive appeal of Kripke semantics led to its dominance, with topological semantics receiving less attention. Kripke frames are often easier to visualize, while topological spaces can appear abstract. The work by Awodey et al. [20] significantly advanced topological semantics by developing topos-based models for higher-order modal logic. Compared to neighborhood semantics, topological semantics offers a unified, spatially intuitive framework, especially for reasoning about continuity and approximation. Classical topological models represent vagueness through open sets and neighborhood-based truth, which allow for localized reasoning and partial observability. However, this imprecision remains qualitative and lacks a mechanism to express varying degrees of truth. This limits its applicability in modeling vagueness, partial belief, or graded phenomena. Fuzzy topological spaces [21] integrate fuzzy set theory with topology, redefining topological concepts in terms of degrees [22]. Unlike classical topologies, fuzzy topologies allow graded openness, partial membership, and imprecise spatial relations, making them ideal for modeling real-world scenarios where information is vague or incomplete. For example this formula is true somewhere around here. But for our fuzzy topological case this example could be seen as this formula is 0.7 true here, and 0.4 true nearby.

In this work, we propose an alternative semantics for fuzzy modal logic that addresses the limitations and gaps mentioned above, specifically by developing a fuzzy topological semantics through an extension of traditional topological semantics. We introduce fuzzy topological semantics, employing Gödel algebra to provide both local (pointwise) and global interpretations. We work with Gödel algebra, since unlike Łukasiewicz or Product algebras, its order theoretic and idempotent structure aligns naturally with the intersection based behavior of fuzzy open sets.

This approach is especially relevant for two main reasons. The first is that many-valued modal logic, as a formal system, requires a solid interpretative foundation. The second is that fuzzy topological spaces improve spatial reasoning by formally handling imprecise regions, continuity, and connectedness. We interpret propositions as fuzzy subsets of a fuzzy topological space [23] and we evaluate modal operators through fuzzy interior and closure operators. We establish the connection between fuzzy Kripke semantics and fuzzy topological semantics through the construction of fuzzy Alexandroff spaces induced by reflexive and transitive fuzzy Kripke frames. We demonstrate that the modal logic $S4$ is sound with respect to the class of all fuzzy topological spaces. This generalization not only preserves the intuitive spatial interpretation of classical topological semantics but also accommodates degrees of truth. The result is a more expressive and flexible semantic framework, suitable for applications in artificial intelligence, epistemic reasoning, and spatial logic. For example, imagine a robot navigating a grid, with Kripke semantics, it either knows the goal is reachable or not. With topological semantics, it can express "I am close to the goal". With fuzzy topological semantics, it can express "I am 80% sure I am near the goal".

Outline of the Paper. The paper is organized as follows. Section 2 introduces the background concepts relevant to this work, focusing on modal logic and alternative semantic frameworks, including Kripke semantics and topological semantics. Additionally, fundamental notions of fuzzy topological spaces and the concept of fuzzy modal logic with fuzzy Kripke semantics are presented in this section. Section 3 then extends classical topological semantics to fuzzy topological semantics, showing that the axioms of the modal logic $S4$ remain valid over the entire class of fuzzy topological spaces. In addition, this section defines the logic corresponding to discrete fuzzy topologies and to fuzzy extremally disconnected spaces. The paper concludes with final remarks and directions for future research.

2 Background

In this section, we present foundational concepts and results required for our study. We briefly review essential notions of modal logic, including Kripke and topological semantics, followed by an overview of fuzzy topological spaces.

Modal Logic. Modal logic was first discussed in a systematic way by Aristotle in *De Interpretatione* [24]. Aristotle noticed not simply that necessity implies possibility (and not the other way around), but also that the notions of necessity and possibility can be defined in terms of one another. The most familiar types of modal statements are those that say something must be true (necessarily true) or could be true (possibly true). The expressions "it is possible that" and "it is necessary that" are called 'modal' operators, because they specify the manner or mode in which the rest of the proposition is claimed to be true. The more known semantics for modal logic is the Kripke semantics [10] (also called relational semantics). Here we follow the presentation in [25].

Definition 1 (Modal Language). *For a given set of propositional letters Prop, we define modal formulas, by the following grammar,*

$$\varphi := p \mid \bot \mid \varphi \to \psi \mid \Box\varphi, \quad p \in Prop.$$

As usual, the remaining boolean operators and the possibility modality $\Diamond$ can be introduced by definition.

Definition 2 (Kripke model). *A* Kripke frame *is a pair* (W, R)*, where* W *is a non-empty set of worlds and* R *is a binary relation on* W*. A* Kripke model *is a triple* (W, R, V)*, where* (W, R) *is Kripke frame and* $V : Prop \to \mathcal{P}(W)$ *is a valuation function.*

Definition 3. *Let* $\mathcal{M} = (W, R, V)$ *be a Kripke model, the truth value of a modal formula* φ *at a world* $w \in W$*, denoted by* $\mathcal{M}, w \models \varphi$*, is defined recursively as follows:*

i) $\mathcal{M}, w \models p$ *iff* $w \in V(p)$*, for* $p \in Prop$
ii) $\mathcal{M}, w \models \bot$ *never*
iii) $\mathcal{M}, w \models \varphi \to \psi$ *iff not* $\mathcal{M}, w \models \varphi$ *or* $\mathcal{M}, w \models \psi$
iv) $\mathcal{M}, w \models \Box\varphi$ *iff for all* $v \in W$ *with* wRv*, we have* $\mathcal{M}, v \models \varphi$*.*

While Kripke semantics has played a central role in shaping the modern understanding of modal logic, it is not the only interpretative framework available. Over time, some alternative semantics have been proposed each offering unique perspectives and tools to handle modalities, especially in contexts where relational structures may be inadequate or too restrictive.

Topological semantics of modal logic extends relational semantics by introducing geometric and spatial interpretations. In this framework, worlds are represented as points in a topological space, and the accessibility relation is interpreted using open sets or neighborhoods, providing a more continuous and structured understanding of modal logic. This extension allows for a richer semantic landscape, where the relations between worlds are understood in terms of proximity and spatial properties. Notably, certain Kripke incomplete logics, such as provability logic (GLP), become topologically complete, meaning that they can be fully realized within topological spaces [26].

We recall that a *topological space* is a pair (X, τ), where X is a set and $\tau \subseteq \mathcal{P}(X)$ is such that $X, \emptyset \in \tau$, τ is closed under finite intersections and τ is closed under arbitrary unions.

Definition 4 (Topological Model - [27]). *A* topological model $M = (X, \tau, V)$ *is a tuple where* (X, τ) *is a topological space and* $V : Prop \to \mathcal{P}(X)$ *is a valuation.*

Definition 5. *Let* $\mathcal{M} = (X, \tau, V)$ *be topological model, the truth value of a formula* φ *at a point* $x \in X$*, denoted* $\mathcal{M}, x \models \varphi$*, is defined recursively as follows:*

i) $\mathcal{M}, x \models p$ *iff* $x \in V(p)$*,* *for* $p \in Prop$*,*
ii) $\mathcal{M}, x \models \bot$ *never,*
iii) $\mathcal{M}, x \models \varphi \to \psi$ *iff not* $\mathcal{M}, x \models \varphi$ *or* $\mathcal{M}, x \models \psi$
iv) $\mathcal{M}, x \models \Box\varphi$ *iff* $x \in \mathrm{Int}(\{y \in X : \mathcal{M}, y \models \varphi)\}$*,*

We should observe that for the possibility operator the truth becomes:

$$\mathcal{M}, w \models \Diamond\varphi \text{ iff } x \in \mathrm{Cl}(\{y \in X : \mathcal{M}, y \models \varphi)\}).$$

A formula φ is *valid in a topological model* $\mathcal{M}$ written $\mathcal{M} \models \varphi$ if for all points $x \in X$, we have $\mathcal{M}, x \models \varphi$. And a formula is *valid in a topological space* (X, τ), if φ is valid in any topological model $\mathcal{M}$ over (X, τ).

To build the connections between topological semantics and Kripke semantics, we consider reflexive and transitive frames and show that they come in one-to-one correspondence with the special class of topological spaces, called *Alexandroff spaces.* Moreover, as established in [27], every topological model satisfies the axioms of the modal logic $S4$. The modal logic $S4$ is the minimal collection of modal formulas that includes all classical tautologies, the following axioms, (N) $\Box\top$, (T) $\Box p \rightarrow p$, (R) $\Box(p \wedge q) \leftrightarrow \Box p \wedge \Box q$ and (4) $\Box p \rightarrow \Box\Box p$ and is closed under modus ponens, substitution and monotonicity (from $\varphi \rightarrow \psi$ derive $\Box\varphi \rightarrow \Box\psi$). This is not exactly the standard way of defining $S4$, it is an equivalent axiomatization of $S4$. For a closer fit to topological reasoning, it is better to work with these (see [28] and [29]).

Definition 6 (Quasi-Ordered Set). *A* quasi-ordered set (qo-set) *is a pair* (W, R) *where* W *is a non-empty set of worlds and* $R \subseteq W \times W$ *is a reflexive and transitive relation.*

Definition 7 (Alexandroff Space). *An* Alexandroff space *is a topological space* (W, τ) *where every point* $w \in W$ *has a smallest open neighborhood* N_w *and arbitrary intersections of open sets are open.*

Alexandroff spaces correspond naturally to quasi-ordered sets by defining the open sets as upward closed subsets with respect to the order. This correspondence provides a bridge between Kripke semantics and topological semantics. Every Kripke frame based on a quasi-order corresponds to an Alexandroff space, and conversely, every Alexandroff space gives rise to such a Kripke frame. As a result, axioms and inference rules of modal logics like $S4$, which are valid on reflexive and transitive Kripke frames, are also valid on all Alexandroff spaces. This structural alignment enables the transfer of completeness results between the two semantic frameworks.

Another alternative semantics is Neighborhood semantics, a more general framework for understanding modal logics, first introduced by Dana Scott and Richard Montague (independently in [14] and [13]). It is particularly advantageous for non-normal modal logics, in which structural principles such as the necessity of tautologies or the distributivity of necessity over conjunction may fail to hold.

Fuzzy Topological Spaces. Fuzzy set theory was first introduced by Lotfi Zadeh in 1965 [30], laying the foundation for fuzzy topology. Building on this, C.L. Chang introduced the concept of fuzzy topological spaces in 1968 [21], defining a fuzzy topology on a set X as a family $\mathbf{T} \subseteq I^X$ (with $I = [0, 1]$) satisfying certain axioms, where each element of $\mathbf{T}$ is regarded as a fuzzy open set. He also extended this framework to membership functions defined over arbitrary complete distributive lattices.

Definition 8 (Fuzzy Topological Space). *A family* $\mathbf{T}$ *of fuzzy sets in* X *is known as* fuzzy topology *on a nonempty set* X *if it satisfies the following axioms:*

(T1) $0_X, 1_X \in \mathbf{T}$

(T2) $G_1 \bigcap G_2 \in \mathbf{T}$, *for any* $G_1, G_2 \in \mathbf{T}$
(T3) $\bigcup G_j \in \mathbf{T}$, *for any arbitrary family* $\{G_j : G_j \in \mathbf{T}, j \in J\}$

The elements of T *are called* fuzzy open sets.

We consider the usual functions 0_X and 1_X, where $0_X(x) = 0$ for all $x \in X$, known as the *fuzzy empty set* or the fuzzy set assigning zero membership to every element, and $1_X(x) = 1$ for all $x \in X$, which represents the fuzzy set with maximum (full) membership value for each element. The standard fuzzy set operations of union and intersection are defined pointwise using the maximum and minimum operators, corresponding to the operations of the Gödel algebra. The complement of a fuzzy set A is defined by $A^c(x) = 1 - A(x)$. A fuzzy set is *fuzzy closed* if its complement is fuzzy open, the collection of all such sets is denoted $T^c = \{A : A^c \in T\}$.

Example 1. Let $X = \{l, m, n\}$, $\lambda_1 = \{(l,\ 0.3),\ (m,\ 0.5),\ (n,\ 0.4)\}$, $\lambda_2 = \{(l,\ 0.2),\ (m,\ 0.5),\ (n,\ 0.4)\}$ and $\lambda_3 = \{(l,\ 0.1),\ (m,\ 0.3),\ (n,\ 0.2)\}$ be fuzzy sets of X, then $\mathbf{T} = \{1, 0, \lambda_1, \lambda_3\}$ is the fuzzy topology on X.

Example 2. The concept of fuzzy metric, introduced by Kramosil and Michalek [31] provides a way to quantify the degree of nearness between two points. Formally, for a set X, a fuzzy metric is a mapping $X \times X \times [0, \infty) \to [0, 1]$, where the value for points $x, y \in X$ and parameter $t \geq 0$ can be intuitively understood as the degree that the distance between x and y is less than t. Building on this Aygünoglu et al. [32] demonstrated how to construct a topology from a fuzzy metric space. Their method produces a family of classical, parameterized topologies $\{T_\alpha\}$ where $\alpha \in [0, \infty)$ on the underlying set X. From this family, a stratified fuzzy topology T^M can be defined. A fuzzy set λ in X belongs to T^M if and only if for $\alpha \in [0, \infty)$, its α-level λ^α is an open set in the corresponding classical topology T_α. This specific construction

$$T^M = \{\lambda : X \to [0, 1] \mid \lambda \in T_\alpha \forall \alpha \in [0, \infty)\}$$

results in a stratified fuzzy topology, a well known type of fuzzy topology originally introduced by Lowen [22].

The degree based framework define topological concepts by membership values, enabling us to study spatial relationships when the boundaries of the spaces are not sharply defined. Below are some important definitions of fuzzy topological notions that will be needed throughout this work to formalize and analyze our proposed semantic framework.

Definition 9 (Fuzzy Interior and Fuzzy Closure). *The union of all open subsets contained in a fuzzy set* λ *is known as the* interior of a fuzzy set λ. *Formally, it is given by* $Int(\lambda) = Sup\{P : P \leq \lambda, P \in \mathbf{T}\}$ *and it is denoted as* $Int(\lambda)$.

The fuzzy closure $Cl(\lambda)$ *is the intersection of all closed sets which contain* λ *and it is given by* $Cl(\lambda) = Inf\{G : G^c \in \mathbf{T}, \lambda \leq G\}$.

Definition 10 (Base and Subbase - [33]). *Let X be a fuzzy topology. A subfamily B of X is a* base for X *if and only if each and every member of X can be written as a union of some members of B.*

A subfamily S of X is a subbase for X *iff the family of finite intersections of members of S form a base for X.*

To relate the work of Cintula et al. [5] to our fuzzy topological semantics, it is essential to examine the notions of the fuzzy neighbourhood of a point and of the fuzzy neighbourhood of a set. In our setting, these neighbourhoods are not given directly but are induced by fuzzy open sets, so that every fuzzy topological model can be viewed as a fuzzy neighbourhood model with additional closure conditions. This explains how the two approaches are related, while also showing that our semantics is grounded in the structure of fuzzy topology.

Definition 11 (Fuzzy Neighborhood of a Point and a Set - [34]). *A fuzzy set λ in a fuzzy topological space $(X, \mathbf{T})$ is* fuzzy neighborhood of a point $x \in X$ *iff there exist $g \in \mathbf{T}$ such that $g \leq \lambda$ and $\lambda(x) = g(x) > 0$. A neighborhood of point x is usually denoted by λ_x. A neighborhood λ_x is called an* open neighborhood of x *iff $\lambda_x \in \mathbf{T}$.*

Let $(X, \mathbf{T})$ be a fuzzy topological space. A fuzzy set μ in X is fuzzy neighborhood of a fuzzy set α *in X iff there exist $g \in \mathbf{T}$ such that $\alpha \leq g \leq \mu$.*

The definitions described above provide a basic grasp of fuzzy topological spaces, laying the groundwork for their use in semantic modeling. These structures provide a flexible and refined approach to spatial thinking, particularly in situations when traditional topological frameworks fall short due to ambiguity or partial truth. For a more detailed and comprehensive study of fuzzy topological spaces, the reader is encouraged to consult the key works of Chang [21] and Lowen [22].

Fuzzy Modal Logic. In the field of fuzzy modal logics, several systems have been developed by combining the principles of fuzzy logic with modal reasoning. These logics differ primarily in the choice of the underlying fuzzy logic, often determined by the selection of a specific t-norm, which governs the interpretation of conjunction and implication in the many-valued setting. In this section, we introduce some basic definitions that will be used throughout the paper. In particular, we introduce the notions of a t-norm and a t-conorm (also called an s-norm), which play a central role in defining a semantics based on fuzzy values. These concepts are inspired by *fuzzy algebras* [35].

A *t-norm* is a binary operation $* : [0,1] \times [0,1] \to [0,1]$ satisfying the following conditions: (i) $*$ is *commutative* and *associative*; (ii) $*$ is *non-decreasing* in both arguments; and (iii) $1 * x = x$ and $0 * x = 0$ for all $x \in [0,1]$. A *continuous t-norm* is a t-norm that is a continuous mapping from $[0,1] \times [0,1]$ into $[0,1]$ (in the usual sense). Next we present some examples

Example 3. The following are some widely used examples of t-norms that serve as fundamental operations in fuzzy logic systems:

i) **Łukasiewicz t-norm:** $x * y = \max(0, x + y - 1)$.
ii) **Gödel t-norm:** $x * y = \min(x, y)$.
iii) **Product t-conorm:** $x * y = x \cdot y$ (product of reals).

An important notion when we intend to use fuzzy semantics to interpret propositional sentences is the choice of the most appropriate implication (see [36]). In CPL, the implication $A \to B$ is true if and only if the truth-value of A is less than or equal to the truth-value of B. On $[0, 1]$, the value of $x \to y$ should be *non-increasing* in x and *non-decreasing* in y. This leads to the definition of the residuum $\Rightarrow$ as: $x * z \leq y$ iff $z \leq (x \Rightarrow y)$.

It follows that $x \Rightarrow y$ is the maximal z satisfying $x * z \leq y$. The following lemma guarantees the existence of such a maximal element (see [35] for details).

Lemma 1 ([35]). *Let $*$ be a continuous t-norm. Then there is a unique operation $x \Rightarrow y$ satisfying, for all $x, y, z \in [0, 1]$, the condition $x * z \leq y$ iff $z \leq (x \Rightarrow y)$, namely $x \Rightarrow y = \max\{z \mid x * z \leq y\}$.*

The operation $x \Rightarrow y$ from Lemma 1 is called the *residuum* of the t-norm. The context will make clear which t-norm is being used whenever we refer to a particular residuum $\Rightarrow$. In the sequel, we will make use of the following facts: (i) $x \leq y$ iff $(x \Rightarrow y) = 1$ and (ii) $(1 \Rightarrow) = x$.

For the previous t-norms we get the following residuum/implications

i) $a \Rightarrow b = \min(1, 1 - a + b)$ **Łukasiewicz implication**

ii) $a \Rightarrow b = \begin{cases} 1 & \text{if } a \leq b, \\ b & \text{otherwise.} \end{cases}$ **Gödel implication**

iii) $a \Rightarrow b = \begin{cases} 1 & \text{if } a \leq b, \\ \frac{b}{a} & \text{otherwise.} \end{cases}$ **Product implication**

The residuum $\Rightarrow$ defines in a natural way a negation $\ominus x := (x \Rightarrow 0)$. For the examples above, we have:

i) $\ominus x = 1 - x$ **Łukasiewicz negation**
ii) $\ominus 0 = 1, \ominus x = 0$ for $x > 0$ **Gödel negation**
iii) The product negation coincides with Gödel negation.

Most fuzzy modal logics have semantics based on continuous t-norms and its residuum. In what follows, we consider a generic continuous t-norm T and its residuum $\Rightarrow$. We begin by defining the syntax of modal language. Here, as we will see, we need to include $\wedge$ and $\Diamond$ as primitive symbols.

Definition 12. *Let Prop be a set of atomic propositions, the set of modal formulas* Form *is given by:*

$$\varphi, \psi ::= p \mid \perp \mid \varphi \wedge \psi \mid \varphi \to \psi \mid \Diamond\varphi \mid \Box\varphi, \textit{ where } p \in \textit{Prop}.$$

Kripke semantics is generalized to the fuzzy context in a natural way. More precisely, we introduce the notion of "fuzzy Kripke model", a generalization of standard Kripke models where instead of relations between elements we have degrees of membership between them see [1,3,37].

Definition 13 (Fuzzy Models). *A* fuzzy frame *is a tuple* $\mathfrak{F} = (W, R)$ *where* W *is a non empty set of states and* $R : W^2 \to [0,1]$ *is a membership function indicating the membership degree of each edge.*

A fuzzy Kripke model *is a triple* $\mathcal{M} = (W, R, V)$ *where* (W, R) *is a fuzzy frame and* $V : Prop \times W \to [0,1]$ *is the valuation function.*

Definition 14. *Let* $\mathcal{M} = (W, R, V)$ *be a fuzzy Kripke model,* $w \in W$*, and let* T *be a continuous t-norm. The semantics of a formula* φ *is given recursively by a function* $[\cdot]^w_{\mathcal{M}} : \mathsf{Form} \to [0,1]$ *as:*

i) $[p]^w_{\mathcal{M}} = v(p, w)$
ii) $[\bot]^w_{\mathcal{M}} = 0$
iii) $[\varphi \wedge \psi]^w_{\mathcal{M}} = T([\varphi]^w_{\mathcal{M}}, [\psi]^w_{\mathcal{M}})$
iv) $[\varphi \to \psi]^w_{\mathcal{M}} = [\varphi]^w_{\mathcal{M}} \Rightarrow [\psi]^w_{\mathcal{M}}$
v) $[\Diamond\varphi]^w_{\mathcal{M}} = \sup_{w' \in W} \left(T(R(w, w'), [\varphi]^{w'}_{\mathcal{M}})\right)$
vi) $[\Box\varphi]^w_{\mathcal{M}} = \inf_{w' \in W} \left(R(w, w') \Rightarrow [\varphi]^{w'}_{\mathcal{M}}\right)$

Note that the negation is not necessarily involutive. Moreover, in general $\Diamond\varphi$ is not equivalent to $\neg\Box\neg\varphi$. So, the modal operator $\Diamond$ has to be considered as a primitive symbol. It is essential to recognize that there is no singular fuzzy modal logic instead, multiple fuzzy modal logics exist, each parameterized by the choice of an underlying fuzzy algebra (i.e., a specific algebraic structure defining logical operations like conjunction, implication, and negation). This algebraic framework dictates the behavior of the logic, leading to significant variations in properties such as truth functionality, axiomatic completeness, and adherence to classical laws (see [3,7,37]). Henceforth, we restrict attention to Gödel modal logic, with semantics defined by the operations of the Gödel algebra.

Definition 15. *A fuzzy Kripke frame* $\mathcal{F} = (W, R)$ *is* reflexive and transitive *if* R *is such that,* $\forall w \in W, R(w, w) = 1$ *and* $\forall u, v, w \in W, T(R(u, v), R(v, w)) \leq R(u, w)$*, respectively.*

If we consider only reflexive and transitive fuzzy Kripke frames all axioms of S4 are valid. The axioms (T) and (4) come from reflexivity and transitivity, respectively.

Theorem 1 (Soundness for Fuzzy Kripke Frames). *All axioms of fuzzy S4 are valid in every reflexive and transitive fuzzy Kripke frame.*

3 Fuzzy Topological Semantics

In this section we introduce fuzzy topological semantics for fuzzy modal logic. In this work, we adopt Gödel algebra as the underlying many-valued framework for our fuzzy topological semantics. This choice is motivated by several compelling theoretical and practical considerations. Gödel algebra, based on the minimum t-norm, offers a natural alignment with the intersection operator of the fuzzy spaces, we discussed. Its residuated structure ensures well behaved implications, which align closely with classical modal reasoning when extended to fuzzy settings. Moreover, Gödel algebra preserves monotonicity and continuity properties that are particularly suited to the semantics of open sets and interior operators in fuzzy topology. In contrast, Łukasiewicz and Product algebras allow additive or multiplicative effects that do not match the intersection based behavior of fuzzy open sets, making Gödel algebra the most natural choice for our semantics. The key insight is that accessibility in fuzzy Kripke semantics corresponds to continuity in fuzzy topological terms, allowing modal formulas to be interpreted as graded conditions over fuzzy open sets.

Definition 16 (Fuzzy Topological Model). *A* fuzzy topological model *is a tuple* $\mathcal{M} = (X, \mathbf{T}, V)$ *where* $(X, \mathbf{T})$ *is a fuzzy topological space and* V *is a valuation, i.e., a map* $V : Prop \to [0,1]^X$.

Definition 17. *Let* $\mathcal{M} = (X, \mathbf{T}, V)$ *be a fuzzy topological model and* $x \in X$. *The semantics of a formula* φ *at a point* $x \in X$, *in symbols* $[\![\varphi]\!]^x_{\mathcal{M}} \in [0,1]$, *is recursively defined by*

i) $[\![p]\!]^x_{\mathcal{M}} = V(p)(x), p \in Prop$
ii) $[\![\bot]\!]^x_{\mathcal{M}} = 0$
iii) $[\![\varphi \wedge \psi]\!]^x_{\mathcal{M}} = \min([\![\varphi]\!]^x_{\mathcal{M}}, [\![\psi]\!]^x_{\mathcal{M}})$ *(Gödel t-norm)*
iv) $[\![\varphi \to \psi]\!]^x_{\mathcal{M}} = \begin{cases} 1 & \text{if } [\![\varphi]\!]^x_{\mathcal{M}} \leq [\![\psi]\!]^x_{\mathcal{M}} \\ [\![\psi]\!]^x_{\mathcal{M}} & \text{otherwise} \end{cases}$ *(Gödel residuum)*
v) $[\![\Diamond\varphi]\!]^x_{\mathcal{M}} = \inf_{\mu^c \in \mathbf{T}}\{\mu(x) : [\![\varphi]\!]_{\mathcal{M}} \leq \mu\}$
vi) $[\![\Box\varphi]\!]^x_{\mathcal{M}} = \sup_{\mu \in \mathbf{T}}\{\mu(x) : \mu \leq [\![\varphi]\!]_{\mathcal{M}}\}$.

The value $[\![\varphi]\!]^x_{\mathcal{M}}$ will be read as "the degree of truth of the formula φ in the model $\mathcal{M}$ at the point x". In some cases it is useful to define the semantics of a formula in a model globally.

Definition 18. *Let* $\mathcal{M} = (X, \mathbf{T}, V)$ *be a topological model,* $w \in X$. *The semantics of a formula* φ *in* M *is a function* $[\![\varphi]\!]_{\mathcal{M}} : X \to [0,1]$ *defined recursively by*

i) $[\![p]\!]_{\mathcal{M}} = V(p), p \in Prop$
ii) $[\![\bot]\!]_{\mathcal{M}} = 0_X$
iii) $[\![\varphi \wedge \psi]\!]_{\mathcal{M}} = [\![\varphi]\!]_{\mathcal{M}} \cap [\![\psi]\!]_{\mathcal{M}}$
iv) $[\![\varphi \to \psi]\!]_{\mathcal{M}} = [\![\varphi]\!]_{\mathcal{M}} \Rightarrow [\![\psi]\!]_{\mathcal{M}}$,
where $([\![\varphi]\!]_{\mathcal{M}} \Rightarrow [\![\psi]\!]_{\mathcal{M}})(x) = \begin{cases} 1 & \text{if } [\![\varphi]\!]_{\mathcal{M}}(x) \leq [\![\psi]\!]_{\mathcal{M}}(x) \\ [\![\psi]\!]_{\mathcal{M}}(x) & \text{otherwise} \end{cases}$
v) $[\![\Box\varphi]\!]_{\mathcal{M}} = int([\![\varphi]\!]_{\mathcal{M}})$
vi) $[\![\Diamond\varphi]\!]_{\mathcal{M}} = cl([\![\varphi]\!]_{\mathcal{M}})$.

A formula φ is said to be *true in a model* $\mathcal{M} = (X, \mathbf{T}, V)$ if $[\![\varphi]\!]^x_{\mathcal{M}} = 1$ for every $x \in X$. φ is *valid in a fuzzy topological space* $(X, \mathbf{T})$ if it is true in every model based on $(X, \mathbf{T})$. Finally, φ is said to be *valid in a class* K of fuzzy topological spaces if it is valid in every member of K. It is not difficult to see that $[\![\varphi]\!]_{\mathcal{M}}(x) = [\![\varphi]\!]^x_{\mathcal{M}}$. Hence, φ is valid in $\mathcal{M}$ iff $[\![\varphi]\!]_{\mathcal{M}} = 1_X$. Consequently, from the properties of residuum, we have that $\varphi \to \psi$ is valid in $\mathcal{M}$ iff $[\![\varphi]\!]_{\mathcal{M}} \leq [\![\psi]\!]_{\mathcal{M}}$. These insights are essential for analyzing modal formulas via fuzzy topological operators, particularly the fuzzy interior and closure, as explored in the following proposition.

Proposition 1 ([22]). *Let $(X, \mathbf{T})$ be a fuzzy topological space, Int and Cl the fuzzy interior and closure operators defined by fuzzy topology, then the following holds for any fuzzy subset of X,*

(I1) $Int(1_X) = 1_X$	***(C1)*** $Cl(0_X) = 0_X$
(I2) $Int(\lambda) \leq \lambda$	***(C2)*** $\lambda \leq Cl(\lambda)$
(I3) $Int(\lambda \wedge \mu) = Int(\lambda) \wedge Int(\mu)$	***(C3)*** $Cl(\lambda \vee \mu) = Cl(\lambda) \vee Cl(\mu)$
(I4) $Int(Int(\lambda)) = Int(\lambda)$	***(C4)*** $Cl(Cl(\lambda)) = Cl(\lambda)$

It is easy to observe that the conditions $(I1)$ through $(I4)$, when interpreted with the $\Box$ operator in place of the fuzzy interior operator Int, resemble the axioms of the modal logic system $S4$. Similarly, the dual conditions $(C1)$ through $(C4)$, when expressed using the $\Diamond$ (diamond) operator instead of the fuzzy closure operator Cl, also mirror the axioms of $S4$. In fact, as we will demonstrate shortly, every fuzzy topological space gives rise to a model that validates all $S4$ axioms. The following proposition states this result.

Proposition 2 (Fuzzy Topological Validity of $S4$). *Let* FTop *be the class of all fuzzy topological spaces. The axioms of $S4$ are valid in every fuzzy topological model over* FTop.

Proof. To prove the validity of $S4$, we prove the validity of axioms of $S4$ one by one. In each case we consider an arbitrary fuzzy topological model $\mathcal{M} = (X, \mathbf{T}, V)$ and any $x \in X$.

- For (N): We have that $[\![\Box\top]\!]^x_{\mathcal{M}} = \sup\{\mu(x) \mid \mu \in \mathbf{T},\ \mu \leq 1_X\}$. Since $1_X \in \mathbf{T}$ and clearly $1_X \leq 1_X$, the supremum contains the value $1_X(x) = 1$ and therefore equals 1. Hence $\Box\top$ is valid.
- For (T): By definition of $\Box$,

$$[\![\Box p]\!]^x_{\mathcal{M}} = \sup\{\mu(x) \mid \mu \in \mathbf{T},\ \mu \leq [\![p]\!]_{\mathcal{M}}\}.$$

  For every $\mu \in \mathbf{T}$ with $\mu \leq [\![p]\!]_{\mathcal{M}}$ we have the point-wise inequality $\mu(x) \leq [\![p]\!]^x_{\mathcal{M}}$. Hence the supremum itself satisfies $[\![\Box p]\!]^x_{\mathcal{M}} \leq [\![p]\!]^x_{\mathcal{M}}$. Therefore,

$$[\![\Box p \to p]\!]^x_{\mathcal{M}} = 1.$$

- For (4): $\Box p \to \Box\Box p$. We aim to prove $[\![\Box p]\!]^x_{\mathcal{M}} \leq [\![\Box\Box p]\!]^x_{\mathcal{M}}$. By definition:

$$[\![\Box p]\!]^x_{\mathcal{M}} = \sup\{\mu(x) \mid \mu \in \mathbf{T},\ \mu \leq [\![p]\!]_{\mathcal{M}}\}$$

Let $\mu \in \mathbf{T}$ be such that $\mu \leq [\![p]\!]_{\mathcal{M}}$. Since $[\![\Box p]\!]_{\mathcal{M}}$ is defined as:

$$[\![\Box p]\!]_{\mathcal{M}} = \sup\{\nu \mid \nu \in \mathbf{T},\ \nu \leq [\![p]\!]_{\mathcal{M}}\}$$

and μ is one such ν, we have:

$$\mu \leq [\![\Box p]\!]_{\mathcal{M}}$$

Hence, every $\mu \in \mathbf{T}$ satisfying $\mu \leq [\![p]\!]_{\mathcal{M}}$ also satisfies $\mu \leq [\![\Box p]\!]_{\mathcal{M}}$. Therefore, every such μ is a candidate in the supremum that defines:

$$[\![\Box\Box p]\!]^x_{\mathcal{M}} = \sup\{\lambda(x) \mid \lambda \in \mathbf{T},\ \lambda \leq [\![\Box p]\!]_{\mathcal{M}}\}$$

In particular, since $\mu \in \mathbf{T}$ and $\mu \leq [\![\Box p]\!]_{\mathcal{M}}$, we get:

$$\mu(x) \leq [\![\Box\Box p]\!]^x_{\mathcal{M}}$$

This holds for all $\mu \in \mathbf{T}$ with $\mu \leq [\![p]\!]_{\mathcal{M}}$, so taking supremum over all such μ gives:

$$[\![\Box p]\!]^x_{\mathcal{M}} \leq [\![\Box\Box p]\!]^x_{\mathcal{M}}$$

Thus,

$$[\![\Box p \to \Box\Box p]\!]^x_{\mathcal{M}} = 1$$

- For (R): $\Box(p \wedge q) \leftrightarrow \Box p \wedge \Box q$. To show that R is valid we need show its validation in both directions, first to prove $\Box(p \wedge q) \to \Box p \wedge \Box q$ we need to show that,

$$[\![\Box(p \wedge q)]\!]^x_{\mathcal{M}} \leq \min([\![\Box p]\!]^x_{\mathcal{M}},\ [\![\Box q]\!]^x_{\mathcal{M}})$$

By definition of $\Box$:

$$[\![\Box(p \wedge q)]\!]^x_{\mathcal{M}} = \sup\{\mu(x) \mid \mu \in \mathbf{T},\ \mu \leq [\![p \wedge q]\!]_{\mathcal{M}}\}$$

Since:

$$[\![p \wedge q]\!]^y_{\mathcal{M}} = \min([\![p]\!]^y_{\mathcal{M}},\ [\![q]\!]^y_{\mathcal{M}}) \quad \text{for all } y \in X,$$

we have that any μ satisfying $\mu \leq [\![p \wedge q]\!]_{\mathcal{M}}$ also satisfies:

$$\mu \leq [\![p]\!]_{\mathcal{M}} \quad \text{and} \quad \mu \leq [\![q]\!]_{\mathcal{M}}$$

Thus:

$$\mu(x) \leq [\![\Box p]\!]^x_{\mathcal{M}},\quad \mu(x) \leq [\![\Box q]\!]^x_{\mathcal{M}} \Rightarrow \mu(x) \leq \min([\![\Box p]\!]^x_{\mathcal{M}},\ [\![\Box q]\!]^x_{\mathcal{M}})$$

Taking supremum over all such μ yields:

$$[\![\Box(p \wedge q)]\!]^x_{\mathcal{M}} \leq \min([\![\Box p]\!]^x_{\mathcal{M}},\ [\![\Box q]\!]^x_{\mathcal{M}})$$

To prove $\Box p \wedge \Box q \to \Box(p \wedge q)$ we need to show that:

$$\min([\![\Box p]\!]^x_{\mathcal{M}},\ [\![\Box q]\!]^x_{\mathcal{M}}) \leq [\![\Box(p \wedge q)]\!]^x_{\mathcal{M}}$$

Let $\mu_1, \mu_2 \in \mathbf{T}$ such that:

$$\mu_1 \leq [\![p]\!]_{\mathcal{M}}, \quad \mu_2 \leq [\![q]\!]_{\mathcal{M}}, \quad \mu_1(x) = [\![\Box p]\!]^x_{\mathcal{M}}, \quad \mu_2(x) = [\![\Box q]\!]^x_{\mathcal{M}}$$

Define $\mu = \mu_1 \cap \mu_2 \in \mathbf{T}$. Then:

$$\mu \leq \min\left([\![p]\!]_{\mathcal{M}},\ [\![q]\!]_{\mathcal{M}}\right) = [\![p \wedge q]\!]_{\mathcal{M}} \quad \Rightarrow \quad \mu \leq [\![p \wedge q]\!]_{\mathcal{M}}$$

So:

$$[\![\Box(p \wedge q)]\!]^x_{\mathcal{M}} \geq \mu(x) = \min\left(\mu_1(x),\ \mu_2(x)\right) = \min\left([\![\Box p]\!]^x_{\mathcal{M}},\ [\![\Box q]\!]^x_{\mathcal{M}}\right)$$

Hence,

$$[\![\Box(p \wedge q)]\!]^x_{\mathcal{M}} = \min\left([\![\Box p]\!]^x_{\mathcal{M}},\ [\![\Box q]\!]^x_{\mathcal{M}}\right) = [\![\Box p \wedge \Box q]\!]^x_{\mathcal{M}}$$

- For Modus Ponens: Fix any $x \in X$ such that $[\![\varphi]\!]^x_{\mathcal{M}} = 1$ and $[\![\varphi \to \psi]\!]^x_{\mathcal{M}} = 1$. Since $1 \Rightarrow 1 = 1$, $[\![\varphi \to \psi]\!]^x_{\mathcal{M}} = 1$
- For Monotonicity: Assume $\varphi \to \psi$ is valid, i.e., $[\![\varphi \to \psi]\!]_{\mathcal{M}} = 1$. Hence $[\![\varphi]\!]^y_{\mathcal{M}} \leq [\![\psi]\!]^y_{\mathcal{M}}$, for every $y \in X$.
  The interior operator $\mathrm{Int}\colon [0,1]^X \to [0,1]^X$ is monotone: $A \leq B \Rightarrow \mathrm{Int}\, A \leq \mathrm{Int}\, B$. Applying this pointwise inequality to $A = [\![\varphi]\!]_{\mathcal{M}}$ and $B = [\![\psi]\!]_{\mathcal{M}}$ yields

$$[\![\Box\varphi]\!]^y_{\mathcal{M}} = \mathrm{Int}\big([\![\varphi]\!]_{\mathcal{M}}\big)(y) \ \leq\ \mathrm{Int}\big([\![\psi]\!]_{\mathcal{M}}\big)(y) = [\![\Box\psi]\!]^y_{\mathcal{M}} \quad \text{for all } y \in X.$$

  Now fix any $x \in X$. By Gödel implication again, $[\![\Box\varphi \to \Box\psi]\!]^x_{\mathcal{M}} = 1$. Since $\mathcal{M}$ and x were arbitrary, the formula is valid.

We have seen that the modal logic S4 with respect to fuzzy topological semantics is sound in every fuzzy topological model. Moreover, we shown the validity of these principles pointwise, confirming that $S4$ formulas are semantically preserved under the structure of fuzzy open sets. Now, we turn to establish the relationship between fuzzy topological and fuzzy Kripke models. For this purpose, we need to consider a particular kind of fuzzy topology, namely the fuzzy Alexandroff topology. We establish a one-to-one correspondence between these fuzzy Alexandroff spaces and fuzzy Kripke frames. This correspondence mirrors the classical scenario where Alexandroff spaces correspond to quasi-ordered sets (qo-sets), thereby highlighting a natural extension of classical modal logic concepts into the fuzzy context.

Definition 19 (Fuzzy Alexandroff Space [38]). *Let W be a set and $\tau \subseteq [0,1]^W$ a fuzzy topology on W. We say that (W, τ) is a* fuzzy Alexandroff space *if, τ is closed under arbitrary intersections (infima) (τ is called a* fuzzy Alexandroff topology*).*

Definition 20. *Given a reflexive and transitive fuzzy Kripke frame $\mathcal{F} = (W, R)$ where W is a non-empty set of worlds and $R : W \times W \to [0,1]$ is a fuzzy accessibility relation, the* fuzzy topology *τ_R is defined as the collection of upward-closed fuzzy sets:*

$$\lambda \in \tau_R \iff \forall w, v \in W : \min(R(w,v), \lambda(w)) \leq \lambda(v).$$

As τ_R is defined, it holds that it contains 0, 1 and it is closed under arbitrary unions and intersections.

Proposition 3. *For any reflexive and transitive fuzzy Kripke frame $\mathcal{F} = (W, R)$, the induced topology τ_R is a fuzzy Alexandroff topology.*

Definition 21 (Interior and Closure in Fuzzy Alexandroff Space). *Given a fuzzy Kripke frame $\mathcal{F} = (W, R)$ with R reflexive and transitive, define the* interior operator *Int on fuzzy subset A of W by,*

$$Int(A)(w) = \inf_{v \in W} (R(w, v) \Rightarrow A(v))$$

for any fuzzy set $A : W \longrightarrow [0, 1]$, where $\Rightarrow$ is the Gödel residuum. Similarly define closure operator *Cl on fuzzy subset A of W by,*

$$Cl(A)(w) = \sup_{v \in W} \big(R(w, v) \wedge A(v)\big)$$

for any fuzzy set $A : W \longrightarrow [0, 1]$ where $\wedge$ is the Gödel minimum.

We choose Gödel residuum due to its natural alignment with fuzzy topological semantics, ensuring the required monotonicity and continuity conditions of the interior operator are satisfied, thereby simplifying and clarifying the fuzzy topological interpretation.

Proposition 4. *Let (W, τ_R) be a fuzzy Alexandroff space induced by a reflexive and transitive fuzzy relation $R : W \times W \longrightarrow [0, 1]$. Then for every fuzzy set $A : W \longrightarrow [0, 1]$ and each $w \in W$, the fuzzy interior and fuzzy closure operators are equivalent to the interior and closure operators defined by Gödel residuum and minimum:*

i) $\sup\{\mu(w) \mid \mu \in \tau_R,\ \mu \leq A\} = \inf_{v \in W} \big(R(w, v) \Rightarrow A(v)\big)$;
ii) $\inf\{\eta(w) \mid \eta \text{ is closed, } A \leq \eta\} = \sup_{v \in W} \big(R(w, v) \wedge A(v)\big)$.

Now, if we consider the interior operator for fuzzy Alexendroff space as defined in Definition 21, the interpretation of $\Box$ operator defined in fuzzy Kripke frames and fuzzy topological semantics coincide (using Gödel algebra). So it is not difficult to see that,

Lemma 2. *Let $\mathcal{F} = (W, R)$ be a fuzzy Kripke frame with R reflexive and transitive and let (W, τ_R) be the corresponding fuzzy Alexandroff space. For any fuzzy valuation V and modal formula φ, we have:*

$$[\varphi]^w_{\mathcal{M}^K} = [\![\varphi]\!]^w_{\mathcal{M}^T},$$

where $\mathcal{M}^K = (W, R, V)$ and $\mathcal{M}^T = (W, \tau_R, V)$.

Proof. The base case for propositional variables is immediate by definition. The inductive steps for the Boolean connectives $\bot$, $\wedge$, and $\rightarrow$ are straightforward, as their interpretations are defined identically in both $\mathcal{M}^K$ and $\mathcal{M}^T$.

Inductive Step: To prove for formulas $\Box\psi$ assume that $[\psi]_{\mathcal{M}^K} = [\![\psi]\!]_{\mathcal{M}^T}$. By definition in fuzzy kripke model

$$[\Box\psi]^w_{\mathcal{M}^K} = \inf_{v\in W}\{R(w,v) \Rightarrow [\psi]^v_{\mathcal{M}^K}\}$$

and in fuzzy topological model, $\Box\psi$ is interpreted as,

$$[\![\Box\psi]\!]^w_{\mathcal{M}^T} = \sup_{\mu\in\mathcal{T}}\{\mu(w) : \mu \leq [\![\psi]\!]_{\mathcal{M}^T}\}.$$

From Proposition 4, $\sup_{\mu\in\mathcal{T}}\{\mu(w) : \mu \leq [\![\psi]\!]_{\mathcal{M}^T}\} = \inf_{v\in W}\{R(w,v) \Rightarrow [\![\psi]\!]^v_{\mathcal{M}^T}\}$, by the induction hypothesis, $[\psi]_{\mathcal{M}^K} = [\![\psi]\!]_{\mathcal{M}^T}$, so we have,

$$[\![\Box\psi]\!]^w_{\mathcal{M}^T} = \inf_{v\in W}\{R(w,v) \Rightarrow [\psi]^v_{\mathcal{M}^K}\} = [\Box\psi]^w_{\mathcal{M}^K}.$$

Using Proposition 4 and induction hypothesis, a similar result can be obtained for the case where $\varphi = \Diamond\psi$.

The next proposition provides two modal characterizations of when a fuzzy topological space is discrete.

Proposition 5 (Discrete Space Characterization). *Let $(X, \mathbf{T})$ be a fuzzy topological space. The following are equivalent:*

i) $p \rightarrow \Box p$ is valid in $(X, \mathbf{T})$;
ii) $\Diamond p \rightarrow p$ is valid in $(X, \mathbf{T})$;
iii) $\mathbf{T} = [0,1]^X$ (i.e., $(X, \mathbf{T})$ is the discrete fuzzy topology).

Proof. *i)* $\Leftrightarrow$ *iii)*: Assume $p \rightarrow \Box p$ is valid in $(X, \mathbf{T})$. For any fuzzy set $A\colon X \rightarrow [0,1]$, consider V_A such that $V_A(p) = A$. Let $\mathcal{M} := (X, \mathbf{T}, V_A)$. Then for all $x \in X$:

$$[\![p \rightarrow \Box p]\!]_{\mathcal{M}}(x) = 1.$$

Hence, $A(x) \leq Int(A)(x)$. Since $Int(A) \leq A$ universally, this means $Int(A) = A$. Thus by Proposition 1, $\mathbf{T} = [0,1]^X$.

Conversely, if $\mathbf{T} = [0,1]^X$, then $Int(A) = A$ for all A. For any $\mathcal{M} = (X, \mathbf{T}, V)$:

$$[\![p \rightarrow \Box p]\!]_{\mathcal{M}}(x) = V(p)(x) \Rightarrow Int(V(p))(x) = V(p)(x) \Rightarrow V(p)(x) = 1.$$

Hence $\mathcal{M} \models p \rightarrow \Box p$.

ii) $\Leftrightarrow$ *iii)*: Assume $\Diamond p \rightarrow p$ is valid in $(X, \mathbf{T})$. For any A, consider V_A such that $V_A(p) = A$. Let $\mathcal{M} := (X, \mathbf{T}, V_A)$. For all x:

$$[\![\Diamond p \rightarrow p]\!]_{\mathcal{M}}(x) = 1.$$

This implies $Cl(A)(x) \leq A(x)$. Since $A \leq Cl(A)$ universally, $Cl(A) = A$. By Proposition 1, $\mathbf{T} = [0,1]^X$.

Conversely, if $\mathbf{T} = [0,1]^X$, then $Cl(A) = A$. For any $\mathcal{M} = (X, \mathbf{T}, V)$:

$$[\![\Diamond p \to p]\!]_{\mathcal{M}}(x) = Cl(V(p))(x) \Rightarrow V(p)(x) = V(p)(x) \Rightarrow V(p)(x) = 1.$$

Hence $\mathcal{M} \models \Diamond p \to p$.

Next we consider another property of fuzzy topological spaces, namelly extremally disconnected.

Definition 22 (Fuzzy Extremally Disconnected Space - [39]). *A fuzzy topological space* $(X, \mathbf{T})$ *is* extremally disconnected *if for every open fuzzy set* $\mu \in \mathbf{T}$*, the closure* $Cl(\mu)$ *is open.*

Extremal disconnectedness can be characterized using the interior and closure operators. The next theorem provides this characterization for fuzzy extremally disconnected spaces.

Theorem 2. *A fuzzy topology* $\mathbf{T}$ *on a set* X *is extremally disconnected if and only if for every fuzzy set* $A\colon X \to [0,1]$, $Cl(Int(A)) \leq Int(Cl(A))$.

Using this result, one can characterize extremally disconnected fuzzy topological spaces by the validity of the modal formula $\Diamond\Box p \to \Box\Diamond p$. This is stated in next proposition.

Proposition 6. *Let* $(X, \mathbf{T})$ *be a fuzzy topological space. The following statements are equivalent:*

i) $\Diamond\Box p \to \Box\Diamond p$ *is valid in* $(X, \mathbf{T})$*;*
ii) $(X, \mathbf{T})$ *is extremally disconnected fuzzy topological space.*

Proof. Assume that the formula $\Diamond\Box p \to \Box\Diamond p$ is valid in $(X, \mathbf{T})$. To show that $(X, \mathbf{T})$ is fuzzy extremally disconnected, take any open fuzzy set $\mu \in \mathbf{T}$, and define a valuation $V(p) = \mu$. Then,

$$[\![\Box p]\!]_{\mathcal{M}} = Int(\mu), \qquad [\![\Diamond\Box p]\!]_{\mathcal{M}} = Cl(Int(\mu)),$$
$$[\![\Diamond p]\!]_{\mathcal{M}} = Cl(\mu), \qquad [\![\Box\Diamond p]\!]_{\mathcal{M}} = Int(Cl(\mu)).$$

By assumption, the axiom $\Diamond\Box p \to \Box\Diamond p$ is valid, so for every $x \in X$,

$$[\![\Diamond\Box p \to \Box\Diamond p]\!]_{\mathcal{M}}(x) = Cl(Int(\mu))(x) \Rightarrow Int(Cl(\mu))(x) = 1.$$

By properties of Gödel algebra, $a \Rightarrow b = 1$ if and only if $a \leq b$. Hence, we obtain,

$$Cl(Int(\mu))(x) \leq Int(Cl(\mu))(x) \quad \text{for all } x \in X.$$

Therefore, $Cl(Int(\mu)) \leq Int(Cl(\mu))$. Since this holds for all fuzzy sets μ, and by definition the interior $Int(\mu) \in \mathbf{T}$ for each such μ, we conclude that,

$$Cl(Int(\mu)) \leq Int(Cl(\mu)) \quad \text{for all } \mu\colon X \to [0,1],$$

by Theorem 2, this implies that $(X, \mathbf{T})$ is fuzzy extremally disconnected.

Now assume that $(X, \mathbf{T})$ is fuzzy extremally disconnected then for every fuzzy set A, $Cl(Int(A)) \leq Int(Cl(A))$. Let V be any valuation such that $V(p) = A$. Then using the interpretations of the modal operators:

$$[\![\Diamond\Box p]\!]_{\mathcal{M}} = Cl(Int(A)), \quad [\![\Box\Diamond p]\!]_{\mathcal{M}} = Int(Cl(A)).$$

By the hypothesis, $Cl(Int(A)) \leq Int(Cl(A))$, so for every $x \in X$,

$$Cl(Int(A))(x) \leq Int(Cl(A))(x).$$

Therefore $[\![\Diamond\Box p \to \Box\Diamond p]\!]_{\mathcal{M}}(x) = Cl(Int(A))(x) \Rightarrow Int(Cl(A))(x) = 1$.

4 Conclusion and Further Work

In this paper, we introduced a fuzzy topological semantics for fuzzy modal logic and described its relationship with fuzzy Kripke semantics. The proposed framework provides a natural extension that brings the spatial intuition of classical topological semantics into the realm of uncertainty characteristic of fuzzy logics. This extension yields a suitable reasoning structure for contexts involving imprecision or graded information. Similar results can also be obtained by considering alternative fuzzy topologies, in which intersection and union are defined using different t-norms and t-conorms. However, the corresponding definition of topological semantics would need to be adapted accordingly.

One natural direction for future work is to establish completeness theorems for fuzzy topological semantics. In addition, we intend to develop a theory of fuzzy bisimulations within fuzzy topological models, which would allow for a refined analysis of model equivalence and invariance under modal formulas. Another promising line of research involves exploring A-lattice-based topological semantics. Such an extension would allow the integration of more general logics and provide a foundation for fuzzy modal reasoning with greater expressive power. On the other hand, investigating whether fuzzy topological semantics can be captured within a coalgebraic framework also appears highly fruitful, as it would unify these semantics within a broader and well-established setting. These research directions are likely to deepen the connections between fuzzy logic, topology, and modal reasoning, while also opening new avenues for applications in areas such as approximate reasoning, formal epistemology, and systems dealing with uncertain information.

Acknowledgement. The present study was developed in the scope of the Project "Agenda ILLIANCE" [C644919832-00000035 | Project n° 46], financed by PRR - Plano de Recuperação e Resiliência under the Next Generation EU from the European Union, by CIDMA under the Portuguese Foundation for Science and Technology (FCT) Multi-Annual Financing Program for R & D Units, grants UID/4106/2025 and UID/PRR/4106/ 2025 and by the Project Banksy, with reference compete2020-feder-00892000.

References

1. Fitting, M.C.: Many-valued modal logics. Fund. Inform. **15**(3–4), 235–254 (1991). https://doi.org/10.3233/FI-1991-153-404
2. Benevides, M., Madeira, A., Martins, M.A.: Graded epistemic logic with public announcement. J. Log. Algebraic Methods Program. **125**, 100732 (2022). https://doi.org/10.1016/j.jlamp.2021.100732
3. Jain, M., Madeira, A., Martins, M.A.: A fuzzy modal logic for fuzzy transition systems. Electron. Notes Theor. Comput. Sci. **348**, 85–103 (2020). https://doi.org/10.1016/j.entcs.2020.02.006
4. Bou, F., et al.: Characterizing fuzzy modal semantics by fuzzy multimodal systems with crisp accessibility relations. In: IFSA/EUSFLAT Conference, pp. 1541–1546 (2009)
5. Cintula, P., Noguera, C.: Neighborhood semantics for modal many-valued logics. Fuzzy Sets Syst. **345**, 99–112 (2018). https://doi.org/10.1016/j.fss.2017.10.009
6. Freire, A.R., Martins, M.A.: Modality across different logics. Logic J. IGPL **33**(3), jzae082 (2024). https://doi.org/10.1093/jigpal/jzae082
7. Vidal, A., Esteva, F., Godo, L.: On modal extensions of product fuzzy logic. J. Logic Comput. **27**(1), 299–336 (2017). https://doi.org/10.1093/logcom/exv046
8. Madeira, A., Neves, R., Martins, M.A.: An exercise on the generation of many-valued dynamic logics. J. Log. Algebraic Methods Program. **85**(5), 1011–1037 (2016). https://doi.org/10.1016/J.JLAMP.2016.03.004
9. Madeira, A., Neves, R., Martins, M.A., Barbosa, L.S.: A dynamic logic for every season. In: Braga, C., Martí-Oliet, N. (eds.) SBMF 2014. LNCS, vol. 8941, pp. 130–145. Springer, Cham (2015). https://doi.org/10.1007/978-3-319-15075-8_9
10. Kripke, S.A.: Semantical analysis of intuitionistic logic I. Stud. Logic Found. Math. **40**, 92–130 (1965). https://doi.org/10.2307/2270547
11. Blackburn, P., De Rijke, M., Venema, Y.: Modal Logic: Graph. Darst, vol. 53. Cambridge University Press (2001)
12. Hansoul, G., Teheux, B.: Extending Łukasiewicz logics with a modality: algebraic approach to relational semantics. Stud. Logica **101**, 505–545 (2013). https://doi.org/10.1007/s11225-012-9396-9
13. Montague, R., et al.: Universal grammar. 1974, pp. 222–246 (1970)
14. Scott, D.: Advice on modal logic. In: Philosophical Problems in Logic: Some Recent Developments, pp. 143–173. Springer (1970). https://doi.org/10.1007/978-94-010-3272-8_7
15. Rodriguez, R.O., Godo, L.: Modal uncertainty logics with fuzzy neighborhood semantics. In: WL4AI-2013, p. 79 (2013)
16. Rodriguez, R.O., Godo, L.: On the fuzzy modal logics of belief KD45 (A) and Prob (Łn): axiomatization and neighbourhood semantics. In: WL4AI-2015, p. 64 (2015)
17. Mendelson, B.: Introduction to topology. Courier Corporation (1990)
18. McKinsey, J.C.C., Tarski, A.: The algebra of topology. Ann. Math. **45**(1), 141–191 (1944). https://doi.org/10.2307/1969080
19. Sustretov, D.: Topological semantics and decidability. arXiv preprint math/0703106 (2007). https://doi.org/10.48550/arXiv.math/0703106
20. Awodey, S., Kishida, K., Kotzsch, H.-C.: Topos semantics for higher-order modal logic. Logique et Analyse **57**(228), 591–636 (2014). http://www.jstor.org/stable/44085305
21. Chang, C.-L.: Fuzzy topological spaces. J. Math. Anal. Appl. **24**(1), 182–190 (1968). https://doi.org/10.1016/0022-247X(68)90057-7

22. Lowen, R.: Fuzzy topological spaces and fuzzy compactness. J. Math. Anal. Appl. **56**(3), 621–633 (1976). https://doi.org/10.1016/0022-247X(76)90029-9
23. Shostak, A.P.: Two decades of fuzzy topology: basic ideas, notions, and results. Russ. Math. Surv. **44**(6), 125 (1989). https://doi.org/10.1070/RM1989v044n06ABEH002295
24. Protin, C.L.: A logic for Aristotle's modal syllogistic. Hist. Philos. Logic **44**(3), 225–246 (2023). https://doi.org/10.1080/01445340.2022.2107382
25. Goldblatt, R.: Logics of Time and Computation. 2nd. CSLI Lecture Notes 7. CSLI Publications, Stanford (1992). ISBN: 9780937073940
26. Bezhanishvili, N., Grilletti, G., Holliday, W.H.: Correction to: algebraic and topological semantics for inquisitive logic via choice-free duality. In: Iemhoff, R., Moortgat, M., de Queiroz, R. (eds.) WoLLIC 2019. LNCS, vol. 11541, pp. C1–C1. Springer, Heidelberg (2019). https://doi.org/10.1007/978-3-662-59533-6_41
27. van Benthem, J., Bezhanishvili, G.: Modal logics of space. In: Handbook of Spatial Logics, pp. 217–298. Springer (2007)
28. Gabelaia, D.: Modal Definability in Topology. ILLC Master of Logic Thesis. Master's thesis. Institute for Logic, Language and Computation (ILLC), University of Amsterdam, Amsterdam, The Netherlands (2001). https://eprints.illc.uva.nl/id/document/1688
29. Aiello, M., Van Benthem, J., Bezhanishvili, G.: Reasoning about space: the modal way. J. Logic Comput. **13**(6), 889–920 (2003). https://doi.org/10.1093/logcom/13.6.889
30. Zadeh, L.A.: Fuzzy sets. Inf. Control **8**(3), 338–353 (1965). https://doi.org/10.1016/S0019-9958(65)90241-X
31. Kramosil, I., Michálek, J.: Fuzzy metrics and statistical metric spaces. Kybernetika **11**(5), 336–344 (1975). http://dml.cz/dmlcz/125556
32. Aygünoğlu, A., Aydoğdu, E., Aygün, H.: Construction of fuzzy topology by using fuzzy metric. Filomat **34**(2), 433–441 (2020). https://doi.org/10.2298/FIL2002433A
33. Cheng-Ming, H.: Fuzzy topological spaces. J. Math. Anal. Appl. **110**(1), 141–178 (1985). https://doi.org/10.1016/0022-247x(85)90340-3
34. Warren, R.H.: Neighborhoods, bases and continuity in fuzzy topological spaces. Rocky Mountain J. Math. 459–470 (1978). https://doi.org/10.1216/RMJ-1978-8-3-459
35. Hájek, P.: Metamathematics of fuzzy logic. English. Vol. 4. Trends Log. Stud. Log. Libr. Kluwer Academic Publishers, Dordrecht (1998). ISBN: 0-7923-5238-6
36. Baczyński, M., Jayaram, B.: Fuzzy implications. English. Vol. 231. Stud. Fuzziness Soft Comput. Springer, Berlin (2008). ISBN: 978-3-540-69080-1. https://doi.org/10.1007/978-3-540-69082-5
37. Esteva, F., Godo, L.: Monoidal t-norm based logic: towards a logic for left-continuous t-norms. Fuzzy Sets Syst. **124**(3), 271–288 (2001). https://doi.org/10.1016/S0165-0114(01)00098-7
38. Bosi, G., Panozzo, C., Zuanon, M.E.: Fuzzy alexandrov topologies associated to fuzzy interval orders. Asian Res. J. Math. **6**(10), 1–6 (2020). https://doi.org/10.9734/ARJOM/2020/v16i1030227
39. Ghosh, B.: Fuzzy extremally disconnected spaces. Fuzzy Sets Syst. **46**(2), 245–250 (1992). https://doi.org/10.1016/0165-0114(92)90137-S

Dynamic Logic of Product Relation Changers

Ryo Hatano[1(✉)], Pimolluck Jirakunkanok[2], and Katsuhiko Sano[3]

[1] Tokyo University of Science, 2641 Yamazaki, Noda, Chiba, Japan
r-hatano@rs.tus.ac.jp
[2] Mueang Suphan Buri District, Suphan Buri Province, Thailand
[3] Hokkaido University, Nishi 7, Kita 10, Kita, Sapporo, Hokkaido, Japan
v-sano@let.hokudai.ac.jp

Abstract. This paper integrates the notions of relation changers and action models within a unified framework, proposing a dynamic logic of product relation changers. This logic generalizes dynamic logic of relation changers by van Benthem and Liu, arrow update logic and generalized arrow update logic by Kooi and Renne. A key feature is that we allow multiple pre- and postconditions in each action and the resulting product models do not require state deletion. We present a sound and complete Hilbert system for dynamic logic of product relation changers, wherein semantic completeness is demonstrated via recursion axioms. Furthermore, we provide a cut-free labelled sequent calculus for the proposed logic.

Keywords: Dynamic epistemic logic · Action models · Arrow updates · Relation changers · Recursion axioms · Cut-free labelled sequent calculus

1 Introduction

In a standard setting of dynamic epistemic logic, the knowledge (or belief) of an agent is defined in terms of a relational structure over possible worlds (or states) called a Kripke model, and an agent's knowledge change is defined as a *model transformation* triggered by a dynamic operator. To represent such a model transformation, we may focus on two primary principles that originate from different approaches for public announcement logic [13,34] (cf. [7,24,25]). The first principle is the state deletion (or domain restriction) of a Kripke model, due to Plaza [34]. The principle led to the development of the line of work of *action models* or *product updates* due to Baltag, Moss and Solecki [3,4,10]. The second principle is the link deletion (or change) of the accessibility relation of a Kripke model, due to Gerbrandy and Groeneveld [13]. This principle led not only to the line of work of *arrow update logic* (**AUL**) due to Kooi and Renne [23,24], but also to *dynamic logic of relation changers* (**DLRC**) due to van Benthem and Liu [7,16,17,25]. Note that these principles directly affect an agent's epistemic (or doxastic) attitude. Intuitively, for example, the link deletion enables an agent to have *regrets* with respect to non-φ worlds, whereas domain restriction does not (cf. [7,25]).

P. Jirakunkanok—Independent Researcher

J. Wang et al. (Eds.): DaLí 2025, LNCS 16472, pp. 134–150, 2026.
https://doi.org/10.1007/978-3-032-22626-6_8

Although relation changers and arrow updates share the same principle, the relationship between these operations has not been explored. Jirakunkanok et al. [20] modified Lorini's [27] modal framework for reasoning about an agent's belief and signed information, proposing private announcements and permissions based on particular action models in order to analyze a judge's change of belief in a court case. In her Ph.D. thesis [19], Jirakunkanok, the second author of the paper, employed the notion of action models that is combined with the idea of relation changers to capture the two private actions given in [20]. However, these action models were insufficient to capture relation changers and generalized arrow updates as special cases.

In this study, we integrate the notions of relation changers and action models within a unified framework, proposing a *dynamic logic of product relation changers* (**PRC**). This logic generalizes dynamic logic of relation changers by van Benthem and Liu [7, 25], arrow update logics by Kooi and Renne [23,24]. A key feature is that we allow multiple pre- and postconditions in each action (this is a difference from [19]) and the resulting product models do not require state deletion. We present a sound and complete Hilbert system for **PRC**, wherein semantic completeness is demonstrated via recursion axioms. Moreover, we provide a cut-free labelled sequent calculus for the logic.

Several authors have conducted proof-theoretic studies of dynamic epistemic logic in terms of sequent calculi. For the public announcement logic, there have been proposed both labelled sequent calculi [28,31,39] and non-labelled ones [26]. For the action model logic, labelled sequent calculi were proposed by Nomura et al. [30], and non-labelled sequent calculus was proposed in [12,38]. The first and third authors of this paper also proposed a labelled sequent calculus for the dynamic logic of relation changers in [16, 18] and a non-labelled sequent calculus for a constructive variant of the dynamic logic of relation changers in [17]. As far as the authors know, there is no sequent calculus for generalized arrow update logic. This paper proposes a cut-free labelled sequent calculus for **PRC** that equipollent with the Hilbert system.

The rest of the paper is organised as follows: Sect. 2 provides the necessary background for dynamic logic of relation changers **DLRC**. Section 3 proposes our dynamic logic of product relation changers **PRC** with several examples. We also present a Hilbert system for our logic and demonstrate its soundness and completeness using recursion axioms. Section 4 moves on to a labelled sequent calculus of **PRC** and establishes that it is equipollent with the Hilbert system and that the cut rule can be eliminated. Section 5 concludes the paper by outlining further research directions.

2 Dynamic Logic of Relation Changers

Let $\mathsf{PROP} = \{ p, q, \dots \}$ be a countably infinite set of *propositional variables* and $\mathsf{AP} = \{ a, b, \dots \}$ a finite set of *atomic programs*. Let us denote by $\mathcal{L}$ the syntax of the iteration-free propositional dynamic logic. We define the set $\mathsf{FORM}_{\mathcal{L}}$ of all formulas of the syntax $\mathcal{L}$ and the set $\mathsf{PR}_{\mathcal{L}}$ of all programs of the syntax $\mathcal{L}$ simultaneously by

$$\mathsf{FORM}_{\mathcal{L}} \ni \varphi ::= p \mid \neg\varphi \mid (\varphi \to \varphi) \mid [\alpha]\varphi, \quad \mathsf{PR}_{\mathcal{L}} \ni \alpha ::= a \mid (\alpha \cup \alpha) \mid (\alpha; \alpha) \mid ?\varphi,$$

where $p \in \mathsf{PROP}$ and $a \in \mathsf{AP}$. We also define $\top, \bot, \varphi \land \psi, \varphi \leftrightarrow \psi$ as the abbreviations of $\varphi \to \varphi, \neg\top, \neg(\varphi \to \neg\psi)$ and $(\varphi \to \psi) \land (\psi \to \varphi)$, respectively. As the standard reading of formulas involving programs, we follow [15]. We read a formula of the form $[\alpha]\varphi$ as "after executing α, it is necessary that φ," a program $\alpha \cup \beta$ "choose either α or β non-deterministically and execute the chosen one," a program $\alpha;\beta$ "execute α, then execute β" and a program $?\varphi$ "test φ, proceed if φ is true, fail otherwise," respectively.

Then we add the relation changers $[\mathsf{r}]$ to our syntax $\mathcal{L}$ to define the expanded syntax $\mathcal{L}^{+}$. We define the sets $\mathsf{FORM}_{\mathcal{L}^+}$, $\mathsf{PR}_{\mathcal{L}^+}$ and $\mathsf{RC}_{\mathcal{L}^+}$ of all formulas, programs and relation changers, respectively, simultaneously by

$$\begin{aligned} \mathsf{FORM}_{\mathcal{L}^+} \ni \varphi &::= p \mid \neg\varphi \mid (\varphi \to \varphi) \mid [\alpha]\varphi \mid [\mathsf{r}]\varphi, \\ \mathsf{PR}_{\mathcal{L}^+} \ni \alpha &::= a \mid (\alpha \cup \alpha) \mid (\alpha;\alpha) \mid ?\varphi, \\ \mathsf{RC}_{\mathcal{L}^+} \ni \mathsf{r} &::= (a := \alpha)_{a \in \mathsf{AP}}. \end{aligned}$$

Since a relation changer $\mathsf{r} = (a := \alpha_a)_{a \in \mathsf{AP}}$ (which rewrites each atomic program a with a (new possibly compounded) program α_a) can be regarded as a function from AP to $\mathsf{PR}_{\mathcal{L}^+}$, we also use the notation $\mathsf{r}(a)$ to mean α_a. A formula of the form $[\mathsf{r}]\varphi$ stands for "after changing the accessibility relation for each program a by the binary relation interpretating α_a, φ holds" when $\mathsf{r} = (a := \alpha_a)_{a \in \mathsf{AP}}$. Let $X = \{\alpha_1, \dots, \alpha_n\}$ be a finite set of programs. Then $\bigcup X := \alpha_1 \cup \dots \cup \alpha_n$ denotes the non-deterministic choice on all programs in X, where $\bigcup \emptyset := ?\bot$.

Let us move to Kripke semantics. A *Kripke model* $\mathfrak{M}$ is a tuple $(W, (R_a)_{a \in \mathsf{AP}}, V)$ where W is a non-empty set of *possible worlds* (or *states*), $R_a \subseteq W \times W$ is an *accessibility relation* ($a \in \mathsf{AP}$), and V is a *valuation* from PROP to the power set of W. For $(w, v) \in R_a$, we also use the infix notation $wR_a v$. We define the *satisfaction relation* $\mathfrak{M}, w \models \varphi$ and the *interpretation* $wR_\alpha v$ *of programs* α by simultaneous induction, as usual, except

$$\begin{array}{lll} \mathfrak{M}, w \models [\alpha]\varphi & \text{iff} & \text{for all } v \in W (wR_\alpha v \text{ implies } \mathfrak{M}, v \models \varphi), \\ wR_{\alpha \cup \beta} v & \text{iff} & wR_\alpha v \text{ or } wR_\beta v, \\ wR_{\alpha;\beta} v & \text{iff} & \text{for some } u \in W (wR_\alpha u \text{ and } uR_\beta v), \\ wR_{?\varphi} v & \text{iff} & w = v \text{ and } \mathfrak{M}, v \models \varphi, \\ \mathfrak{M}, w \models [\mathsf{r}]\varphi & \text{iff} & \mathfrak{M}^{\mathsf{r}}, w \models \varphi, \end{array}$$

where $\mathfrak{M}^{\mathsf{r}} = (W, (R^{\mathsf{r}}_a)_{a \in \mathsf{AP}}, V)$ and R^{r}_a is defined by

$$R^{\mathsf{r}}_a := R_{\mathsf{r}(a)} = R_{\alpha_a} \text{ when } \mathsf{r} = (a := \alpha_a)_{a \in \mathsf{AP}}.$$

The above definition of R^{r}_a reflects intuitive meaning of a relation changer r that only rewrites an accessibility relation R_a of $\mathfrak{M}$ by a (new) program $\mathsf{r}(a)$ ($= \alpha_a$). We define the *truth set* $[\![\varphi]\!]_{\mathfrak{M}}$ of a formula φ in a model $\mathfrak{M}$ by $[\![\varphi]\!]_{\mathfrak{M}} := \{ w \in W \mid \mathfrak{M}, w \models \varphi \}$. When the underlying model $\mathfrak{M}$ is clear from the context, we simply write $[\![\varphi]\!]$ instead of writing $[\![\varphi]\!]_{\mathfrak{M}}$. We say that φ is *valid on a model* $\mathfrak{M}$ (notation: $\mathfrak{M} \models \varphi$) if $[\![\varphi]\!]_{\mathfrak{M}} = W$. We also say that φ is *valid in a class* $\mathbb{M}$ *of models* (notation: $\mathbb{M} \models \varphi$) if $\mathfrak{M} \models \varphi$ for all $\mathfrak{M} \in \mathbb{M}$. We denote *the class of all models* by $\mathbb{M}_{\mathrm{all}}$.

Example 1. 1. *Public preference upgrade* [7,25]: van Benthem and Liu introduced the notion of link-cutting public preference upgrade as suggestion $[\sharp\varphi]$, whose intuitive meaning is "let's take φ." After publicly suggesting φ, we remove the links from φ-worlds to $\neg\varphi$-worlds but keep existing links to φ-worlds. Then the corresponding relation changer is given as follows: $\mathsf{r}_{\sharp\varphi} = (a := (?\neg\varphi; a) \cup (a; ?\varphi))_{a \in \mathsf{AP}}$.

2. *Arrow update* [23]: Kooi and Renne proposed the notion of arrow update, which is a generalization of the notion of Gerbrandy and Groeneveld's public announcements [13] and eliminates epistemic access of agents due to some $(\varphi, a, \psi) \in U \subseteq \mathsf{FORM}_{\mathcal{L}} \times \mathsf{AP} \times \mathsf{FORM}_{\mathcal{L}}$ (U is a given finite non-empty set) where φ and ψ are regarded as source and target conditions. Then the corresponding relation changer is $\mathsf{r}_{*U} = (a := \bigcup\{(?\varphi; a; ?\psi) \mid (\varphi, a, \psi) \in U\})_{a \in \mathsf{AP}}$.
3. *Atomic programs exchange* [18]: When $\mathsf{AP} = \{a, b\}$, agents' relation exchange can be defined by a relation changer $(a := b, b := a)$. By extending this idea, Hatano et al. [18] presented such a relation changer that rewrites the link of each agent a by that of b if φ holds; otherwise c. The corresponding relation changer is $(a := (?\varphi; b) \cup (?\neg\varphi; c))_{a \in \mathsf{AP}}$.

Note that item 1 above is definable by arrow updates (item 2), because, e.g., $a; ?\varphi$ in a relation changer can be replaced equivalently by $?\top; a; ?\varphi$. However, it is impossible to define item 3 above by an arrow update (item 2). Thus, the syntax for relation changers is more expressive than that for arrow updates [23].

Table 1 provides a Hilbert system **HDLRC** of dynamic logic of relation changers. We denote by $\mathbf{HPDL}^-$ a Hilbert system consisting of axioms and rules of propositional logic and additional axioms and rules of the iteration-free propositional dynamic logic in Table 1. To define **HDLRC**, we add the set of four recursion axioms together with the necessitation rule for $[\mathsf{r}]$ to the axiomatization $\mathbf{HPDL}^-$. These recursion axioms allow us to reduce the semantic completeness of **HDLRC** to that of $\mathbf{HPDL}^-$. Applying the "inside-out" strategy to establish semantic completeness requires the necessitation rule for $[\mathsf{r}]$. A *derivation* in **HDLRC** is a finite list of formulas consisting of an instance of an axiom of **HDLRC** or the result of applying an inference rule of **HDLRC** to formulas that occur earlier. A formula φ is a *theorem* of **HDLRC** (notation: $\vdash_{\mathbf{HDLRC}} \varphi$) if it occurs in the last of a derivation in **HDLRC**. The same convention applied to $\mathbf{HPDL}^-$.

Table 1. Hilbert-style Axiomatizations $\mathbf{HPDL}^-$ and **HDLRC**

Axioms and Rules for $\mathbf{HPDL}^-$			
(**Taut**)	All instances of propositional tautologies	(**MP**)	From φ and $\varphi \to \psi$, infer ψ
($\mathbf{K}_{[\alpha]}$)	$[\alpha](\varphi \to \psi) \to ([\alpha]\varphi \to [\alpha]\psi)$	($[\cup]$)	$[\alpha \cup \beta]\varphi \leftrightarrow [\alpha]\varphi \wedge [\beta]\varphi$
($[;]$)	$[\alpha;\beta]\varphi \leftrightarrow [\alpha][\beta]\varphi$	($[?]$)	$[?\psi]\varphi \leftrightarrow (\psi \to \varphi)$
($\mathbf{Nec}_{[\alpha]}$)	From φ, infer $[\alpha]\varphi$		
Additional Axioms and Rules to $\mathbf{HPDL}^-$ for **HDLRC**			
($[\mathsf{r}]$**at**)	$[\mathsf{r}]p \leftrightarrow p$	($[\mathsf{r}]\neg$)	$[\mathsf{r}]\neg\varphi \leftrightarrow \neg[\mathsf{r}]\varphi$
($[\mathsf{r}] \to$)	$[\mathsf{r}](\varphi \to \psi) \leftrightarrow ([\mathsf{r}]\varphi \to [\mathsf{r}]\psi)$	($[\mathsf{r}][a]$)	$[\mathsf{r}][a]\varphi \leftrightarrow [\mathsf{r}(a)][\mathsf{r}]\varphi$
($\mathbf{Nec}_{[\mathsf{r}]}$)	From φ, infer $[\mathsf{r}]\varphi$		

Fact 2. ([7,25]) *For any formula* φ, $\mathbb{M}_{\text{all}} \models \varphi$ iff $\vdash_{\mathbf{HDLRC}} \varphi$.

3 Dynamic Logic of Product Relation Changers

In order to generalize the notion of relation changer in terms of an appropriate notion of action model, our starting point is "product upgrade" given in [7]:

> The output for product upgrade on epistemic preference models are again the above epistemic models $\mathcal{M} \times \mathcal{E}$. But this time, we keep all world/action pairs (s, a) represented, as these are the non-realized options that we can still have regrets about. [7, p.18]

There is, however, a contradiction between this quotation and [7, Fact 26] given immediately after the quotation. [7, Fact 26] states that $\mathcal{M}^{\sharp\varphi}$ is isomorphic to $\mathcal{M} \times \mathcal{E}^{\sharp\varphi}$, where $\mathcal{M}^{\sharp\varphi}$ is a model $\mathcal{M}$ preference-upgraded by φ (recall item 1 of Example 1) and "the event model $\mathcal{E}^{\sharp\varphi}$ has two events "seeing that φ" (event 1), "seeing that not-φ" (event 2), with event 2 $\leq$ event 1" [7, p.18]. When the cardinality of a domain of the model $\mathcal{M}$ is finite, say $n \geqslant 1$, then the domain of $\mathcal{M}^{\sharp\varphi}$ has n elements (since the preference upgrade is a link-deletion) but the domain of $\mathcal{M} \times \mathcal{E}^{\sharp\varphi}$ has $2n$ elements (by following the quotation above) hence both models never be isomorphic. We respect the original idea in the quotation above to generalize the notion of relation changer and resolve the contradiction above in a natural manner (see Example 3).

3.1 Syntax and Kripke Semantics

To our syntax $\mathcal{L}$ of the iteration-free propositional dynamic logic, we add action models $\mathbb{E}$ and the corresponding dynamic operators $[\mathbb{E}, \mathsf{e}]$ to define the expanded syntax $\mathcal{L}^{\otimes}$, where e is an element (called *event* or *action point*) in an action model $\mathbb{E}$.

We define the sets $\mathsf{FORM}_{\mathcal{L}^{\otimes}}$, $\mathsf{PR}_{\mathcal{L}^{\otimes}}$ and $\mathsf{AM}_{\mathcal{L}^{\otimes}}$ of all formulas, programs and action models, respectively, by simultaneous induction as follows.

$$\begin{aligned} \mathsf{FORM}_{\mathcal{L}^{\otimes}} \ni \varphi &::= p \mid \neg\varphi \mid (\varphi \to \varphi) \mid [\alpha]\varphi \mid [\mathbb{E}, \mathsf{e}]\varphi, \\ \mathsf{PR}_{\mathcal{L}^{\otimes}} \ni \alpha &::= a \mid (\alpha \cup \alpha) \mid (\alpha; \alpha) \mid ?\varphi, \\ \mathsf{AM}_{\mathcal{L}^{\otimes}} \ni \mathbb{E} &::= (\mathtt{E}, (\mathtt{Q}_a)_{a\in\mathsf{AP}}, \mathtt{CND}, (\mathtt{rel}_a)_{a\in\mathsf{AP}}), \end{aligned}$$

where $p \in \mathsf{PROP}$, $a \in \mathsf{AP}$ and $\mathbb{E}$ is an *action model* to be defined as a tuple $\mathbb{E} = (\mathtt{E}, (\mathtt{Q}_a)_{a\in\mathsf{AP}}, \mathtt{CND}, (\mathtt{rel}_a)_{a\in\mathsf{AP}})$ such that

- $\mathtt{E}$ is a finite set of action points (or events),
- $\mathtt{Q}_a \subseteq \mathtt{E} \times \mathtt{E}$ is an accessibility relation ($a \in \mathsf{AP}$),
- $\mathtt{CND}$ is a family $(\mathtt{CND}(\mathsf{e}))_{\mathsf{e}\in\mathtt{E}}$ of finite lists of formulas such that, for each $\mathsf{e} \in \mathbb{E}$, $\mathtt{CND}(\mathsf{e}) := (\mathtt{cnd}_1(\mathsf{e}), \ldots, \mathtt{cnd}_n(\mathsf{e}))$ (n depends on e), where we assume that the length of $\mathtt{CND}(\mathsf{e})$ can be zero. The list $(\mathtt{cnd}_1(\mathsf{e}), \ldots, \mathtt{cnd}_n(\mathsf{e}))$ represents *associated* conditions for action e, where *associated* conditions mean formulas that can be used as a precondition or a postcondition for action e. So $\mathtt{CND}(\mathsf{e})$ is a generalization of a precondition of an ordinary action model [10].
- $\mathtt{rel}_a$ is a function sending each $(\mathsf{e}, \mathsf{f}) \in \mathtt{Q}_a$ to a program $\alpha_{(\mathsf{e},\mathsf{f})}$ defined as follows:

$$\alpha_{(\mathsf{e},\mathsf{f})} ::= b \mid ?\mathtt{cnd}_i(\mathsf{e}) \mid ?\mathtt{cnd}_j(\mathsf{f}) \mid (\alpha \cup \alpha) \mid (\alpha; \alpha),$$

  where $b \in \mathsf{AP}$, $\mathtt{cnd}_i(\mathsf{e}) \in \mathtt{CND}(\mathsf{e})$, and $\mathtt{cnd}_j(\mathsf{f}) \in \mathtt{CND}(\mathsf{f})$. In what follows, we often denote $\mathtt{rel}_a(\mathsf{e}, \mathsf{f})$ by $\alpha_{(a,\mathsf{e},\mathsf{f})}$.

Let us move to Kripke semantics. Given a Kripke model $\mathfrak{M} = (W, (R_a)_{a\in\mathsf{AP}}, V)$, we add the following new satisfaction clause:

$$\mathfrak{M}, w \models [\mathbb{E}, \mathtt{e}]\varphi \text{ iff } \mathfrak{M}^{\otimes\mathbb{E}}, (w, \mathtt{e}) \models \varphi,$$

where the updated model $\mathfrak{M}^{\otimes\mathbb{E}} = (W^{\otimes\mathbb{E}}, (R_a^{\otimes\mathbb{E}})_{a\in\mathsf{AP}}, V^{\otimes\mathbb{E}})$ of $\mathfrak{M}$ by the action model $\mathbb{E} = (\mathtt{E}, (\mathtt{Q}_a)_{a\in\mathsf{AP}}, \mathtt{CND}, (\mathtt{rel}_a)_{a\in\mathsf{AP}})$ is defined as follows:

$$\begin{aligned} W^{\otimes\mathbb{E}} &:= W \times \mathtt{E}, \\ (w, \mathtt{e})R_a^{\otimes\mathbb{E}}(v, \mathtt{f}) &\text{ iff } wR_{\alpha_{(a,\mathtt{e},\mathtt{f})}}v \text{ and } \mathtt{e}\mathtt{Q}_a\mathtt{f}, \\ (w, \mathtt{e}) \in V^{\otimes\mathbb{E}}(p) &\text{ iff } w \in V(p), \end{aligned}$$

where recall that $\alpha_{(a,\mathtt{e},\mathtt{f})} := \mathtt{rel}_a(\mathtt{e}, \mathtt{f})$. We define the notions of *truth set* $[\![\varphi]\!]_{\mathfrak{M}}$ and validity $\mathfrak{M} \models \varphi$ of a formula on a model as before.

Example 3 (Relation Changers). Let r be a relation changer. It is noted that there is a set $\{?\varphi_1, \dots, ?\varphi_n\}$ of test operators such that, for all $a \in \mathsf{AP}$, the set of all test operators occurring in r_a is contained in the set $\{?\varphi_1, \dots, ?\varphi_n\}$. We assume that all φ_is are from $\mathcal{L}$. (This is the reason why we set up "CND" as a finite list of formulas.) Define $\mathbb{E}^{\mathsf{r}} = (\mathtt{E}, (\mathtt{Q}_a)_{a\in\mathsf{AP}}, \mathtt{CND}, (\mathtt{rel}_a)_{a\in\mathsf{AP}})$ as follows:

- $\mathtt{E} = \{\mathtt{e}\}$ and $\mathtt{Q}_a = \{(\mathtt{e}, \mathtt{e})\}$ $(a \in \mathsf{AP})$,
- $\mathtt{CND}(\mathtt{e}) := (\varphi_1, \dots, \varphi_n)$ and $\mathtt{rel}_a(\mathtt{e}, \mathtt{e}) = \mathsf{r}_a$ $(a \in \mathsf{AP})$,

where note that $\mathtt{rel}_a(\mathtt{e}, \mathtt{e})$ is well-defined by $\mathtt{CND}(\mathtt{e}) := (\varphi_1, \dots, \varphi_n)$. Then it is easy to see that $\mathfrak{M}^{\mathsf{r}} \cong \mathfrak{M}^{\otimes\mathbb{E}^{\mathsf{r}}}$. This corrects the flaw of [7, Fact 26] in the level of relation changers. For a formula φ in $\mathcal{L}$, the relation changer $\mathsf{r}_{\sharp\varphi}$ for the public preference upgrade [7, 25] is captured by the action model $\mathbb{E}^{\sharp\varphi} := (\{\mathtt{e}\}, (\{(\mathtt{e}, \mathtt{e})\})_{a\in\mathsf{AP}}, \mathtt{CND}, (\mathtt{rel}_a)_{a\in\mathsf{AP}})$ where

- $\mathtt{CND}(\mathtt{e}) := (\varphi, \neg\varphi)$ and $\mathtt{rel}_a(\mathtt{e}, \mathtt{e}) := (?\neg\varphi; a) \cup (a; ?\varphi)$, $(a \in \mathsf{AP})$.

Note that $\mathsf{cnd}_2(\mathtt{e}) := \neg\varphi$ is employed as a source condition before executing program a and $\mathsf{cnd}_1(\mathtt{e}) := \varphi$ is used as a target condition after executing a.

Example 4 (Generalized Arrow Update [24]). A (generalized) arrow update U in [9, Definition 1][1] is defined as a tuple $U := (\mathtt{E}, (\mathtt{A}_a^U)_{a\in\mathsf{AP}})$ where $\mathtt{A}_a^U \subseteq (\mathtt{E}\times\mathsf{Form})\times(\mathtt{E}\times\mathsf{Form})$ is finite. Define $\mathbb{E}^U = (\mathtt{E}, (\mathtt{Q}_a)_{a\in\mathsf{AP}}, \mathtt{CND}, (\mathtt{rel}_a)_{a\in\mathsf{AP}})$ as follows:

- $\mathtt{Q}_a = \mathtt{E} \times \mathtt{E}$ $(a \in \mathsf{AP})$,
- $\mathtt{CND}(\mathtt{e}) := (\varphi_1, \dots, \varphi_n, \psi_1, \dots, \psi_m)$ such that $(\mathtt{e}, \varphi_i)$s are all the elements in the domain of $\mathtt{A}_a^U$ and $(\mathtt{e}, \psi_j)$s are all the elements in the range of $\mathtt{A}_a^U$.
- $\mathtt{rel}_a(\mathtt{e}, \mathtt{f})$ is defined as: $\bigcup\left\{ ?\varphi_i; a; ?\psi_j \mid (\mathtt{e}, \varphi_i)\mathtt{A}_a^U(\mathtt{f}, \psi_j) \right\}$.

Example 5 (Private Announcement). A private announcement $!^a_\varphi$ of φ to agent a is a well-studied notion. Let $\mathsf{AP} = \{a, b\}$, $\mathfrak{M}$ be a model where $W = \{w, v\}$, $R_a = R_b = W \times W$ and $V(p) = \{w\}$ (see Fig. 1). We define $\mathbb{E}^{!^a_p} = (\{ !^a_p, \top \}, (\mathtt{Q}_c)_{c\in\mathsf{AP}}, \mathtt{CND}, (\mathtt{rel}_c)_{c\in\mathsf{AP}})$ where $(\mathtt{Q}_c)_{c\in\mathsf{AP}}$ is defined as in Fig. 1 and

[1] In [24, p.206, Definition 2.1], an *arrow function* was defined as a mapping $\mathtt{a} : \mathsf{AP} \times \mathtt{E} \to \mathsf{Form} \times \mathtt{E} \times \mathsf{Form}$ (in terms of our notation) but this should be corrected as a mapping $\mathtt{a} : \mathsf{AP} \times \mathtt{E} \to \wp(\mathsf{Form} \times \mathtt{E} \times \mathsf{Form})$, which is equivalent to an *arrow relation* in [9, Definition 1].

– $\mathtt{CND}(!^a_p) := p$ and $\mathtt{CND}(\top) := \top$, $\mathtt{rel}_c(\mathtt{e}, \mathtt{f}) := c; ?\mathtt{cnd}_1(\mathtt{f})$ for all $(\mathtt{e}, \mathtt{f}) \in \mathsf{Q}_c$.

Here, $\mathtt{cnd}_1(\mathtt{f})$ is used as a target condition after executing c. In particular if $c = a$ and $\mathtt{f} = !^a_p$, then $\mathtt{rel}_a(\mathtt{e}, !^a_p) := a; ?p$ and if $c = b$ and $\mathtt{f} = \top$, then $\mathtt{rel}_b(\mathtt{e}, \top) := b; ?\top$. Then an updated model $\mathfrak{M}^{\otimes \mathbb{E}^{!^a_p}} = (W^{\otimes \mathbb{E}^{!^a_p}}, (R_a^{\otimes \mathbb{E}^{!^a_p}})_{a \in \mathsf{AP}}, V^{\otimes \mathbb{E}^{!^a_p}})$ is depicted as in Fig. 1 where $V^{\otimes \mathbb{E}^{!^a_p}}(p) = \{(w, !^a_p), (w, \top)\}$.

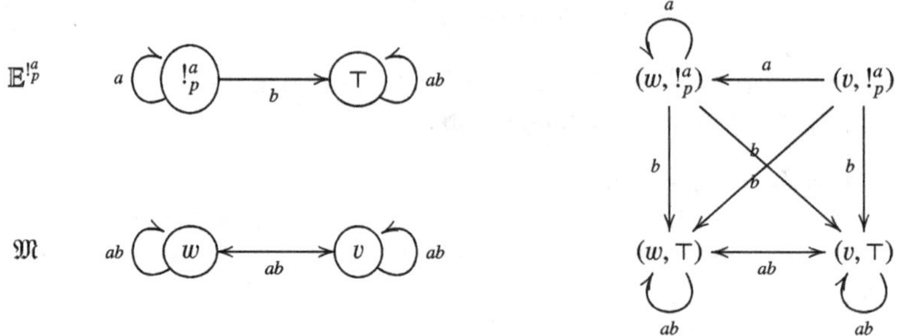

Fig. 1. Model update $\mathfrak{M}^{\otimes \mathbb{E}^{!^a_p}}$ (right) by private announcement $\mathbb{E}^{!^a_p}$.

3.2 Hilbert-style Axiomatization

Table 2 provides a Hilbert system **HPRC** of dynamic logic of product relation changers consisting of axioms and rules of **HPDL**$^-$ in Table 1 plus additional seven axioms and the necessitation rule for $[\mathbb{E}, \mathtt{e}]$, where we define $\mathsf{Q}_a(\mathtt{e}) := \{\mathtt{f} \in \mathtt{E} \mid \mathtt{e}\mathsf{Q}_a\mathtt{f}\}$ and recall that $\alpha_{(a,\mathtt{e},\mathtt{f})} := \mathtt{rel}_a(\mathtt{e}, \mathtt{f})$. We note that axioms ($[\mathbb{E}, \mathtt{e}][\cup]$), ($[\mathbb{E}, \mathtt{e}][;]$), and ($[\mathbb{E}, \mathtt{e}][?]$) in Table 2 are derivable from ($[\cup]$), ($[;]$) and ($[?]$) of Table 1, respectively, in terms of ($\mathtt{Nec}_{[\mathbb{E},\mathtt{e}]}$) and ($[\mathbb{E}, \mathtt{e}] \to$). However, we keep them for our purpose of proving semantic completeness (see Lemma 12). To establish semantic completeness via the inside-out strategy, as in **HDLRC**, one needs the necessitation rule for $[\mathbb{E}, \mathtt{e}]$. In addition, we use the same convention to define a *derivation* and a *theorem* in **HPRC** as in **HDLRC**.

Table 2. Additional Axioms and Rule to **HPDL**$^-$ for **HPRC**

($[\mathbb{E}, \mathbf{e}]\mathbf{at}$) $[\mathbb{E}, \mathbf{e}]p \leftrightarrow p$	($[\mathbb{E}, \mathbf{e}]\neg$) $[\mathbb{E}, \mathbf{e}]\neg\varphi \leftrightarrow \neg[\mathbb{E}, \mathbf{e}]\varphi$
($[\mathbb{E}, \mathbf{e}] \to$) $[\mathbb{E}, \mathbf{e}](\varphi \to \psi) \leftrightarrow ([\mathbb{E}, \mathbf{e}]\varphi \to [\mathbb{E}, \mathbf{e}]\psi)$	($[\mathbb{E}, \mathbf{e}][a]$) $[\mathbb{E}, \mathbf{e}][a]\varphi \leftrightarrow \bigwedge_{\mathtt{f} \in \mathsf{Q}_a(\mathtt{e})} [\alpha_{(a,\mathtt{e},\mathtt{f})}][\mathbb{E}, \mathbf{f}]\varphi$
($[\mathbb{E}, \mathbf{e}][\cup]$) $[\mathbb{E}, \mathbf{e}][\alpha \cup \beta]\varphi \leftrightarrow [\mathbb{E}, \mathbf{e}][\alpha]\varphi \wedge [\mathbb{E}, \mathbf{e}][\beta]\varphi$	($[\mathbb{E}, \mathbf{e}][;]$) $[\mathbb{E}, \mathbf{e}][\alpha; \beta]\varphi \leftrightarrow [\mathbb{E}, \mathbf{e}][\alpha][\beta]\varphi$
($[\mathbb{E}, \mathbf{e}][?]$) $[\mathbb{E}, \mathbf{e}][?\psi]\varphi \leftrightarrow [\mathbb{E}, \mathbf{e}](\psi \to \varphi)$	($\mathrm{Nec}_{[\mathbb{E},\mathbf{e}]}$) From φ, infer $[\mathbb{E}, \mathbf{e}]\varphi$

Proposition 6. If $\vdash_{\mathbf{HPRC}} \varphi$, then $\mathbb{M}_{\text{all}} \models \varphi$, for any formula φ.

Proof. We focus on the validity of ($[\mathbb{E}, \mathtt{e}][a]$) here. Fix any model $\mathfrak{M}$ and any w. It suffices to show that $\mathfrak{M}^{\otimes\mathbb{E}}, (w, \mathtt{e}) \models [a]\varphi$ iff $\mathfrak{M}, w \models \bigwedge_{\mathtt{f} \in \mathtt{Q}_a(\mathtt{e})}[\alpha_{(a,\mathtt{e},\mathtt{f})}][\mathbb{E}, \mathtt{f}]\varphi$. We proceed as follows: $\mathfrak{M}^{\otimes\mathbb{E}}, (w, \mathtt{e}) \models [a]\varphi$ iff

$$
\begin{aligned}
&\text{for all } (v, \mathtt{f}) \in W^{\otimes\mathbb{E}} \ ((w, \mathtt{e})R_a^{\otimes\mathbb{E}}(v, \mathtt{f}) \text{ implies } \mathfrak{M}^{\otimes\mathbb{E}}, (v, \mathtt{f}) \models \varphi),\\
\text{iff } &\text{for all } \mathtt{f} \in \mathbb{E} \ (\mathtt{e}\mathtt{Q}_a\mathtt{f} \text{ implies (for all } v \in W \ (wR_{\alpha_{(a,\mathtt{e},\mathtt{f})}}v \text{ implies } \mathfrak{M}, v \models [\mathbb{E}, \mathtt{f}]\varphi))),\\
\text{iff } &\text{for all } \mathtt{f} \in \mathbb{E} \ (\mathtt{e}\mathtt{Q}_a\mathtt{f} \text{ implies } \mathfrak{M}, w \models [\alpha_{(a,\mathtt{e},\mathtt{f})}][\mathbb{E}, \mathtt{f}]\varphi),
\end{aligned}
$$

The last line is equivalent to $\mathfrak{M}, w \models \bigwedge_{\mathtt{f} \in \mathtt{Q}_a(\mathtt{e})}[\alpha_{(a,\mathtt{e},\mathtt{f})}][\mathbb{E}, \mathtt{f}]\varphi$.

Example 7. Recall Example 4. For generalized arrow update logic, a key recursion axiom is the following in terms of our notation (cf. [24, Axiom U4, Table 1]):

$$[\mathbb{E}^U, \mathtt{e}][a]\gamma \leftrightarrow \bigwedge \left\{ (\varphi_i \to [a](\psi_j \to [\mathbb{E}^U, \mathtt{f}]\gamma)) \mid (\mathtt{e}, \varphi_i)\mathtt{A}_a^U(\mathtt{f}, \psi_j) \right\}.$$

This can be computed from our axiom ($[\mathbb{E}, \mathtt{e}][a]$) as follows:

$$
\begin{aligned}
[\mathbb{E}^U, \mathtt{e}][a]\gamma &\leftrightarrow \bigwedge \left\{ [\alpha_{(a,\mathtt{e},\mathtt{f})}][\mathbb{E}^U, \mathtt{f}]\gamma \mid \mathtt{f} \in \mathtt{Q}_a(\mathtt{e}) \right\}\\
&\leftrightarrow \bigwedge \left\{ \left[\bigcup \left\{ ?\varphi_i; a; ?\psi_j \mid (\mathtt{e}, \varphi_i)\mathtt{A}_a^U(\mathtt{f}, \psi_j) \right\}\right] [\mathbb{E}^U, \mathtt{f}]\gamma \mid \mathtt{f} \in \mathtt{E} \right\} \text{ by definition of } \mathbb{E}^U\\
&\leftrightarrow \bigwedge \left\{ \left[?\varphi_i; a; ?\psi_j\right] [\mathbb{E}^U, \mathtt{f}]\gamma \mid (\mathtt{e}, \varphi_i)\mathtt{A}_a^U(\mathtt{f}, \psi_j) \right\} \text{ by axiom } [\cup]\\
&\leftrightarrow \bigwedge \left\{ [?\varphi_i][a][?\psi_j][\mathbb{E}^U, \mathtt{f}]\gamma \mid (\mathtt{e}, \varphi_i)\mathtt{A}_a^U(\mathtt{f}, \psi_j) \right\} \text{ by axiom } [;]\\
&\leftrightarrow \bigwedge \left\{ \left(\varphi_i \to [a]\left(\psi_j \to [\mathbb{E}^U, \mathtt{f}]\gamma\right)\right) \mid (\mathtt{e}, \varphi_i)\mathtt{A}_a^U(\mathtt{f}, \psi_j) \right\} \text{ by axiom } [?].
\end{aligned}
$$

We establish the semantic completeness of **HPRC** with respect to recursion axioms by reducing it to the completeness of **HPDL**$^-$. Specifically, we translate a formula of $\mathcal{L}^{\otimes}$ into $\mathcal{L}$ in such a way that the formula and its translation are equivalent. To this end, we introduce the notions of length and weight for formulas, programs, and action models, respectively.

Definition 8. Define the *length* ℓ of a formula, a program and an action model by

$$
\begin{aligned}
\ell(p) &= 1, & \ell(a) &= 1,\\
\ell(\neg\varphi) &= \ell(\varphi) + 1, & \ell(\alpha \cup \beta) &= \ell(\alpha) + \ell(\beta) + 1,\\
\ell(\varphi \to \psi) &= \ell(\varphi) + \ell(\psi) + 1, & \ell(\alpha;\beta) &= \ell(\alpha) + \ell(\beta) + 1,\\
\ell([\alpha]\varphi) &= \ell(\alpha) + \ell(\varphi), & \ell(?\varphi) &= \ell(\varphi) + 1,\\
\ell([\mathbb{E}, \mathtt{e}]\varphi) &= (\ell(\mathbb{E}) + 1)\ell(\varphi), & \ell(\mathbb{E}) &= \sum\{\ell(\alpha_{(a,\mathtt{e},\mathtt{f})}) \mid (a, (\mathtt{e}, \mathtt{f})) \in \mathsf{AP} \times \mathtt{Q}_a\},
\end{aligned}
$$

where $\alpha_{(a,\mathtt{e},\mathtt{f})} := \mathtt{rel}_a(\mathtt{e}, \mathtt{f})$ of an action model $\mathbb{E} = (\mathtt{E}, (\mathtt{Q}_a)_{a \in \mathtt{AP}}, \mathtt{CND}, (\mathtt{rel}_a)_{a \in \mathtt{AP}})$.

The notion of length is modified from the notion of complexity in [10, p.195, Definition 7.38] for an action model logic. The following is the corresponding to [10, Lemma 7.39].

Lemma 9. The following hold, where $i \in \{1, 2\}$ in (iii), (v) and (vii):

(i) $\ell([\mathbb{E}, \mathtt{e}]p) > \ell(p)$	(v) $\ell([\mathbb{E}, \mathtt{e}][\alpha_1 \cup \alpha_2]\varphi) > \ell([\mathbb{E}, \mathtt{e}][\alpha_i]\varphi)$
(ii) $\ell([\mathbb{E}, \mathtt{e}]\neg\varphi) > \ell([\mathbb{E}, \mathtt{e}]\varphi)$	(vi) $\ell([\mathbb{E}, \mathtt{e}][\alpha_1;\alpha_2]\varphi) > \ell([\mathbb{E}, \mathtt{e}][\alpha_1][\alpha_2]\varphi)$
(iii) $\ell([\mathbb{E}, \mathtt{e}](\varphi_1 \to \varphi_2)) > \ell([\mathbb{E}, \mathtt{e}]\varphi_i)$	(vii) $\ell([\mathbb{E}, \mathtt{e}][?\varphi_1]\varphi_2) > \ell([\mathbb{E}, \mathtt{e}]\varphi_i)$
(iv) $\ell([\mathbb{E}, \mathtt{e}][a]\varphi) > \ell([\alpha_{(a,\mathtt{e},\mathtt{f})}][\mathbb{E}, \mathtt{f}]\varphi)$	(viii) $\ell([\mathbb{E}, \mathtt{e}][\mathbb{E}', \mathtt{e}']\varphi) > \ell([\mathbb{E}', \mathtt{e}']\varphi)$

Definition 10. The *weight* w (the nesting depth of action model operators) of a formula, a program and an action model is defined as follows:

$$\begin{array}{rlrl} w(p) &= 0, & w(a) &= 0, \\ w(\neg\varphi) &= w(\varphi), & w(\alpha \cup \beta) &= \max(w(\alpha), w(\beta)), \\ w(\varphi \to \psi) &= \max(w(\varphi), w(\psi)), & w(\alpha;\beta) &= \max(w(\alpha), w(\beta)), \\ w([\alpha]\varphi) &= \max(w(\alpha), w(\varphi)), & w(?\varphi) &= w(\varphi), \\ w([\mathbb{E}, \mathtt{e}]\varphi) &= w(\mathbb{E}) + w(\varphi) + 1, & w(\mathbb{E}) &= \max\{w(\alpha_{(a,\mathtt{e},\mathtt{f})}) \mid (a, (\mathtt{e}, \mathtt{f})) \in \mathsf{AP} \times \mathsf{Q}_a\}. \end{array}$$

Definition 11. The (inside-out) *translation* $t^{io} : \mathsf{FORM}_{\mathcal{L}^{\otimes}} \to \mathsf{FORM}_{\mathcal{L}}$ is defined by

$$\begin{array}{rlrl} t^{io}(p) &= p, & t^{io}([\mathbb{E}, \mathtt{e}]p) &= p, \\ t^{io}(\neg\varphi) &= \neg t^{io}(\varphi), & t^{io}([\mathbb{E}, \mathtt{e}]\neg\varphi) &= \neg t^{io}([\mathbb{E}, \mathtt{e}]\varphi), \\ t^{io}(\varphi \to \psi) &= t^{io}(\varphi) \to t^{io}(\psi), & t^{io}([\mathbb{E}, \mathtt{e}](\varphi \to \psi)) &= t^{io}([\mathbb{E}, \mathtt{e}]\varphi) \to t^{io}([\mathbb{E}, \mathtt{e}]\psi), \\ t^{io}([a]\varphi) &= [a]t^{io}(\varphi), & t^{io}([\mathbb{E}, \mathtt{e}][a]\varphi) &= \bigwedge_{\mathtt{f} \in \mathsf{Q}_a(\mathtt{e})} t^{io}([\alpha_{(a,\mathtt{e},\mathtt{f})}][\mathbb{E}, \mathtt{f}]\varphi), \\ t^{io}([\alpha \cup \beta]\varphi) &= t^{io}([\alpha]\varphi) \wedge t^{io}([\beta]\varphi), & t^{io}([\mathbb{E}, \mathtt{e}][\alpha \cup \beta]\varphi) &= t^{io}([\mathbb{E}, \mathtt{e}][\alpha]\varphi) \wedge t^{io}([\mathbb{E}, \mathtt{e}][\alpha]\varphi), \\ t^{io}([\alpha;\beta]\varphi) &= t^{io}([\alpha][\beta]\varphi), & t^{io}([\mathbb{E}, \mathtt{e}][\alpha;\beta]\varphi) &= t^{io}([\mathbb{E}, \mathtt{e}][\alpha][\beta]\varphi), \\ t^{io}([?\psi]\varphi) &= t^{io}(\psi) \to t^{io}(\varphi), & t^{io}([\mathbb{E}, \mathtt{e}][?\psi]\varphi) &= t^{io}([\mathbb{E}, \mathtt{e}]\psi) \to t^{io}([\mathbb{E}, \mathtt{e}]\varphi), \\ & & t^{io}([\mathbb{E}, \mathtt{e}][\mathbb{E}', \mathtt{e}']\varphi) &= t^{io}([\mathbb{E}, \mathtt{e}]t^{io}([\mathbb{E}', \mathtt{e}']\varphi)). \end{array}$$

Lemma 12. For any formula φ, $\vdash_{\mathbf{HPRC}} \varphi \leftrightarrow t^{io}(\varphi)$.

Proof. The proof is shown by induction on $m \geqslant w(\varphi)$ for a formula φ. Let $m = 0$. That is, φ does not contain any action model operators, and so it is easy to show that $\vdash \varphi \leftrightarrow t^{io}(\varphi)$. For inductive step, suppose that $\vdash \varphi \leftrightarrow t^{io}(\varphi)$ for all formulas φ such that $w(\varphi) \leqslant m$. We show that $\vdash \varphi \leftrightarrow t^{io}(\varphi)$ for all formulas φ such that $w(\varphi) \leqslant m + 1$. We can show this statement by the second induction on $n \geqslant \ell(\varphi)$. □

Theorem 1. If $\mathbb{M}_{\text{all}} \models \varphi$, then $\vdash_{\mathbf{HPRC}} \varphi$, for any formula φ.

4 Labelled Sequent Calculus for Product Relation Changers

4.1 Labelled Formalism

Let $\mathsf{VAR} = \{ x, y, z, ... \}$ be a countably infinite set of *variables* and

$$L = ((\mathbb{E}_1, \mathtt{e}_1), (\mathbb{E}_2, \mathtt{e}_2), ..., (\mathbb{E}_n, \mathtt{e}_n))$$

be a (possibly empty) finite list of pointed action models. If L is empty, the symbol ε denotes such an empty list. Given an action model $\mathbb{E} = (\mathtt{E}, (\mathsf{Q}_a)_{a\in\mathtt{AP}}, \mathtt{CND}, (\mathtt{rel}_a)_{a\in\mathtt{AP}})$, we define Q_α for any $\alpha \in \mathsf{PR}_{\mathcal{L}^{\otimes}}$ as follows:

$$\mathsf{Q}_{\alpha\cup\beta} = \mathsf{Q}_\alpha \cup \mathsf{Q}_\beta, \quad \mathsf{Q}_{\alpha;\beta} = \mathsf{Q}_\alpha \circ \mathsf{Q}_\beta, \quad \mathsf{Q}_{?\varphi} = \{ (\mathtt{e}_1, \mathtt{e}_2) \mid \mathtt{e}_1 = \mathtt{e}_2 \},$$

where $\circ$ is a relational composition. Given two lists $L = ((\mathbb{E}_1, \mathtt{e}_1), (\mathbb{E}_2, \mathtt{e}_2), \ldots, (\mathbb{E}_n, \mathtt{e}_n))$ and $L' = ((\mathbb{F}_1, \mathtt{f}_1), (\mathbb{F}_2, \mathtt{f}_2), \ldots, (\mathbb{F}_m, \mathtt{f}_m))$, an expression of the form $(x, L)R_\alpha(y, L')$ is a *relational atom* for program α if $m = n$, $\mathbb{E}_i = \mathbb{F}_i = (\mathtt{E}^{(i)}, (\mathsf{Q}^{(i)}_a)_{a\in\mathtt{AP}}, \mathtt{CND}^{(i)}, (\mathtt{rel}^{(i)}_a)_{a\in\mathtt{AP}})$ and $(\mathtt{e}_i, \mathtt{f}_i) \in \mathsf{Q}^{(i)}_\alpha$ for all $1 \leqslant i \leqslant m$. A relational atom $(x, L)R_\alpha(y, L')$ means that "after the successive updates of action models in L and L', y is accessible from x by an execution of program α." We define *labelled expressions* by

$$A ::= x :^{L} \varphi \mid (x, L)R_{\alpha}(y, L') \mid x = y,$$

where $x, y \in \mathsf{VAR}$, L and L' are lists of pointed action models, $\varphi \in \mathsf{FORM}_{\mathcal{L}^{\otimes}}$, $\alpha \in \mathsf{PR}_{\mathcal{L}^{\otimes}}$ and $(x, L)R_{\alpha}(y, L')$ is a relational atom. When L is an empty list ε, we often simply write $x : \varphi$ instead of $x :^{\varepsilon} \varphi$. We also say that $x :^{L} \varphi$ is a *labelled formula* and $x = y$ an *equality atom*. A labelled formula $x :^{L} \varphi$ stands for "after the successive updates of pointed action models in L, a formula φ holds at state x," and an equality atom $x = y$ "state x is equal to state y." We define a *labelled sequent* $\Gamma \Rightarrow \Delta$ by a pair (Γ, Δ) of finite multisets of labelled expressions and it stands for "if all of Γ hold, then some of Δ holds." Then we define a *labelled sequent calculus* $\mathsf{G}(\mathbf{PRC})$ by all initial sequents and all inference rules in Tables 3 and 4. We also define $\mathsf{G}^{-}(\mathbf{PRC})$ by $\mathsf{G}(\mathbf{PRC})$ without the cut rule (Cut). For each rule except (Cut), a labelled expression other than Γ, Δ in the lower sequent (or the conclusion) of the rule is said to be *principal*. A *derivation* $\mathfrak{D}$ in $\mathsf{G}(\mathbf{PRC})$ is inductively defined as a finite tree generated by initial sequents and inference rules in Tables 3 and 4. We say that a labelled sequent $\Gamma \Rightarrow \Delta$ is *derivable* in $\mathsf{G}(\mathbf{PRC})$ (notation: $\vdash_{\mathsf{G}(\mathbf{PRC})} \Gamma \Rightarrow \Delta$) if there exists a derivation $\mathfrak{D}$ of $\mathsf{G}(\mathbf{PRC})$ such that the root node of $\mathfrak{D}$ is $\Gamma \Rightarrow \Delta$. A formula φ is a *theorem* of $\mathsf{G}(\mathbf{PRC})$ if the labelled sequent $\vdash_{\mathsf{G}(\mathbf{PRC})} \Rightarrow x :^{\varepsilon} \varphi$ for any $x \in \mathsf{VAR}$. These definitions are also used for $\mathsf{G}^{-}(\mathbf{PRC})$.

Table 3. Labelled sequent calculus $\mathsf{G}(\mathbf{PRC})$: basic rules

(Initial sequents) $A \Rightarrow A$

(Structural rules)

$$\frac{\Gamma \Rightarrow \Delta}{\Gamma \Rightarrow \Delta, A}\,(Rw) \quad \frac{\Gamma \Rightarrow \Delta}{A, \Gamma \Rightarrow \Delta}\,(Lw) \quad \frac{\Gamma \Rightarrow \Delta, A, A}{\Gamma \Rightarrow \Delta, A}\,(Rc) \quad \frac{A, A, \Gamma \Rightarrow \Delta}{A, \Gamma \Rightarrow \Delta}\,(Lc) \quad \frac{\Gamma \Rightarrow \Delta, A \quad A, \Sigma \Rightarrow \Pi}{\Gamma, \Sigma \Rightarrow \Delta, \Pi}\,(Cut)$$

(Logical rules)

$$\frac{x :^{L} \varphi, \Gamma \Rightarrow \Delta}{\Gamma \Rightarrow \Delta, x :^{L} \neg\varphi}\,(R\neg) \quad \frac{\Gamma \Rightarrow \Delta, x :^{L} \varphi}{x :^{L} \neg\varphi, \Gamma \Rightarrow \Delta}\,(L\neg)$$

$$\frac{x :^{L} \varphi_1, \Gamma \Rightarrow \Delta, x :^{L} \varphi_2}{\Gamma \Rightarrow \Delta, x :^{L} \varphi_1 \rightarrow \varphi_2}\,(R\rightarrow) \quad \frac{\Gamma \Rightarrow \Delta, x :^{L} \varphi_1 \quad x :^{L} \varphi_2, \Gamma \Rightarrow \Delta}{x :^{L} \varphi_1 \rightarrow \varphi_2, \Gamma \Rightarrow \Delta}\,(L\rightarrow)$$

$$\frac{\left\{(x, L)R_{\alpha}(y, L'), \Gamma \Rightarrow \Delta, y :^{L'} \varphi \mid (\mathsf{e}_i, \mathsf{f}_i) \in \mathsf{Q}_{\alpha}^{(i)} \text{ for all } 1 \leqslant i \leqslant n\right\}}{\Gamma \Rightarrow \Delta, x :^{L} [\alpha]\varphi}\,(R[\alpha])^{\dagger}$$

$\dagger$: y is fresh, $L \equiv ((\mathbb{E}_1, \mathsf{e}_1), \ldots, (\mathbb{E}_n, \mathsf{e}_n))$ and $L' \equiv ((\mathbb{E}_1, \mathsf{f}_1), \ldots, (\mathbb{E}_n, \mathsf{f}_n))$.

$$\frac{\Gamma \Rightarrow \Delta, (x, L)R_{\alpha}(y, L') \quad y :^{L'} \varphi, \Gamma \Rightarrow \Delta}{x :^{L} [\alpha]\varphi, \Gamma \Rightarrow \Delta}\,(L[\alpha])$$

(Equality rules)

$$\frac{}{\Rightarrow x = x}\,(R=) \quad \frac{\Gamma[x/w] \Rightarrow \Delta[x/w]}{x = y, \Gamma[y/w] \Rightarrow \Delta[y/w]}\,(L=_1) \quad \frac{\Gamma[y/w] \Rightarrow \Delta[y/w]}{x = y, \Gamma[x/w] \Rightarrow \Delta[x/w]}\,(L=_2)$$

An underlying idea of our labelled sequent calculus is that $\mathsf{G}(\mathbf{PRC})$ can be regarded as a formalized version of Kripke semantics. Let us comment on the inference rules on $\mathsf{G}(\mathbf{PRC})$. Initial sequents, structural rules, logical rules, program rules, and relational atom rules except $(R[\alpha])$ and $(R;)$ can be regarded as a natural generalization of the

Table 4. Labelled sequent calculus G(**PRC**): additional rules

(Action model rules)

$$\frac{\Gamma \Rightarrow \Delta, x :^{L} p}{\Gamma \Rightarrow \Delta, x :^{L,(\mathbb{E},\mathsf{e})} p}\ (Rat) \qquad \frac{x :^{L} p, \Gamma \Rightarrow \Delta}{x :^{L,(\mathbb{E},\mathsf{e})} p, \Gamma \Rightarrow \Delta}\ (Lat)$$

$$\frac{\Gamma \Rightarrow \Delta, x :^{L,(\mathbb{E},\mathsf{e})} \varphi}{\Gamma \Rightarrow \Delta, x :^{L} [\mathbb{E},\mathsf{e}]\varphi}\ (R[\mathbb{E},\mathsf{e}]) \qquad \frac{x :^{L,(\mathbb{E},\mathsf{e})} \varphi, \Gamma \Rightarrow \Delta}{x :^{L} [\mathbb{E},\mathsf{e}]\varphi, \Gamma \Rightarrow \Delta}\ (L[\mathbb{E},\mathsf{e}])$$

(Relational atom rules)

$$\frac{\Gamma \Rightarrow \Delta, (x, L)R_{\alpha(a,\mathsf{e},\mathsf{f})}(y, L')}{\Gamma \Rightarrow \Delta, (x, L, (\mathbb{E},\mathsf{e}))R_{a}(y, L', (\mathbb{E},\mathsf{f}))}\ (Rrel) \qquad \frac{(x, L)R_{\alpha(a,\mathsf{e},\mathsf{f})}(y, L'), \Gamma \Rightarrow \Delta}{(x, L, (\mathbb{E},\mathsf{e}))R_{a}(y, L', (\mathbb{E},\mathsf{f})), \Gamma \Rightarrow \Delta}\ (Lrel)$$

(Program rules)

$$\frac{\Gamma \Rightarrow \Delta, (x, L)R_{\alpha_i}(y, L')}{\Gamma \Rightarrow \Delta, (x, L)R_{\alpha_1 \cup \alpha_2}(y, L')}\ (R\cup) \qquad \frac{(x, L)R_{\alpha_1}(y, L'), \Gamma \Rightarrow \Delta \quad (x, L)R_{\alpha_2}(y, L'), \Gamma \Rightarrow \Delta}{(x, L)R_{\alpha_1 \cup \alpha_2}(y, L'), \Gamma \Rightarrow \Delta}\ (L\cup)$$

$$\frac{\Gamma \Rightarrow \Delta, (x, L)R_{\alpha_1}(z, L'') \quad \Gamma \Rightarrow \Delta, (z, L'')R_{\alpha_2}(y, L')}{\Gamma \Rightarrow \Delta, (x, L)R_{\alpha_1;\alpha_2}(y, L')}\ (R;)$$

$$\frac{\left\{(x, L)R_{\alpha_1}(z, L''), (z, L'')R_{\alpha_2}(y, L'), \Gamma \Rightarrow \Delta \mid (\mathsf{e}_i, \mathsf{d}_i) \in \mathsf{Q}^{(i)}_{\alpha_1} \text{ and } (\mathsf{d}_i, \mathsf{f}_i) \in \mathsf{Q}^{(i)}_{\alpha_2} \text{ for all } 1 \leqslant i \leqslant n.\right\}}{(x, L)R_{\alpha_1;\alpha_2}(y, L'), \Gamma \Rightarrow \Delta}\ (L;)^{\ddagger}$$

‡: z is fresh, $L \equiv ((\mathbb{E}_1, \mathsf{e}_1), \dots, (\mathbb{E}_n, \mathsf{e}_n))$, $L' \equiv ((\mathbb{E}_1, \mathsf{f}_1), \dots, (\mathbb{E}_n, \mathsf{f}_n))$, and $L'' \equiv ((\mathbb{E}_1, \mathsf{d}_1), \dots, (\mathbb{E}_n, \mathsf{d}_n))$.

$$\frac{\Gamma \Rightarrow \Delta, x = y \quad \Gamma \Rightarrow \Delta, x :^{L} \varphi}{\Gamma \Rightarrow \Delta, (x, L)R_{?\varphi}(y, L')}\ (R?) \qquad \frac{x = y, \Gamma \Rightarrow \Delta}{(x, L)R_{?\varphi}(y, L'), \Gamma \Rightarrow \Delta}\ (L?_1) \qquad \frac{x :^{L} \varphi, \Gamma \Rightarrow \Delta}{(x, L)R_{?\varphi}(y, L'), \Gamma \Rightarrow \Delta}\ (L?_2)$$

labelled sequent calculus **GDLRC** for dynamic logic of relation changers in [18] where the list L in a labelled formula and a relational atom is adapted to handle pointed action models instead of relation changers. Similarly, action model rules are also introduced in the same form as relation changer rules, as in [18]. The logical rule $(R[\alpha])$ and the program rule $(R;)$ are generalized to handle multiple premises. It is noted that

$$\{\,(x, L)R_{\alpha}(y, L'), \Gamma \Rightarrow \Delta, y :^{L'} \varphi \mid (\mathsf{e}_i, \mathsf{f}_i) \in \mathsf{Q}^{(i)}_{\alpha} \ \textit{for all}\ 1 \leqslant i \leqslant n\,\}$$

in $(R[\alpha])$ stands for a finite set of premises of the rule. The number of premises is the cardinality $\# \prod_{1 \leqslant i \leqslant n} \mathsf{Q}^{(i)}_{\alpha}(\mathsf{e}_i)$. The same can be applied to the rule $(L;)$. A similar notations are also used in a labelled sequent calculus for dynamic epistemic logic in [30]. Our equality rules are adopted from a sequent calculus of hybrid logic by Seligman [35] (its origines can be traced back to [6,8,21]), and these rules are required to handle a derivation containing applications of test rules, similarly to our previous work in [16] for a labelled sequent calculus for dynamic logic of relation changers.

Definition 13. A substitution $A[y/x]$ (the result of substituting x in A uniformly with y) is defined by the following clauses:

$$z[y/x] \equiv \begin{cases} z \text{ if } x \neq z, \\ y \text{ if } x = z, \end{cases} \qquad \begin{aligned} (z :^{L} \varphi)[y/x] &\equiv z[y/x] :^{L} \varphi, \\ (z, L)R_{\alpha}(w, L')[y/x] &\equiv (z[y/x], L)R_{\alpha}(w[y/x], L'), \\ (w = z)[y/x] &\equiv w[y/x] = z[y/x]. \end{aligned}$$

where $w, x, y, z \in \mathsf{VAR}$, $\varphi \in \mathsf{Form}_{\mathcal{L}^{\otimes}}$, $\alpha \in \mathsf{PR}_{\mathcal{L}^{\otimes}}$, L and L' are the lists of pointed action models, and A is a labelled expression. A substitution $\Gamma[y/x]$ is also defined by $\Gamma[y/x] := \{\,A[y/x] \mid A \in \Gamma\,\}$ where Γ is a finite multiset of labelled expressions.

The following is easy to establish by induction on derivation.

Lemma 14. If $\mathfrak{D}$ is a derivation of $\Gamma \Rightarrow \Delta$ in G(**PRC**) then there exists a derivation $\mathfrak{D}'$ of $\Gamma[y/x] \Rightarrow \Delta[y/x]$ such that $\mathfrak{D}'$ has the same height as $\mathfrak{D}$.

Theorem 2. If $\vdash_{\mathbf{HPRC}} \varphi$, then $\vdash_{\mathsf{G}(\mathbf{PRC})} \Rightarrow x :^{\varepsilon} \varphi$, for any $\varphi \in \mathsf{FORM}_{\mathcal{L}^{\otimes}}$ and any $x \in$ VAR.

Proof. We only consider the axiom $([\mathbb{E}, \mathtt{e}][a])$ and rules $(\mathbf{Nec}_{[\mathbb{E},\mathtt{e}]})$ and (**MP**). It is noted that Lemma 14 is necessary to simulate $\mathbf{Nec}_{[\alpha]}$. As for the right-to-left direction of $([\mathbb{E}, \mathtt{e}][a])$, it suffices to prove the following:

$$\dfrac{\dfrac{\dfrac{\dfrac{\dfrac{xR_{\alpha(a,\mathtt{e},\mathtt{f})}y \Rightarrow xR_{\alpha(a,\mathtt{e},\mathtt{f})}y}{xR_{\alpha(a,\mathtt{e},\mathtt{f})}y \Rightarrow (x,(\mathbb{E},\mathtt{e}))R_a(y,(\mathbb{E},\mathtt{f}))}\,(Rrel) \quad y :^{(\mathbb{E},\mathtt{f})} \varphi \Rightarrow y :^{(\mathbb{E},\mathtt{f})} \varphi}{xR_{\alpha(a,\mathtt{e},\mathtt{f})}y, x :^{(\mathbb{E},\mathtt{e})} [a]\varphi \Rightarrow y :^{(\mathbb{E},\mathtt{f})} \varphi}\,(L[a]),(w)}{xR_{\alpha(a,\mathtt{e},\mathtt{f})}y, x :^{(\mathbb{E},\mathtt{e})} [a]\varphi \Rightarrow y : [\mathbb{E},\mathtt{f}]\varphi}\,(R[\mathbb{E},\mathtt{f}])}{x :^{(\mathbb{E},\mathtt{e})} [a]\varphi \Rightarrow x : [\alpha_{(a,\mathtt{e},\mathtt{f})}][\mathbb{E},\mathtt{f}]\varphi}\,(R[\alpha_{(a,\mathtt{e},\mathtt{f})}])\ y\text{: fresh}}{x : [\mathbb{E},\mathtt{e}][a]\varphi \Rightarrow x : [\alpha_{(a,\mathtt{e},\mathtt{f})}][\mathbb{E},\mathtt{f}]\varphi}\,(L[\mathbb{E},\mathtt{e}])$$

As for the left-to-right direction of $([\mathbb{E}, \mathtt{e}][a])$, we proceed as follows:

$$\dfrac{\dfrac{\dfrac{\left\{(x,(\mathbb{E},\mathtt{e}))R_a(y,(\mathbb{E},\mathtt{f})), [\alpha_{(a,\mathtt{e},\mathtt{f})}][\mathbb{E},\mathtt{f}]\varphi \Rightarrow y :^{(\mathbb{E},\mathtt{f})} \varphi \mid \mathtt{e}\mathsf{Q}_a\mathtt{f}\right\}}{\left\{(x,(\mathbb{E},\mathtt{e}))R_a(y,(\mathbb{E},\mathtt{f})), x : \bigwedge_{\mathtt{f}\in\mathsf{Q}_a(\mathtt{e})}[\alpha_{(a,\mathtt{e},\mathtt{f})}][\mathbb{E},\mathtt{f}]\varphi \Rightarrow y :^{(\mathbb{E},\mathtt{f})} \varphi \mid \mathtt{e}\mathsf{Q}_a\mathtt{f}\right\}}\,(L\neg),(R\to)}{x : \bigwedge_{\mathtt{f}\in\mathsf{Q}_a(\mathtt{e})}[\alpha_{(a,\mathtt{e},\mathtt{f})}][\mathbb{E},\mathtt{f}]\varphi \Rightarrow x :^{(\mathbb{E},\mathtt{e})} [a]\varphi}\,(R[a])\ y\text{: fresh}}{x : \bigwedge_{\mathtt{f}\in\mathsf{Q}_a(\mathtt{e})}[\alpha_{(a,\mathtt{e},\mathtt{f})}][\mathbb{E},\mathtt{f}]\varphi \Rightarrow x : [\mathbb{E},\mathtt{e}][a]\varphi}\,(R[\mathbb{E},\mathtt{e}]),$$

where the topmost sequent is derivable for each $\mathtt{f}$ such that $\mathtt{e}\mathsf{Q}_a\mathtt{f}$ as follows:

$$\dfrac{\dfrac{xR_{\alpha(a,\mathtt{e},\mathtt{f})}y \Rightarrow xR_{\alpha(a,\mathtt{e},\mathtt{f})}y}{(x,(\mathbb{E},\mathtt{e}))R_a(y,(\mathbb{E},\mathtt{f})) \Rightarrow xR_{\alpha(a,\mathtt{e},\mathtt{f})}y}\,(Lrel) \quad \dfrac{y :^{(\mathbb{E},\mathtt{f})} \varphi \Rightarrow y :^{(\mathbb{E},\mathtt{f})} \varphi}{y : [\mathbb{E},\mathtt{f}]\varphi \Rightarrow y :^{(\mathbb{E},\mathtt{f})} \varphi}\,(L[\mathbb{E},\mathtt{f}])}{(x,(\mathbb{E},\mathtt{e}))R_a(y,(\mathbb{E},\mathtt{f})), [\alpha_{(a,\mathtt{e},\mathtt{f})}][\mathbb{E},\mathtt{f}]\varphi \Rightarrow y :^{(\mathbb{E},\mathtt{f})} \varphi}\,(L[\alpha_{(a,\mathtt{e},\mathtt{f})}]),(w)$$

For the modus ponens, we proceed as follows (the cut rule is necessary here):

$$\dfrac{\Rightarrow x : \varphi \to \psi \quad \dfrac{\Rightarrow x : \varphi \quad x : \psi \Rightarrow x : \psi}{x : \varphi \to \psi \Rightarrow x : \psi}\,(L\to)}{\Rightarrow x : \psi}\,(Cut)$$

As for the rule $(\mathtt{Nec}_{[\mathbb{E},\mathtt{e}]})$, let $\mathfrak{D}$ be a derivation of $\Rightarrow x :^{\varepsilon} \varphi$. Then we can insert $(\mathbb{E}, \mathtt{e})$ in the innermost position of all the "stacks" L in the derivation $\mathfrak{D}$, which gives us a derivation of $\Rightarrow x :^{(\mathbb{E},\mathtt{e})} \varphi$. By the rule $(R[\mathbb{E}, \mathtt{e}])$, we conclude that $\Rightarrow x :^{\varepsilon} [\mathbb{E}, \mathtt{e}]\varphi$. □

4.2 Soundness of G(PRC) for Kripke Semantics

Definition 15. Let $\mathfrak{M} = (W, (R_a)_{a \in \mathsf{AP}}, V)$ be a model and $g : \mathsf{VAR} \to W$ be an *assignment* on $\mathfrak{M}$. Given any model $\mathfrak{M}$, any assignment g on $\mathfrak{M}$ and any labelled expression A, we define $\mathfrak{M}, g \models A$ by the following clause:

$$
\begin{array}{ll}
\mathfrak{M}, g \models x :^{(\mathbb{E}_1, \mathsf{e}_1), \ldots, (\mathbb{E}_n, \mathsf{e}_n)} \varphi & \text{iff } \mathfrak{M}^{\otimes \mathbb{E}_1 \otimes \cdots \otimes \mathbb{E}_n}, (g(x), \mathsf{e}_1, \ldots, \mathsf{e}_n) \models \varphi, \\
\mathfrak{M}, g \models x = y & \text{iff } g(x) = g(y), \\
\mathfrak{M}, g \models (x, L) R_\alpha (y, L') & \text{iff } (g(x), \mathsf{e}_1, \ldots, \mathsf{e}_n) R_\alpha^{\otimes \mathbb{E}_1 \otimes \cdots \otimes \mathbb{E}_n} (g(y), \mathsf{d}_1, \ldots, \mathsf{d}_n)),
\end{array}
$$

where $L \equiv ((\mathbb{E}_1, \mathsf{e}_1), \ldots, (\mathbb{E}_n, \mathsf{e}_n))$ and $L' \equiv ((\mathbb{E}_1, \mathsf{d}_1), \ldots, (\mathbb{E}_n, \mathsf{d}_n))$. In addition, for any labelled sequent $\Gamma \Rightarrow \Delta$, we write $\mathfrak{M}, g \models \Gamma \Rightarrow \Delta$ if, whenever $\mathfrak{M}, g \models A$ for all $A \in \Gamma$, it follows that $\mathfrak{M}, g \models C$ for some $C \in \Delta$.

Proposition 16. Let $\mathfrak{M} = (W, (R_a)_{a \in \mathsf{AP}}, V)$. Given any lists $L \equiv ((\mathbb{E}_1, \mathsf{e}_1), \ldots, (\mathbb{E}_n, \mathsf{e}_n))$ and $L' \equiv ((\mathbb{E}_1, \mathsf{d}_1), \ldots, (\mathbb{E}_n, \mathsf{d}_n))$, $(w, \mathsf{e}_1, \ldots, \mathsf{e}_n) R_\alpha^{\otimes \mathbb{E}_1 \otimes \cdots \otimes \mathbb{E}_n} (v, \mathsf{d}_1, \ldots, \mathsf{d}_n)$ implies $\mathsf{e}_i Q_\alpha^{(i)} \mathsf{d}_i$ for all $1 \leqslant i \leqslant n$ where $\mathbb{E}_i = (\mathsf{E}^{(i)}, (\mathsf{Q}_a^{(i)})_{a \in \mathsf{AP}}, \mathtt{CND}^{(i)}, (\mathtt{rel}_a^{(i)})_{a \in \mathsf{AP}})$.

Proof. By double induction on program α and the length n of L (and L'). □

Theorem 3 (Soundness). For any Γ and Δ, if $\vdash_{\mathsf{G}(\mathbf{PRC})} \Gamma \Rightarrow \Delta$, then $\mathfrak{M}, g \models \Gamma \Rightarrow \Delta$. Therefore, if $\vdash_{\mathsf{G}(\mathbf{PRC})} \Rightarrow x :^{\varepsilon} \varphi$ for all $x \in \mathsf{VAR}$, then $\mathbb{M}_{\text{all}} \models \varphi$.

4.3 Cut Elimination of G(PRC)

Definition 17. The length ℓ of a labelled expression is defined by

$$
\begin{aligned}
\ell(\varepsilon) &= 0, \\
\ell(L) &= \ell((\mathbb{E}_1, \mathsf{e}_1), \ldots, (\mathbb{E}_n, \mathsf{e}_n)) = \ell(\mathbb{E}_1) + \cdots + \ell(\mathbb{E}_n), \\
\ell(x :^L \varphi) &= \ell(L) + \ell(\varphi), \\
\ell((x, L) R_\alpha (y, L')) &= \ell(L) + \ell(\alpha), \\
\ell(x = y) &= 0,
\end{aligned}
$$

where recall that we have defined the length ℓ of a formula, a program, and an action model in Definition 8.

Theorem 4 (Cut Elimination). If $\vdash_{\mathsf{G}(\mathbf{PRC})} \Gamma \Rightarrow \Delta$, then $\vdash_{\mathsf{G}^-(\mathbf{PRC})} \Gamma \Rightarrow \Delta$.

Proof. We follow the method by Ono and Komori [33] (see also [22,32]). Instead of the cut rule, we eliminate the following *extended cut rule*:

$$
\frac{\Gamma \Rightarrow \Delta, A^m \quad A^n, \Sigma \Rightarrow \Pi}{\Gamma, \Sigma \Rightarrow \Delta, \Pi} \ (Ecut)
$$

where $m, n \geq 0$ and A^i is i times repetition of A, and we say that A is an *Ecut labelled expression*. We define $\mathsf{G}^*(\mathbf{PRC})$ as $\mathsf{G}^-(\mathbf{PRC})$ with the rule of $(Ecut)$.

Claim 1 If $\vdash_{\mathsf{G}^-(\mathbf{PRC})} \Gamma \Rightarrow \Delta, A^m$ and $\vdash_{\mathsf{G}^-(\mathbf{PRC})} A^n, \Sigma \Rightarrow \Pi$ then $\vdash_{\mathsf{G}^-(\mathbf{PRC})} \Gamma, \Sigma \Rightarrow \Delta, \Pi$.

Then the original statement follows, since (*Cut*) is a special case of (*Ecut*).

(Proof of Claim) In what follows, let $m > 0$ and $n > 0$, because otherwise we can obtain the conclusion by weakening rules. By induction on $\mathtt{c}$ (complexity), i.e., the length of the Ecut labelled expression A and subinduction on the sum denoted by $\mathtt{w}$ (weight), of the number of all sequents in a derivation $\mathfrak{L}$ in $\mathsf{G}^{-}(\mathbf{PRC})$ of $\Gamma \Rightarrow \Delta, A^{m}$ and the number of all sequents in a derivation $\mathfrak{R}$ in $\mathsf{G}^{-}(\mathbf{PRC})$ of $A^{n}, \Sigma \Rightarrow \Pi$. Given a derivation $\mathfrak{A}$, we write $\mathtt{rule}(\mathfrak{A})$ to mean the last applied rule including initial sequents (which can be regarded as zero premise rules). One of the points is that the length of a labelled expression A of Definition 17 based on the length of formula in Definition 8 plays a proof-theoretic role of complexity (or the length of a cut formula).[2] We divide our argument into the following cases:

1. $\mathtt{rule}(\mathfrak{L})$ or $\mathtt{rule}(\mathfrak{R})$ is an initial sequent.
2. $\mathtt{rule}(\mathfrak{L})$ or $\mathtt{rule}(\mathfrak{R})$ is a structural rule.
3. $\mathtt{rule}(\mathfrak{L})$ or $\mathtt{rule}(\mathfrak{R})$ is a logical, equality, action model, relational atom or program rules where A is not principal.
4. both $\mathtt{rule}(\mathfrak{L})$ and $\mathtt{rule}(\mathfrak{R})$ are logical, equality, action model, relational atom or program rules where A is principal for both rules.

For Case 2, when $\mathtt{rule}(\mathfrak{L})$ or $\mathtt{rule}(\mathfrak{R})$ is a contraction rule and A is principal, then the use of (*Ecut*) is inevitable. For Case 3, (*Ecut*) rule can be moved up above an application of $\mathtt{rule}(\mathfrak{L})$ or $\mathtt{rule}(\mathfrak{R})$ in a derivation with the same complexity. (For equality rules, we can basically follow an argument given in [35, Theorem 3.1], see also an argument for equality rules given in [16].) For Case 4, we comment on the case where $\mathtt{rule}(\mathfrak{L}) = (R[\alpha])$ or $\mathtt{rule}(\mathfrak{R}) = (L[\alpha])$. For simplicity, we let $L = (\mathbb{E}, \mathsf{e})$ and $m = n = 1$. Then we have the following two derivations:

$$\mathfrak{L} \equiv \frac{\left\{(x,(\mathbb{E},\mathsf{e}))R_{\alpha}(y,(\mathbb{E},\mathsf{f})), \Gamma \Rightarrow \Delta, y :^{(\mathbb{E},\mathsf{f})} \varphi \mid \mathsf{e}\mathsf{Q}_{a}\mathsf{f}\right\}}{\Gamma \Rightarrow \Delta, x :^{(\mathbb{E},\mathsf{e})} [\alpha]\varphi} \ (R[\alpha])$$

$$\mathfrak{R} \equiv \frac{\Sigma \Rightarrow \Pi, (x,(\mathbb{E},\mathsf{e}))R_{\alpha}(z,(\mathbb{E},\mathsf{d})) \quad z :^{(\mathbb{E},\mathsf{d})} \varphi, \Sigma \Rightarrow \Pi}{x :^{(\mathbb{E},\mathsf{e})} [\alpha]\varphi, \Sigma \Rightarrow \Pi} \ (L[\alpha])$$

Since y is fresh, we apply the substitution $[y/z]$ to get a derivation:

$$\frac{\Sigma \Rightarrow \Pi, (x,(\mathbb{E},\mathsf{e}))R_{\alpha}(z,(\mathbb{E},\mathsf{d})) \quad \dfrac{\mathfrak{L}_{\mathsf{d}}[y/z]}{(x,(\mathbb{E},\mathsf{e}))R_{\alpha}(z,(\mathbb{E},\mathsf{d})), \Gamma \Rightarrow \Delta, z :^{(\mathbb{E},\mathsf{d})} \varphi}}{\Gamma, \Sigma \Rightarrow \Delta, \Pi, z :^{(\mathbb{E},\mathsf{d})} \varphi} \ (Cut),$$

where the application of cut is eliminable by induction hypothesis and the length of the cut labelled expression (recall Definition 17) is smaller than $x :^{(\mathbb{E},\mathsf{e})} [\alpha]\varphi$. Again by taking (*Cut*) with the result above and $z :^{(\mathbb{E},\mathsf{d})} \varphi, \Sigma \Rightarrow \Pi$, we obtain the following:

$$\dfrac{\dfrac{\Gamma, \Sigma \Rightarrow \Delta, \Pi, z :^{(\mathbb{E},\mathsf{d})} \varphi \quad z :^{(\mathbb{E},\mathsf{d})} \varphi, \Sigma \Rightarrow \Pi,}{\Gamma, \Sigma, \Sigma \Rightarrow \Delta, \Pi, \Pi} \ (Cut)}{\Gamma, \Sigma \Rightarrow \Delta, \Pi} \ (Rc), (Lc)$$

[2] This is an extension of cut elimination arguments given in [17,26].

where the application of cut is eliminable by induction hypothesis and the length of the cut labelled expression is smaller than $x :^{(\mathbb{E},\mathsf{e})} [\alpha]\varphi$. This implies $\vdash_{\mathsf{G}^{-}(\mathbf{PRC})} \Gamma, \Sigma \Rightarrow \Delta, \Pi$, as desired. ⊣

This finishes our argument for the cut elimination theorem. □

5 Conclusion

The technical results of this paper can be summarized as follows:

Corollary 1. Given any formula $\varphi \in \mathsf{FORM}_{\mathcal{L}^{\otimes}}$, the following are all equivalent: (i) $\mathbb{M}_{\mathrm{all}} \models \varphi$; (ii) $\vdash_{\mathsf{H}\mathbf{PRC}} \varphi$; (iii) $\vdash_{\mathsf{G}(\mathbf{PRC})} \Rightarrow x :^{\varepsilon} \varphi$ for all $x \in \mathsf{VAR}$; (iv) $\vdash_{\mathsf{G}^{-}(\mathbf{PRC})} \Rightarrow x :^{\varepsilon} \varphi$ for all $x \in \mathsf{VAR}$.

Proof. The equivalence between (i) and (ii) is due to Proposition 6 and Theorem 1. A direction from (ii) and (iii) is by Theorem 2. A direction from (iii) to (iv) is due to Theorem 4. Finally the direction from (iv) to (i) is by Theorem 3. □

Let us comment on possible directions for further study. First, we could enrich our base logic with program constructors to capture a wider range of relation changers and action models. Possible candidates include the program intersection ∩ (cf. [1,2]) and the Kleene star. Since completeness proofs for **PDL** with intersection and its iteration-free variant are more involved than in intersection-free cases, we may restrict intersection to atomic programs, which still suffices to capture distributed knowledge [11,40]. It would then be interesting to compare the resulting dynamic extension of our logic, obtained by adding the Kleene star and restricted intersection, with recent work on data exchange and communication among agents [5]. Second, we may also provide a non-labelled sequent calculus for the dynamic logic of product relation changers **PRC**, as was done in [17] for a constructive dynamic logic of relation changers. Third, we may also investigate an alternative semantics for **PRC** by regarding $[\mathbb{E}, \mathsf{e}]$ as a modal operator and defining its accessibility relation in one (huge) Kripke model (the recursion axioms would restrict the accessibility relation). This alternative semantics would allow us to prove the semantic completeness of the axiomatization for the original (Kripke) semantics without a translation strategy based on recursion axioms (cf. [16,17,29,36,37]). Fourth, we may explore the relationship between **PRC** and *general dynamic dynamic logic* (**GDDL**) proposed by Girard et al. [14]. Kooi and Renne [24] proved that generalized arrow updates have the same expressivity as product updates (action models) [4], and the present paper showed that **PRC** captures generalized arrow updates (see Example 4). Given that a product update is known as a special case of a **GDDL** dynamic operator [14], it would be interesting to investigate the relationship between **GDDL** and **PRC**. Fifth, we may extend the approach of Jirakunkanok et al. [19,20] to legal applications by using our framework, which is based on Lorini's [27] modal framework for reasoning about an agent's beliefs and signed information. Since our framework can capture relation changers and generalized arrow updates in terms of action models, it enables us to analyze various belief and reliability changes of a judge in a court case, beyond private announcements and permissions. Finally, we may generalize **PRC** to an intuitionistic/constructive variant as was done for dynamic logic of relation changers in [17].

Acknowledgement. We would like to thank the anonymous reviewers for their helpful comments and suggestions on our manuscript. We would also like to thank Alexandru Baltag and Tim French for their helpful comments and suggestions during the discussions at the DaLi2025 workshop. The first author's work was supported by JSPS KAKENHI Grant-in-Aid for Early-Career Scientists Grant Number JP23K16952. The third author's work was also supported by JSPS KAKENHI Grant-in-Aid for Scientific Research (B) (Grant Number JP23K21869) and (C) (Grant Number JP25K03537).

References

1. Balbiani, P., Vakarelov, D.: Iteration-free PDL with intersection: a complete axiomatization. Fund. Inform. **45**(3), 173–194 (2001)
2. Balbiani, P., Vakarelov, D.: PDL with intersection of programs: a complete axiomatization. J. Appl. Non-Classical Logics **13**(3–4), 231–276 (2003)
3. Baltag, A., Moss, L.S.: Logics for epistemic programs. Synthese **139**(2), 165–224 (2004)
4. Baltag, A., Moss, L.S., Solecki, S.: The logic of public announcements, common knowledge, and private suspicions. In: Proceedings of the 7th Conference on Theoretical Aspects of Rationality and Knowledge (TARK VII), pp. 43–56 (1998)
5. Baltag, A., Smets, S.: Logics for data exchange and communication. In: Ciabattoni, A., Gabelaia, D., Sedlár, I. (eds.) Advances in Modal Logic, AiML 2024, Prague, Czech Republic, 19–23 August (2024), pp. 147–170. College Publications (2024)
6. Barwise, J.: An introduction to first-order logic. In: Barwise, J. (ed.) Handbook of Mathematical Logic, pp. 5–46. North-Holland Publishing Company (1977)
7. van Benthem, J., Liu, F.: Dynamic logic of preference upgrade. J. Appl. Non-Classical Logics **17**(2), 157–182 (2007)
8. Degtyarev, A., Voronkov, A.: Equality reasoning in sequent-based calculi. In: Robinson, A., Voronkov, A. (eds.) Handbook of Automated Reasoning, vol. I, chap. 10, pp. 611–706. Elsevier Science (2001)
9. van Ditmarsch, H., van der Hoek, W., Kooi, B., Kuijer, L.B.: Arrow update synthesis. Inf. Comput. **275**, 104544 (2020)
10. van Ditmarsch, H., van der Hoek, W., Kooi, B.P.: Dynamic epistemic logic, vol. 337. Springer Science and Business Media (2007)
11. Fagin, R., Halpern, J.Y., Moses, Y., Vardi, M.Y.: Reasoning about Knowledge. MIT Press (1995), paperback Edition 2003
12. Frittella, S., Greco, G., Kurz, A., Palmigiano, A., Sikimić, V.: A proof-theoretic semantic analysis of dynamic epistemic logic. J. Log. Comput. **26**(6), 1961–2015 (2014)
13. Gerbrandy, J., Groeneveld, W.: Reasoning about information change. J. Logic Lang. Inform. **6**(2), 147–169 (1997)
14. Girard, P., Seligman, J., Liu, F.: General dynamic dynamic logic. In: Advances in Modal Logic, vol. 9, pp. 239–260. College Publications (2012)
15. Harel, D., Kozen, D., Tiuryn, J.: Dynamic logic. MIT Press (2000)
16. Hatano, R., Sano, K.: Recapturing dynamic logic of relation changers via bounded morphisms. Stud. Logica. **109**, 95–124 (2020)
17. Hatano, R., Sano, K.: Three faces of recursion axioms: the case of constructive dynamic logic of relation changers. J. Log. Comput. **33**(6), 1399–1436 (2022)
18. Hatano, R., Sano, K., Tojo, S.: Cut free labelled sequent calculus for dynamic logic of relation changers. In: Yang, S.C.M., Lee, K.Y., Ono, H. (eds.) Philosophical Logic: Current Trends in Asia, pp. 153–180. Springer Singapore (2017)

19. Jirakunkanok, P.: Logic-based analysis of belief change in judgment. Ph.D. thesis, School of Information Science, Japan Advanced Institute of Science and Technology, Japan (2017)
20. Jirakunkanok, P., Sano, K., Tojo, S.: Dynamic epistemic logic of belief change in legal judgments. Artifi. Intell. Law **26**, 201–249 (2018)
21. Kanger, S.: A simplified proof method for elementary logic. In: Computer Programming and Formal Systems, pp. 87–94. Elsevier (1963)
22. Kashima, R.: Mathematical logic. Asakura Publishing Co., Ltd (in Japanese) (2009)
23. Kooi, B., Renne, B.: Arrow update logic. Rev. Symbolic Logic **4**(4), 536–559 (2011)
24. Kooi, B., Renne, B.: Generalized arrow update logic. In: Proceedings of the 13th Conference on Theoretical Aspects of Rationality and Knowledge, pp. 205–211 (2011)
25. Liu, F.: Reasoning about preference dynamics, vol. 354. Springer Science and Business Media (2011)
26. Liu, S., Sano, K.: Non-labelled sequent calculi of public announcement expansions of K45 and S5. In: Alechina, N., Herzig, A., Liang, F. (eds.) Logic, Rationality, and Interaction. vol. 14329, pp. 190–206. Springer Nature Switzerland (2023)
27. Lorini, E., Perrussel, L., Thévenin, J.-M.: A modal framework for relating belief and signed information. In: Leite, J., Torroni, P., Ågotnes, T., Boella, G., van der Torre, L. (eds.) CLIMA 2011. LNCS (LNAI), vol. 6814, pp. 58–73. Springer, Heidelberg (2011). https://doi.org/10.1007/978-3-642-22359-4_5
28. Maffezzoli, P., Negri, S.: A proof theoretical perspective on public announcement logic. Logic Philos. Sci. **9**, 49–59 (2011)
29. Motoura, S.: A general framework for dynamic epistemic logic: towards canonical correspondences. J. Appl. Non-Classical Logics **27**(1–2), 50–89 (2017)
30. Nomura, S., Ono, H., Sano, K.: A cut-free labelled sequent calculus for dynamic epistemic logic. J. Log. Comput. **30**(1), 321–348 (2020)
31. Nomura, S., Sano, K., Tojo, S.: Revising a sequent calculus for public announcement logic. Structural Analysis of Non-classical Logics: The Proceedings of the Second Taiwan Philosophical Logic Colloquium (TPLC-2014), pp. 131–157 (2015)
32. Ono, H.: Proof theory and algebra in logic. Springer Singapore (2019)
33. Ono, H., Komori, Y.: Logics without the contraction rule. J. Symbolic Logic **50**(01), 169–201 (1985)
34. Plaza, J.: Logics of public communications. In: Emrich, M.L., Pfeifer, M.S., Hadzikadic, M., Ras, Z.W. (eds.) Proceedings of the 4th International Symposium on Methodologies for Intelligent Systems, pp. 201–216 (1989)
35. Seligman, J.: Internalization: the case of hybrid logics. J. Log. Comput. **11**(5), 671–689 (2001)
36. Wang, Y., Aucher, G.: An alternative axiomatization of DEL and its applications. In: Proceedings of the Twenty-Third IJCAI International Joint Conference in Artificial Intelligence, pp. 1139–1146 (2013)
37. Wang, Y., Cao, Q.: On axiomatizations of public announcement logic. Synthese **190**(1), 103–134 (2013)
38. Wirsing, M., Knapp, A.: A reduction-based cut-free gentzen calculus for dynamic epistemic logic. Logic J. IGPL **31**(6), 1047–1068 (2023)
39. Wu, H., van Ditmarsch, H., Chen, J.: A labelled sequent calculus for public announcement logic. Stud. Logic **16**(3), 89–107 (2023)
40. Ågotnes, T., Wáng, Y.N.: Resolving distributed knowledge, Artif. Intell. **252**, 1–21 (2017)

Completeness and Decidability of Protocol-Dependent Knowledge in Gossip

Hans van Ditmarsch[1], Malvin Gattinger[2], and Wouter J. Smit[2(✉)]

[1] CNRS, IRIT, University of Toulouse, Toulouse, France
[2] ILLC, University of Amsterdam, Amsterdam, The Netherlands
wouter@wouterjsmit.nl

Abstract. In the gossip problem, a group of agents aims to efficiently share information using one-to-one communication. This often occurs in decentralised systems, where agents must rely on protocols to efficiently coordinate their communication. Recent work has used epistemic logic to define gossip protocols, including protocol-dependent knowledge modalities: agent knowledge assuming common knowledge that all agents follow said protocol. While axiomatisations exist for various versions of the gossip problem, none of these include protocol-dependent knowledge.

We show that protocol-dependent knowledge is strictly more expressive than standard knowledge, and we provide axiomatisations for four logics of gossip with protocol-dependent knowledge. We show that all four axiomatisations are sound and complete, as well as decidable.

1 Introduction

The gossip problem, introduced in the 1970's as the *telephone problem* [4,19] addresses how to spread information (secrets) among a group of agents by sequential pairwise communication, often called telephone calls. The goal is for all agents to know all secrets. We assume that each agent holds a single secret and that when calling each other, both agents exchange all the secrets they know, but no other information. There are many variations of the gossip problem: *one-directional exchange*, when only one agent in a call informs the other but not vice versa; multiple calls made in *parallel*; restrictions on the *network*, limiting what agents you can reach; *(a)synchronicity* determining whether agents are aware of a global clock when making calls; and more—see [15] for a survey.

Initially, research focused on finding optimal sequences of calls to achieve the goal that all agents know all secrets. Such optimal sequences require a central scheduler to plan calls. Later publications shifted to *distributed* gossip [17]. In the absence of a central scheduler, agents must rely on some protocol to coordinate their calls. Often, they rely on making calls at random. More recent developments focus on *epistemic* gossip protocols which specify pre-conditions that an

---

This article is based on the Master's thesis of the third author [18].

J. Wang et al. (Eds.): DaLí 2025, LNCS 16472, pp. 151–166, 2026.
https://doi.org/10.1007/978-3-032-22626-6_9

agent must know are true before making a call [1,3]. There exist many such distributed epistemic protocols [1,8,9]. Other aspects of gossip protocols may also be epistemic, such as the *termination goal* (when the goal is not merely that all secrets are known by all agents, but that they also know this) [10,11], or the *messages* being sent [7,16]. Another variation is to allow *faulty* messages [5].

To analyse how the information develops over the course of a protocol's execution we can use a dynamic epistemic logic with knowledge modalities for individual agents and a dynamic modality for calls. Various such logics have been axiomatised [13]. In these settings and even when all agents follow a given protocol, they do not assume that the other agents follow the same protocol. That is, their knowledge is not based on other agents also selecting calls on the same grounds as themselves. In [12] the authors propose so-called *protocol-dependent knowledge* (see also [10,14]), which are epistemic modalities formalizing what an agent knows *given that it is common knowledge that all agents follow some protocol.* Strengthening gossip protocols with these modalities makes some protocols more successful [12]. However, the corresponding logics with protocol-dependent knowledge modalities have not yet been axiomatised.

We show that protocol-dependent knowledge is strictly more expressive than standard knowledge and provide axiomatisations for four logics of gossip with protocol-dependent knowledge modalities. To axiomatise dynamic modalities for calls, we use the technique of reduction axioms by which a formula with dynamic modalities is provably equivalent to one without. We show that all four axiomatisations are sound and complete, and that all four logics are decidable.

We proceed as follows. We first recall the definitions of gossip and protocol-dependent knowledge in Sect. 2 before proving our expressivity results in Sect. 3. We then introduce our proof systems. In Sect. 4 we present the logic of initial models in a call-free setting, in Sect. 5 we define call reductions, and in Sect. 6 the logic of gossip models including calls. In the latter we also show that these proof systems are decidable. Section 7 concludes the article.

2 Definitions

We assume a finite set of *agents* $a, b, \ldots \in \mathbb{A}$. Each agent knows a corresponding *secret*, also denoted $a, b, \ldots$ and $\mathbb{S}$ is the set of all secrets. We only express the secret distribution among the agents, not the contents of the secrets.

A *call* is a pair of agents $a \neq b \in \mathbb{A}$. A *call sequence* $\sigma \in \Sigma$ is a list of calls. For n agents there are $n \cdot (n-1)$ different calls and all calls can always be executed. We write ab as shorthand for the call (a, b) and use a period to concatenate calls into a sequence, e.g. $ab.cd$.

Gossip has been studied under various settings [13,15]. We consider synchronous gossip with symmetric calls, i.e. agents always know how many calls took place but not necessarily which, and the caller and callee both share all secrets they know. Agents *inspect-then-merge* the secrets the other agent knows, meaning they first find out what the other agent knew before the call. They share no higher-order epistemic information or other information in a call. Agents form a total network, meaning they can always call anybody else, and calls are one-to-one.

We define the syntax with the following two mutually recursive definitions.

Definition 1 (Language). *Let $a, b \in \mathbb{A}$ be agents and $P \in \mathbb{P}$ a protocol. We define the* language $\mathcal{L}^{\mathbb{P}}$ *as*

$$\varphi ::= S_a b \mid \neg\varphi \mid (\varphi \wedge \varphi) \mid K_a^P \varphi \mid [ab]\varphi.$$

The syntax extends standard epistemic logic. The atom $S_a b$ means that *agent a knows the secret of agent b*. The protocol-dependent knowledge modality $K_a^P \varphi$ says that *assuming it is common knowledge that protocol P is followed, agent a knows that φ*. The dynamic modality $[ab]\varphi$ means that *φ holds after call ab*.

Protocols are essentially (distributed) algorithms that select calls to execute until some goal is achieved, usually that all agents know all secrets. For the purpose of this article it suffices to define a protocol by its protocol conditions, which specifies for each call when it may be executed.

Definition 2 (Protocols). *A* gossip protocol *P is a list of $n \cdot (n-1)$* protocol conditions *$P_{ab} \in \mathcal{L}^{\mathbb{P}}$ for agents $a \neq b \in \mathbb{A}$. Let $\mathbb{P}$ be the set of all protocols.*

Protocols cannot reference themselves, directly or indirectly. This means that P_{ab} may not contain a modality K^Q such that $Q = P$ or Q references P.

The operators $\vee$, $\rightarrow$, $\top$, and $\bot$ are defined as abbreviations in the usual way. We also define the epistemic dual of K_a^P by $\widehat{K}_a^P := \neg K_a^P \neg$. While $\mathcal{L}^{\mathbb{P}}$ does not contain the standard knowledge operator K, we will use it as abbreviation for K^{ANY}, knowledge assuming the trivial protocol ANY defined in Example 3.

For any agent a and set of secrets $R \subseteq \mathbb{S}$ we furthermore define the following abbreviation to describe that a *only knows* the secrets in R.

$$O_a R := \bigwedge_{b \in R} S_a b \wedge \bigwedge_{b \notin R} \neg S_a b \qquad \text{“}a \textit{ only knows the secrets in } R\text{”}$$

We also define variations of this language. Firstly we define the basic language of gossip $\mathcal{L}$ by replacing K^P with the standard knowledge operator K.

The static fragments $\mathcal{L}^{\mathbb{P}}_{-}$ and $\mathcal{L}_{-}$ of these languages omit the dynamic modality $[ab]$. As this modality represents a call in the setting of gossip, we also use the term *call-free*.

Agents can coordinate their calls by means of protocols. A protocol stipulates whether a call is permitted at a certain point, characterised by a protocol condition. There are various well-known gossip protocols, such as *Learn New Secrets* (LNS) where agents are only allowed to make a call to an agent whose secret they do not yet know. We also use the trivial protocol *Any Call* (ANY).

Example 3 (Protocols LNS *and* ANY*).* The protocols LNS and ANY are defined by the protocol conditions $\mathsf{LNS}_{ab} := \neg S_a b$ and $\mathsf{ANY}_{ab} := \top$ for all $a \neq b \in \mathbb{A}$.

We view the protocol conditions P_{ab} as subformulas of the K^P modality, which gives rise to the following definition of modal degree.

Definition 4 (Modal Degree). *The* modal degree $d(\varphi)$ *of a formula* $\varphi \in \mathcal{L}^{\mathbb{P}}$ *is defined recursively by*

$$\begin{aligned} d(S_a b) &:= 0, \\ d(\neg\varphi) &:= d(\varphi), \\ d(\varphi \wedge \psi) &:= \max\{d(\varphi), d(\psi)\}, \\ d(K_a^P \varphi) &:= 1 + \max\{d(\varphi), d(P)\}, \\ d([ab]\varphi) &:= d(\varphi), \end{aligned}$$

where $d(P) := \max\{d(P_{ab}) \mid a \neq b \in \mathbb{A}\}$ *is the* degree of protocol P.

We now define gossip models in two steps: an initial model represents the situation before any calls have happened and is then lifted to a gossip model that includes calls. Initial models with n agents are essentially $\mathbf{S5}_n$ Kripke models.

The following definition by [13] models arbitrary initial settings of gossip without protocol-dependent knowledge. These allow any secret distribution and knowledge thereof, as long as agents at least know their own secret and are aware which secrets they know. Protocols only effectively change the semantics after calls have taken place. We can therefore use the same definition for protocol-dependent knowledge too.

Definition 5 (Initial Models). *An* initial model *is a triple* $I = \langle W_0, R_0, V_0 \rangle$ *where* W_0 *is a set of worlds,* $R_0 : \mathbb{A} \to 2^{W_0 \times W_0}$ *is an equivalence relation for each agent, and* $V_0 : \mathbb{A} \times W_0 \to 2^{\mathbb{S}}$ *is a function mapping agent-world pairs to sets of secrets, such that (i)* $a \in V_0(a, w)$ *for all* $w \in W$*, and (ii) if we have* $(w_1, w_2) \in R_0(a)$ *then* $V_0(a, w_1) = V_0(a, w_2)$*.*

We can lift any initial model to a gossip model by inducing calls [13]. As calls are always possible, this creates for each world in the initial model an $n \cdot (n-1)$ branching tree. The resulting models are therefore forests.

Definition 6 (Gossip States). *Given an initial model* $I = \langle W_0, R_0, V_0 \rangle$*, a* gossip state *is a pair* (w, σ) *where* $w \in W_0$ *and* σ *is a call sequence. The set of gossip states induced from* I *is* $W(I) := W_0 \times \Sigma$*.*

Definition 7 (Valuation). *Given an initial model* $I = \langle W_0, R_0, V_0 \rangle$ *and some gossip state* $(w, \sigma) \in W(I)$*, the* valuation $V_a(w, \sigma)$ *denotes the set of secrets known by agent* a *at state* (w, σ)*. We define* V *recursively as follows.*

$$\begin{aligned} V_a(w, \epsilon) &= V_0(a, w) && \textit{Empty sequence} \\ V_a(w, \sigma.bc) &= V_b(w, \sigma) \cup V_c(w, \sigma) && a \in \{b, c\} \\ V_a(w, \sigma.bc) &= V_a(w, \sigma) && a \notin \{b, c\} \end{aligned}$$

The semantics on gossip models are now defined as follows, with Definitions 8 to 10 being mutually recursive. The effects of protocols emerge in Definition 8 of the epistemic accessibility relation. While calls are always possible, only those calls that are P-permitted are considered by agent a under the relation $\sim_a^P$. The epistemic relation $\sim_a$ for standard gossip as in [13, Definition 3.5] instantiated for synchronous, bi-directional calls with inspect-then-merge can be retrieved by omitting the protocol conditions, marked with $(*)$.

Definition 8 (Epistemic Relation). *Given an initial model $I = \langle W_0, R_0, V_0 \rangle$, we lift the initial relation R_0 for each $P \in \mathbb{P}$ and agent $a \in \mathbb{A}$ to $\sim_a^P \subseteq W(I) \times W(I)$ such that:*

$$
\begin{array}{lll}
(w_i, \epsilon) \sim_a^P (w_j, \epsilon) & \textit{iff } (w_i, w_j) \in R_0(a); & \\
(w_i, \sigma.ab) \sim_a^P (w_j, \tau.ab) & \textit{iff } (w_i, \sigma) \sim_a^P (w_j, \tau) & \\
& \quad \textit{and } V_b(w_i, \sigma) = V_b(w_j, \tau) & \\
& \quad \textit{and } (w_i, \sigma) \models P_{ab} \textit{ and } (w_j, \tau) \models P_{ab}; & (*) \\
(w_i, \sigma.ba) \sim_a^P (w_j, \tau.ba) & \textit{iff } (w_i, \sigma) \sim_a^P (w_j, \tau) & \\
& \quad \textit{and } V_b(w_i, \sigma) = V_b(w_j, \tau) & \\
& \quad \textit{and } (w_i, \sigma) \models P_{ba} \textit{ and } (w_j, \tau) \models P_{ba}; & (*) \\
(w_i, \sigma.bc) \sim_a^P (w_j, \tau.de) & \textit{iff } (w_i, \sigma) \sim_a^P (w_j, \tau) & \\
& \quad \textit{and } a \notin \{b, c, d, e\} & \\
& \quad \textit{and } (w_i, \sigma) \models P_{bc} \textit{ and } (w_j, \tau) \models P_{de}. & (*)
\end{array}
$$

Definition 9 (Gossip Models). *For an initial model I, we define the (induced)* gossip model *$M(I) := \langle W(I), \sim, V \rangle$ with $W(I)$, $\sim$, and V from Definitions 6 to 8. When the initial model is clear from context, we omit it and write M.*

Definition 10 (Semantics). *Let $\varphi, \psi \in \mathcal{L}^{\mathbb{P}}$. We define the relation $\models$ between pointed gossip models and formulas as follows.*

$$
\begin{array}{lcl}
M, (w, \sigma) \models S_a b & \iff & b \in V_a(w, \sigma) \\
M, (w, \sigma) \models \neg \varphi & \iff & M, (w, \sigma) \not\models \varphi \\
M, (w, \sigma) \models \varphi \wedge \psi & \iff & M, (w, \sigma) \models \psi \textit{ and } M, (w, \sigma) \models \psi \\
M, (w, \sigma) \models K_a^P \varphi & \iff & M, (w, \sigma') \models \varphi \textit{ for all } (w, \sigma') \textit{ s.t. } (w, \sigma) \sim_a^P (w, \sigma') \\
M, (w, \sigma) \models [ab] \varphi & \iff & M, (w, \sigma.ab) \models \varphi
\end{array}
$$

When M and w are clear from context, we omit them and write $\sigma \models \varphi$.

We also define semantics for the call-free language on initial models such that $I, w \models_{\mathcal{I}} \varphi$ iff $M(I), (w, \epsilon) \models \varphi$ for all $\varphi \in \mathcal{L}_{-}^{\mathbb{P}}$.

Effectively, the epistemic relation for protocol P is restricted to call sequences that are *P-permitted.*

Definition 11 (P-Permitted Calls). *Given a protocol P, a call ab is P-*permitted *at state $M, (w, \sigma)$ if its protocol condition P_{ab} holds: $M, (w, \sigma) \models P_{ab}$. A call sequence σ is P-permitted if each call is P-permitted. A call or sequence is P-*illegal *if it is not P-permitted. We omit P if the protocol is clear from context.*

Protocol-dependent knowledge only differs from standard knowledge *after* calls happen. In initial models no calls have happened yet, hence knowledge does not yet depend on which protocol an agent assumes.

Lemma 12. *Let P and Q be any two protocols and let a be any agent. Then for any initial model I and world $w \in W(I)$ we have $I, w \models_{\mathcal{I}} K_a^P \varphi$ iff $I, w \models_{\mathcal{I}} K_a^Q \varphi$.*

Proof. By the semantics, in particular note that the ϵ clause of Definition 8 does not depend on the protocol but only on R_0. □

While initial models are **S5**, protocol-dependent gossip models are not: the epistemic relation for a protocol P excludes any states with P-illegal call sequences, breaking reflexivity. They are instead partial equivalence relations. As a side-effect of these semantics, the P-dependent knowledge-base in P-illegal states becomes inconsistent, meaning that for any agent a the state satisfies $K_a^P \bot$. This property is known as the *global alarm* [12].

We can describe any distribution of secrets or knowledge thereof with an initial model. However, the classic gossip problem assumes a specific initial setting: in the *root model* all agents know only their own secret and this is common knowledge among the agents.

Definition 13 (Root Model). *The* root model $I_{\mathcal{R}}$ *is the initial model with* $W_0 = \{w_{\mathcal{R}}\}$, *and* $R_0(a) = \{(w_{\mathcal{R}}, w_{\mathcal{R}})\}$ *and* $V_0(a, w_{\mathcal{R}}) = \{a\}$ *for all agents* a. *We write* $\mathcal{I}$ *for the singleton class of* $I_{\mathcal{R}}$.

We call the gossip model induced from $I_{\mathcal{R}}$ the *tree model*, as it contains a single tree rooted in $w_{\mathcal{R}}$. While other single-tree models exist, we use the name exclusively for this model. It is semantically equivalent to models used in literature that assume the classical setting [2,10–12].

Definition 14 (Tree Model). *The* tree model $M_{\mathcal{T}} := M(I_{\mathcal{R}})$ *is the gossip model induced from the root model. We write* $\mathcal{T}$ *for the singleton class of* $M_{\mathcal{T}}$.

We have now defined four classes of models: $\mathcal{G}$, $\mathcal{I}$, $\mathcal{R}$, and $\mathcal{T}$. We visualise the relation between them in Fig. 1 below. The main class is $\mathcal{G}$. It consists of the induced gossip models of all initial models in $\mathcal{I}$, i.e. the execution trees with an initial model as their root. One particular model is the tree model $M_{\mathcal{T}} \in \mathcal{T}$ induced by the root model $I_{\mathcal{R}} \in \mathcal{R}$.

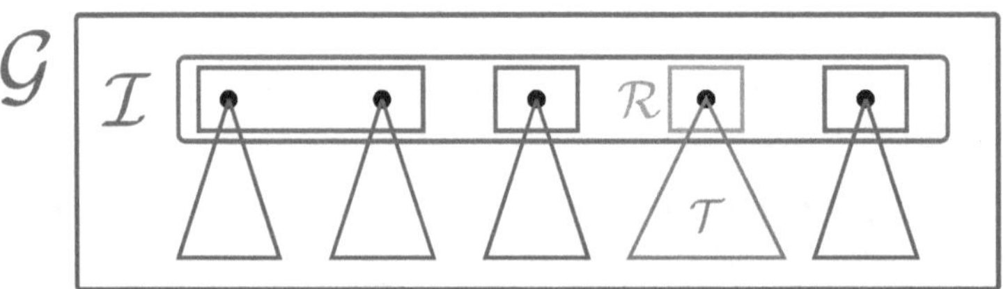

Fig. 1. The classes $\mathcal{I}$, $\mathcal{G}$, $\mathcal{R}$, and $\mathcal{T}$.

3 Expressivity of Protocol-Dependent Knowledge

We now show that $\mathcal{L}^{\mathbb{P}}$ is strictly more expressive than $\mathcal{L}$. To this end, we first show that $\mathcal{L}^{\mathbb{P}}$ can express the length of call sequences using protocol-dependent knowledge modalities. We do so by recursively defining protocols that depend on the violation of the previous.

Definition 15 (Counting Protocols). *For all natural numbers $k \geq 0$ and all agents $a \neq b$, and an arbitrary agent u we define P^k_{ab} recursively as follows.*

$$P^0_{ab} := \bot \qquad \text{“Allow no calls”}$$
$$P^{k+1}_{ab} := \widehat{K}^{P^k}_u \top \qquad \text{“The previous protocol has not been violated”}$$

These protocols leverage the global alarm property. A sequence violates protocol P^n iff it exceeds length n. Recall however that a protocol can only be violated by an illegal call: the lack of a call can never cause violation. A counting protocol therefore provides an upper bound, but not a lower bound.

Lemma 16. *For all agents u, call sequences σ and all $k \geq 0$ we have*

$$M, (w, \sigma) \models \widehat{K}^{P^k}_u \top \textit{ if and only if } |\sigma| \leq k.$$

Proof. Let agent u and call sequence σ be arbitrary. We use induction on k.

Base case. Suppose $k = 0$.

($\Longrightarrow$) Suppose $M, (w, \sigma) \models \widehat{K}^{P^0}_u \top$. Hence σ has not violated the protocol P^0. However, $P^0_{ab} = \bot$ for all calls ab, so no calls are P^0-permitted. Therefore we find that $\sigma = \epsilon$ and conclude that $|\sigma| \leq 0$.

($\Longleftarrow$) Suppose $|\sigma| \leq 0$. Then $\sigma = \epsilon$. The empty call sequence cannot violate any protocol, also not P^0. Hence $M, (w, \sigma) \models \neg K^{P^0}_u \bot$, i.e. $M, (w, \sigma) \models \widehat{K}^{P^0}_u \top$.

Induction Hypothesis. Let k be arbitrary and suppose we have $M, (w, \sigma) \models \widehat{K}^{P^k}_u \top$ if and only if $|\sigma| \leq k$.

Induction Step. Suppose the induction hypothesis holds for k. We show it holds for $k + 1$. Let σ be an arbitrary call sequence such that $|\sigma| = k + 1$ and write it as $\sigma = \tau.ab$. Using the semantics of $[ab]$ and $\widehat{K}_u$ we get the following equivalences:

$$\begin{aligned} M, (w, \tau.ab) \models \widehat{K}^{P^{k+1}}_u \top &\iff M, (w, \tau) \models [ab]\widehat{K}^{P^{k+1}}_u \top \\ &\iff M, (w, \tau) \models \neg [ab] K^{P^{k+1}}_u \bot \end{aligned}$$

We now distinguish three cases whether agent u is one of the agents a and b.

Suppose $u \notin \{a, b\}$. We get the following equivalences. The steps **Pri** and **Fnc** follow from the soundness of the axioms with the same name, that we will show below in Lemma 24. For the step at $(*)$ we use the definition of P^{k+1}_{ab} in both directions. Observe for the backwards direction that P^k is not violated at τ, so neither is P^{k+1}. This means we have $\tau \sim_u^{P^{k+1}} \tau$ which is sufficient to obtain $\tau \models \widehat{K}_u^{P^{k+1}}(P_{ab}^{k+1} \land [ab]\top)$. Hence τ satisfies the disjunct for $de = ab$.

$$\begin{aligned}
&\tau \models \neg[ab]K_u^{P^{k+1}}\bot \\
\iff &\tau \not\models P_{ab}^{k+1} \to \textstyle\bigwedge_{d,e\neq u} K_u^{P^{k+1}}(P_{de}^{k+1} \to [de]\bot) && \textbf{(Pri)} \\
\iff &\tau \models P_{ab}^{k+1} \land \textstyle\bigvee_{d,e\neq u} \neg(K_u^{P^{k+1}}(P_{de}^{k+1} \to [de]\bot)) && \text{(De Morgan)} \\
\iff &\tau \models P_{ab}^{k+1} \land \textstyle\bigvee_{d,e\neq u} \widehat{K}_u^{P^{k+1}}(P_{de}^{k+1} \land \neg[de]\bot) && (\text{Sem.}\widehat{K}_u) \\
\iff &\tau \models P_{ab}^{k+1} \land \textstyle\bigvee_{d,e\neq u} \widehat{K}_u^{P^{k+1}}(P_{de}^{k+1} \land [de]\top) && \textbf{(Fnc)} \\
\iff &\tau \models \widehat{K}_u^{P^k}\top && (*)
\end{aligned}$$

Suppose $u = a$. We get the following equivalences. The step $\mathbf{Obs}_1$ follows from Lemma 24 below. For the backwards direction of the step at (†), observe that there is precisely one set $Q \subseteq \mathbb{S}$ such that $O_b Q$. For all other sets $T \neq Q$ the conjunct is satisfied with $\neg O_b T$. The conjunct for $R = Q$ is satisfied by the knowledge operator: we use again the reflexive relation $\tau \sim^{P^{k+1}} \tau$ to obtain $\tau \models \widehat{K}_u^{P^{k+1}}(P_{ab}^{k+1} \land O_b Q \land [ab]\top)$.

$$\begin{aligned}
&\tau \models \neg[ab]K_u^{P^{k+1}}\bot \\
\iff &\tau \not\models P_{ab} \to \textstyle\bigvee_{R\subseteq\mathbb{S}}(O_b R \land K_u^P(P_{ab} \to (O_b R \to [ab]\bot))) && (\mathbf{Obs}_1) \\
\iff &\tau \models P_{ab} \land \textstyle\bigwedge_{R\subseteq\mathbb{S}} \neg O_b R \lor \neg K_u^P(P_{ab} \to (O_b R \to [ab]\bot)) && \text{(De Morgan)} \\
\iff &\tau \models P_{ab} \land \textstyle\bigwedge_{R\subseteq\mathbb{S}} \neg O_b R \lor \widehat{K}_u^P(P_{ab} \land O_b R \land \neg[ab]\bot) && (\text{Sem.}\widehat{K}_u) \\
\iff &\tau \models P_{ab} \land \textstyle\bigwedge_{R\subseteq\mathbb{S}} \neg O_b R \lor \widehat{K}_u^P(P_{ab} \land O_b R \land [ab]\top) && \textbf{(Fnc)} \\
\iff &\tau \models \widehat{K}_u^{P^k}\top && (\dagger)
\end{aligned}$$

Suppose $u = b$. We repeat the steps for $u = a$ and instead apply $\mathbf{Obs}_2$.

Hence $\tau.ab \models \widehat{K}_u^{P^{k+1}}\top \iff \tau \models \widehat{K}_u^{P^k}\top$. We finish by applying the induction hypothesis to find $\tau \models \widehat{K}_u^{P^k}\top \iff |\tau| \leq k \iff |\sigma| \leq k+1$. □

We negate the global alarm to invert the bound and define *counting formulas.*

Definition 17 (Counting Formulas). *We define* $\varphi_0 := \widehat{K}_u^{P^0}\top$ *and for every other natural number* $k \geq 1$, *let* $\varphi_k := \widehat{K}_u^{P^k}\top \land K_u^{P^{k-1}}\bot$.

Lemma 18. *For every call sequence* σ *and all* $k \in \mathbb{N}$ *we have* $\sigma \models \varphi_k$ *iff* $|\sigma| = k$.

Proof. Immediate by Lemma 16, as $K_u^{P^{k-1}}\bot$ is the negation of $\widehat{K}_u^{P^{k-1}}\top$. □

In order to show that the standard language of gossip $\mathcal{L}$ cannot express the length of call sequences, we give a standard definition of bisimulation. We can apply this definition to protocol-dependent gossip models as $\sim$ is equivalent to $\sim^{\mathsf{ANY}}$. Lemma 20 can be shown in the usual way [6, Theorem 2.20].

Definition 19 (Bisimulation). *Let* $M = \langle W, \sim, V\rangle$ *and* $M' = \langle W', \sim', V'\rangle$ *be gossip models. Two states are* bisimilar, *written* $M, s \leftrightarrow M', s'$, *if a bisimulation* Z *exists with* sZs'. *A relation* $Z \subseteq W \times W'$ *is a* bisimulation *if and only if for all gossip states* s, s' *such that* sZs' *we have:*

1. *(**Atoms**) For every agent a we have $V_a(s) = V'_a(s')$;*
2. *(**Forth**) For each agent a, if $s \sim_a t$ then there is a $t' \in W'$ such that $s' \sim_a t'$ and tZt';*
3. *(**Back**) For each agent a, if $s' \sim_a t'$ then there is a $t \in W$ such that $s \sim_a t$ and tZt'.*

Lemma 20. *If two states are bisimilar, they satisfy the same formulas in $\mathcal{L}$.*

Theorem 21. *$\mathcal{L}^{\mathbb{P}}$ is strictly more expressive than $\mathcal{L}$.*

Proof. We show that there exist two states that satisfy the same formulas in $\mathcal{L}$ but that can be distinguished by a formula in $\mathcal{L}^{\mathbb{P}}$.

Let $I_{\mathcal{R}}$ be the initial root model for two agents a, b and $M_{\mathcal{T}}$ the gossip model induced from it. Consider call sequences $\sigma = ab$ and $\tau = ba.ab$.

We show that $M_{\mathcal{T}}, (w_{\mathcal{R}}, \sigma) \leftrightarrows M_{\mathcal{T}}, (w_{\mathcal{R}}, \tau)$. Let Z be an equivalence relation containing two equivalence classes: one containing only the empty sequence $(w_{\mathcal{R}}, \epsilon)$ and one containing the states for all other call sequences. By definition $(w_{\mathcal{R}}, \sigma)Z(w_{\mathcal{R}}, \tau)$ and we claim that Z is a bisimulation.

Observe that all atoms are satisfied after the first call, whether it is ab or ba. Next, both agents can distinguish each point in the model: they are involved in all calls. The epistemic relations of both agents contain only reflexive relations.

1. (**Atoms**) For all reflexive relations in Z, **Atoms** holds trivially. For all other $(w_{\mathcal{R}}, \sigma')Z(w_{\mathcal{R}}, \tau')$, observe that σ' and τ' must be non-empty sequences, as ϵ only occurs reflexively in Z. All secrets are already shared, so **Atoms** holds.
2. (**Forth/Back**) For each agent, $\sim^{\mathsf{ANY}}$ contains precisely one reflexive relation for each call sequence. Hence **Forth** and **Back** are satisfied.

Hence Z is a bisimulation. By Lemma 20 then $(w_{\mathcal{R}}, \sigma)$ and $(w_{\mathcal{R}}, \tau)$ agree on all $\varphi \in \mathcal{L}$. However, by Lemma 18 we find $\sigma \models \varphi_1$ and $\tau \not\models \varphi_1$ and $\varphi_1 \in \mathcal{L}^{\mathbb{P}}$. □

4 The Logic of Initial Models

We now provide call-free axiomatisations for the class of initial models $\mathcal{I}$ and the singleton class $\mathcal{R}$ of the root model $I_{\mathcal{R}}$. The logic for both is essentially **S5** plus axioms about agents knowing (only) their own secrets. The **PI** axiom ensures protocol invariance, which is required because protocol-dependent knowledge can only differ from standard knowledge after calls have happened.

Definition 22. *The proof system $\vdash_{\mathcal{R}}$ for the call-free language $\mathcal{L}^{\mathbb{P}}_{-}$ is defined as shown in Table 1. The system $\vdash_{\mathcal{I}}$ is obtained by omitting the* **Only** *axiom.*

Theorem 23. *The proof system $\vdash_{\mathcal{I}}$ is sound and complete for the class of all initial models $\mathcal{I}$: for all $\varphi \in \mathcal{L}^{\mathbb{P}}_{-}$ we have $\models_{\mathcal{I}} \varphi$ iff $\vdash_{\mathcal{I}} \varphi$. The system $\vdash_{\mathcal{R}}$ is sound and complete for the root model $I_{\mathcal{R}}$: for all $\varphi \in \mathcal{L}^{\mathbb{P}}_{-}$ we have $\models_{\mathcal{R}} \varphi$ iff $\vdash_{\mathcal{R}} \varphi$.*

Proof. Soundness follows directly from the semantics. Completeness can be shown using standard methods by building a canonical model [6, Section 4.2]. In particular we can define the canonical relation $R_0(a)$ over maximally consistent sets as usual using K^{ANY}. The **PI** axiom ensures that in any maximally consistent set the modalities K^P and K^Q for any two protocols P and Q agree with each other, matching Lemma 12 and therefore the semantics. □

Table 1. Rules and axioms of $\vdash_{\mathcal{R}}$. Omitting **Only** produces $\vdash_{\mathcal{I}}$.

Prop	propositional tautologies	**K**	$K_a^P(\varphi \to \psi) \to (K_a^P\varphi \to K_a^P\psi)$
MP	$\vdash \varphi, \vdash \varphi \to \psi$ imply $\vdash \psi$	**T**	$K_a^P\varphi \to \varphi$
Sub	$\vdash \varphi \leftrightarrow \psi$ implies $\vdash \chi \leftrightarrow \chi[\varphi/\psi]$	**4**	$K_a^P\varphi \to K_a^P K_a^P\varphi$
Own	$S_a a$	**5**	$\neg K_a^P\varphi \to K_a^P\neg K_a^P\varphi$
Only	$O_a a$	**Nec**	$\vdash \varphi$ implies $\vdash K_a^P\varphi$
PFi	$S_a b \to K_a^P S_a b$		
NPi	$\neg S_a b \to K_a^P \neg S_a b$	**PI**	$K^P\varphi \to K^Q\varphi$

5 Call Reductions

While in Sect. 3 we have seen that $\mathcal{L}^{\mathbb{P}}$ is more expressive than $\mathcal{L}$, we now show that the protocol-dependent language still shares a feature with the standard language, namely that the call operator $[ab]$ can be eliminated. We will use the established idea of *reduction axioms* to translate any formula to a call-free formula using the validities shown in Table 2. These validities originate from [13], and only the final three are adapted to protocol-dependent knowledge.

However, we will not actually stipulate these formulas as axioms for the proof systems $\vdash_{\mathcal{G}}$ and $\vdash_{\mathcal{T}}$. We only use them as rewrite rules and rely on their semantic validity to prove the soundness and completeness of $\vdash_{\mathcal{G}}$ and $\vdash_{\mathcal{T}}$.

Table 2. Call Reduction Validities on Gossip Models.

Call Basics		**Call Effects**		
Con	$[ab](\varphi \wedge \psi) \leftrightarrow ([ab]\varphi \wedge [ab]\psi)$	**Eff**	$[ab]S_c d \leftrightarrow (S_a d \vee S_b d)$	$c \in \{a, b\}$
Fnc	$[ab]\neg\varphi \leftrightarrow \neg[ab]\varphi$	**Ext**	$[ab]S_c d \leftrightarrow S_c d$	$c \notin \{a, b\}$
Calls and Protocol-Dependent Knowledge				
Obs₁	$[ab]K_a^P\varphi \leftrightarrow (P_{ab} \to \bigvee_{R \subseteq S}(O_b R \wedge K_a^P(P_{ab} \to (O_b R \to [ab]\varphi))))$			
Obs₂	$[ab]K_b^P\varphi \leftrightarrow (P_{ab} \to \bigvee_{R \subseteq S}(O_a R \wedge K_b^P(P_{ab} \to (O_a R \to [ab]\varphi))))$			
Pri	$[ab]K_c^P\varphi \leftrightarrow (P_{ab} \to \bigwedge_{d,e \neq a} K_c^P(P_{de} \to [de]\varphi))$			$c \notin \{a, b\}$

Lemma 24. *All axioms in Table 2 are valid on the class of gossip models $\mathcal{G}$.*

Proof. The Call Basics are valid because calls are deterministic actions that can always be executed. The Call Effects are valid as they are equal to Definition 7.

To show $\mathbf{Obs}_1$, let (w, σ) be an arbitrary state in some model M and let $\varphi \in \mathcal{L}^{\mathbb{P}}$ be arbitrary. We have the following chains of equivalences. Recall that $V_b(w, \sigma)$ is the set of secrets that agent b knows at state (w, σ). At step $(*)$ we use a disjunct to enumerate all possible sets of secrets $V_b(w,\sigma)$ that agent b might know. There is precisely one set $V_b(w,\sigma) = R \subseteq \mathbb{S}$ such that $O_b R$ holds.

$$\begin{array}{lr}
(w,\sigma) \models [ab]K_a^P\varphi & \\
\iff (w,\sigma.ab) \models K_a^P\varphi & \text{(Sem. } [ab]) \\
\iff \forall(w',\tau.de) \text{ s.t. } (w,\sigma.ab) \sim_a^P (w',\tau.de) : (w',\tau.de) \models \varphi & \text{(Sem. } K_a^P) \\
\iff \forall(w',\tau) \text{ s.t. } (w,\sigma.ab) \sim_a^P (w',\tau.ab) : (w',\tau.ab) \models \varphi & \text{(Def. } \sim_a^P) \\
\iff \forall(w',\tau) \text{ s.t. } (w,\sigma) \sim_a^P (w',\tau) : & \text{(Def. } \sim_a^P) \\
\quad \text{if } (w,\sigma) \models P_{ab} \text{ and } (w',\tau) \models P_{ab} \text{ and } V_b(w,\sigma) = V_b(w',\tau) & \\
\quad \text{then } (w',\tau.ab) \models \varphi & \\
\iff \forall(w',\tau) \text{ s.t. } (w,\sigma) \sim_a^P (w',\tau) : & \text{(Semantics)} \\
\quad \text{If } (w,\sigma) \models P_{ab} \text{ then } (w',\tau) \models P_{ab} \to (O_b V_b(w,\sigma) \to [ab]\varphi) & \\
\iff (w,\sigma) \models P_{ab} \to K_a^P(P_{ab} \to (O_b V_b(w,\sigma) \to [ab]\varphi)) & \text{(Sem. } K_a^P) \\
\iff (w,\sigma) \models P_{ab} \to \bigvee_{R\subseteq\mathbb{S}}(O_b R \wedge K_a^P(P_{ab} \to (O_b R \to [ab]\varphi))) & (*)
\end{array}$$

The proof for **Obs**$_2$ is analogous to **Obs**$_1$. For **Pri** we have the following chain of equivalences.

$$\begin{array}{lr}
(w,\sigma) \models [ab]K_c^P\varphi & \\
\iff (w,\sigma.ab) \models K_c^P\varphi & \text{(Sem. } [ab]) \\
\iff \forall(w',\tau.de) \text{ s.t. } (w,\sigma.ab) \sim_c^P (w',\tau.de) \text{ and } c \neq d,e : & \text{(Sem. } K_c^P) \\
\quad (w',\tau.de) \models \varphi & \\
\iff \forall d,e \neq c : \forall(w',\tau) \text{ s.t. } (w,\sigma) \sim_c^P (w',\tau) : & \text{(Def. } \sim_c^P) \\
\quad \text{if } (w,\sigma) \models P_{ab} \text{ and } \tau \models P_{de} \text{ then } (w',\tau.de) \models \varphi & \\
\iff \forall d,e \neq c : \forall(w',\tau') \text{ s.t. } (w,\sigma) \sim_c^P (w',\tau) : & \text{(Semantics)} \\
\quad \text{if } (w,\sigma) \models P_{ab} \text{ then } (w',\tau) \models P_{de} \to [de]\varphi & \\
\iff \text{If } (w,\sigma) \models P_{ab} \text{ then } \forall d,e \neq c : (w,\sigma) \models K_c^P(P_{de} \to [de]\varphi) & \text{(Sem.} K_c^P) \\
\iff (w,\sigma) \models P_{ab} \to \bigwedge_{d,e\neq c} K_c^P(P_{de} \to [de]\varphi) & \square
\end{array}$$

Given the validity of the call reductions, we now define and prove the following.

Definition 25 (Call Reduction). *For any formula $\varphi \in \mathcal{L}^{\mathbb{P}}$, let $\mathsf{cr}(\varphi) \in \mathcal{L}^{\mathbb{P}}_{-}$ denote the formula obtained by rewriting φ using the validities shown in Table 2 from left to right.*

Note that $\mathsf{cr}(\cdot)$ is well-defined: the rewriting terminates because after each step the subformula under the call is a strict subformula of the previous formula.

Lemma 26. *For every formula $\varphi \in \mathcal{L}^{\mathbb{P}}$ we have that $\models_{\mathcal{G}} \varphi \leftrightarrow \mathsf{cr}(\varphi)$.*

Proof. By induction on the structure of the formula, using Lemma 24. $\square$

Corollary 27. *All axioms in Table 2 are valid on the class of the tree model $\mathcal{T}$ and for every formula $\varphi \in \mathcal{L}^{\mathbb{P}}$ we have that $\models_{\mathcal{T}} \varphi \leftrightarrow \mathsf{cr}(\varphi)$.*

Proof. Immediate from $\mathcal{T} \subseteq \mathcal{G}$ and Lemmas 24 and 26. $\square$

Definition 25 also guarantees that rewriting never increases the modal degree.

Lemma 28. *Let $\varphi \in \mathcal{L}^{\mathbb{P}}$. Its call reduction $\psi \in \mathcal{L}^{\mathbb{P}}_{-}$ has degree $d(\psi) \leq d(\varphi)$.*

Proof. It suffices to show that $d(\mathsf{cr}([ab]\chi)) \leq d([ab]\chi)$. The proof is by induction on χ. We explicitly show the induction step for $\chi = K^P_c \chi'$ with $c = a$.

Induction Hypothesis. Let $\chi \in \mathcal{L}^{\mathbb{P}}$ be arbitrary. For each strict subformula χ' of χ we have $d(\mathsf{cr}(\chi')) \leq d(\chi')$.

Induction Step. Suppose $\chi = K^P_a \chi'$. Thus $\varphi = [ab]K^P_a \chi'$. Let $n = d(\varphi)$.

We have $\mathsf{cr}(\chi) = (\mathsf{cr}(P_{ab}) \to \bigvee_{R \subseteq S}(O_b R \wedge K^P_a(\mathsf{cr}(P_{ab}) \to (O_b R \to \mathsf{cr}(\chi')))))$ by Definition 25. By Definition 4, each of the strict subformulas of χ has a degree at most $n-1$. By the induction hypothesis on χ' we have thus $d(\mathsf{cr}(\chi')) \leq d(\chi') < d(\chi) = n$ and as P_{ab} is too a strict subformula of χ we have $d(\mathsf{cr}(P_{ab})) \leq d(P_{ab}) < d(\chi) = n$. Compute the degree of $\mathsf{cr}(\chi)$ to conclude that $d(\mathsf{cr}(\chi)) \leq n$. □

6 The Logic of Gossip Models

We cannot extend the axiomatisation $\vdash_{\mathcal{I}}$ and standard canonical model construction to $\mathcal{G}$ because **PI** no longer holds after calls have been made. However, we can still define provability in $\mathcal{G}$ in terms of $\mathcal{I}$, in a way that is similar to the *Tree Rule* in [13].

First, observe that it follows directly from the semantics that for any pointed gossip model and formula $\varphi \in \mathcal{L}^{\mathbb{P}}$ we have $M, (w, \sigma) \models \varphi$ iff $M, (w, \epsilon) \models [\sigma]\varphi$. We can therefore check the truth of some formula anywhere in the model already in the root state. Furthermore we can use Definition 25 to find the call reduction of $[\sigma]\varphi$. As the root states are semantically equivalent to their counterpart in the initial model, we can then use $\vdash_{\mathcal{I}}$ to determine validity of $[\sigma]\varphi$ in the roots of the gossip models, and subsequently conclude that φ always holds after sequence σ.

To find out whether φ holds after every call sequence and therefore is a validity, we only need to repeat this process for the other call sequences. However, there are infinitely many. Fortunately, we can bound the number of call sequences that we must check by showing that the number of n-bisimilarity classes is finite.

For this we now define n-bisimulation for protocol-dependent knowledge. Lemma 31 is a standard result about n-bisimulation and can be shown in the usual way, so we omit the proof [6, Proposition 2.31].

Definition 29 (n-bisimulation for Protocol-Dependent Knowledge). *Let $n \in \mathbb{N}$, and $M = \langle W, \sim, V\rangle$ and $M' = \langle W', \sim', V'\rangle$ be protocol-dependent gossip models. Two states $s \in W$ and $s' \in W'$ are* n-bisimilar, *denoted $M, s \leftrightarroweq_n M', s'$, if and only if the following conditions hold.*

1. ***(Atoms)** For every agent a we have $V_a(s) = V'_a(s')$.*

Additionally if $n > 0$ we have for each a, and for all $P \in \mathbb{P}$ with $d(P) < n$ an instance of the following two conditions.

2. ***(Forth)*** *For every* $t \in W$ *we have: if* $s \sim_a^P t$ *then there is a* $t' \in W'$ *such that* $s' \sim_a'^P t'$ *and* $M, t \underline{\leftrightarrow}_{n-1} M', t'$.
3. ***(Back)*** *For every* $t' \in W$ *we have: if* $s' \sim_a^{P'} t'$ *then there is a* $t \in W$ *such that* $s \sim_a^P t$ *and* $M, t \underline{\leftrightarrow}_{n-1} M', t'$.

Definition 30. *Two states are* n-equivalent, *written* $M, (w, \sigma) \equiv_n M', (w', \sigma')$, *iff for all* $\varphi \in \mathcal{L}^{\mathbb{P}}$ *with* $d(\varphi) \leq n$ *we have* $M, (w, \sigma) \models \varphi \iff M', (w', \sigma') \models \varphi$.

Lemma 31. *For any two states* $M, (w, \sigma)$ *and* $M', (w', \sigma')$ *we have*

$$M, (w, \sigma) \underline{\leftrightarrow}_n M', (w', \sigma') \text{ if and only if } M, (w, \sigma) \equiv_n M', (w', \sigma').$$

Lemma 32. *The number of* $\underline{\leftrightarrow}_n$*-equivalence classes* $f(n)$ *is finite.*

Proof. First we claim that there are only finitely many semantically different modal formulas of modal degree up to n. This can be shown by induction on n [6, Proposition 2.29]. For the induction step from n to $n+1$ with $K_a^P \varphi$, note that the protocol P must be defined by formulas of degree n or lower. By the induction hypothesis there are only finitely many semantically different formulas, and thus also only finitely many semantically different protocols. Hence for each degree we have only finitely many semantically different modalities. The rest is standard.

Now towards a contradiction, suppose $f(n)$ is infinite. Then there exist states $M_k, (w_k, \sigma_k)$ for $k \in \mathbb{N}$ that are all not n-bisimilar to each other. By Lemma 31 then there are formulas $\varphi_{i,j}$ that are each true at the state with index i and false at the state with index j. Moreover, infinitely many of these formulas must be semantically different, contradicting the claim. □

We now get the following result, which states that anything true after some sequence of calls is also true after some sequence of calls no longer than the number of n-bisimilarity classes.

Lemma 33. *Let* $\varphi \in \mathcal{L}^{\mathbb{P}}$ *and* $n = d(\varphi)$. *For any state* $(w, \sigma.\tau)$ *in any gossip model* M *that satisfies* φ *there is a sequence* τ' *such that* $|\tau'| \leq f(n)$ *and* $M, (w, \sigma.\tau') \models \varphi$, *where* $f(n)$ *is the number of* n*-bisimilarity classes.*

Proof. Let $M, (w, \sigma.\tau)$ and φ be arbitrary such that $M, (w, \sigma.\tau) \models \varphi$ and $|\tau| > f(n)$. Then τ must have two initial fragments $\tau_1 \neq \tau_2$ such that $(w, \tau_1) \underline{\leftrightarrow}_n (w, \tau_2)$. W.l.o.g. let $|\tau_1| < |\tau_2|$ and $\tau = \tau_2.\tau_3$. Then $M, (w, \sigma.\tau_2) \models [\tau_3]\varphi$. By definition $d([\tau_3]\varphi) = n$, so by n-bisimilarity and Lemma 31 also $M, (w, \sigma.\tau_1) \models [\tau_3]\varphi$ and $M, (w, \sigma.\tau_1.\tau_3) \models \varphi$. Note that $|\tau_1.\tau_3| < |\tau|$ and we can repeat until done. □

Definition 34 (Proof System $\vdash_{\mathcal{G}}$). *Let* $\varphi \in \mathcal{L}^{\mathbb{P}}$ *and* $n = d(\varphi)$. *We define* $\vdash_{\mathcal{G}}$ *as follows, where* $f(n)$ *is the number of* n*-bisimilarity classes.*

$$\vdash_{\mathcal{G}} \varphi \iff \forall \sigma : |\sigma| \leq f(n) \text{ we have } \vdash_{\mathcal{I}} \mathsf{cr}([\sigma]\varphi)$$

Theorem 35 (Soundness and Completeness for $\mathcal{G}$). *For any $\varphi \in \mathcal{L}^{\mathbb{P}}$, we have $\models_{\mathcal{G}} \varphi$ iff $\vdash_{\mathcal{G}} \varphi$.*

Proof (**Soundness**). By contraposition, suppose $\not\models_{\mathcal{G}} \varphi$. We show that $\nvdash_{\mathcal{G}} \varphi$.

By assumption there is some model $M \in \mathcal{G}$ and some state (w, σ) such that $M, (w, \sigma) \models \neg\varphi$, i.e. we have $M, (w, \epsilon) \models \neg[\sigma]\varphi$. By Lemma 33 we can assume that $|\sigma| \leq f(d(\varphi))$ and by Lemma 26 we have the call reduction $\mathsf{cr}([\sigma]\varphi)$.

Now suppose $\vdash_{\mathcal{G}} \varphi$, in order to reach a contradiction. By Definition 34 and from $|\sigma| \leq f(d(\varphi))$ we get $\vdash_{\mathcal{I}} \mathsf{cr}([\sigma]\varphi)$. By soundness of $\vdash_{\mathcal{I}}$ we then have $\models_{\mathcal{I}} \mathsf{cr}([\sigma]\varphi)$ and in particular $M, (w, \epsilon) \models \mathsf{cr}([\sigma]\varphi)$. By validity of the call reductions we get $M, (w, \epsilon) \models [\sigma]\varphi$. This contradicts $M, (w, \epsilon) \models \neg[\sigma]\varphi$, so we find $\nvdash_{\mathcal{G}} \varphi$.

(**Completeness**) By contraposition, suppose $\nvdash_{\mathcal{G}} \varphi$. We show that $\not\models_{\mathcal{G}} \varphi$.

By Definition 34 there must be a call sequence σ such that $\nvdash_{\mathcal{I}} \mathsf{cr}([\sigma]\varphi)$. By completeness of $\vdash_{\mathcal{I}}$ we have some model and state such that $M, (w, \epsilon) \not\models \mathsf{cr}([\sigma]\varphi)$. By validity of the call reductions we get $M, (w, \epsilon) \not\models [\sigma]\varphi$. This implies $M, (w, \sigma) \not\models \varphi$ and thus $\not\models_{\mathcal{G}} \varphi$. □

We were able to use $\vdash_{\mathcal{I}}$ in defining $\vdash_{\mathcal{G}}$ because there is a bijective relation between the two classes: each model in $\mathcal{G}$ is induced by an initial model in $\mathcal{I}$ and vice versa. Recall that $\mathcal{T}$ is the singleton class of the tree model induced by the root model in $\mathcal{R}$. We therefore define $\vdash_{\mathcal{T}}$ analogously to $\vdash_{\mathcal{G}}$, but using $\vdash_{\mathcal{R}}$ instead of $\vdash_{\mathcal{I}}$. The proof of Theorem 37 is analogous to Theorem 35.

Definition 36 (Proof System $\vdash_{\mathcal{T}}$). *Let $\varphi \in \mathcal{L}^{\mathbb{P}}$ and $n = d(\varphi)$. We define $\vdash_{\mathcal{T}}$ as follows, where $f(n)$ is the number of n-bisimilarity classes.*

$$\vdash_{\mathcal{T}} \varphi \iff \forall \sigma : |\sigma| \leq f(n) \text{ we have } \vdash_{\mathcal{R}} \mathsf{cr}([\sigma]\varphi)$$

Theorem 37 (Soundness and Completeness for $\mathcal{T}$). *For any $\varphi \in \mathcal{L}^{\mathbb{P}}$, we have $\models_{\mathcal{T}} \varphi$ iff $\vdash_{\mathcal{T}} \varphi$.*

The definitions of both $\vdash_{\mathcal{G}}$ and $\vdash_{\mathcal{T}}$ bound the number of times they invoke $\vdash_{\mathcal{I}}$ and $\vdash_{\mathcal{R}}$ by the number of n-bisimilarity classes. As this is finite, these systems too are decidable.

Theorem 38. *Proof systems $\vdash_{\mathcal{G}}$ and $\vdash_{\mathcal{T}}$ are decidable.*

Proof. Immediate by Lemma 32. □

7 Conclusion

We have shown that the protocol-dependent knowledge modality is more expressive than the standard epistemic knowledge modality. We have defined four logics for gossip using protocol-dependent knowledge modalities and provided sound and complete proof systems that are decidable. In particular we have provided a proof system for $\mathcal{T}$, the model defined in [12]. These contributions give insight into the use of protocol-dependent knowledge modalities in epistemic logic.

An interesting avenue for future work is to generalise the use and axiomatisation of protocol-dependent knowledge modalities to domains outside of gossip.

References

1. Apt, K.R., Grossi, D., van der Hoek, W.: Epistemic protocols for distributed gossiping. In: Proceedings Fifteenth Conference on Theoretical Aspects of Rationality and Knowledge, TARK 2015. EPTCS, vol. 215, pp. 51–66 (2015). https://doi.org/10.4204/EPTCS.215.5
2. Apt, K.R., Wojtczak, D.: Common knowledge in a logic of gossips. In: Proceedings Sixteenth Conference on Theoretical Aspects of Rationality and Knowledge, TARK 2017. EPTCS, vol. 251, pp. 10–27 (2017). https://doi.org/10.4204/EPTCS.251.2
3. Attamah, M., van Ditmarsch, H., Grossi, D., Hoek, W.: Knowledge and gossip. Front. Artif. Intell. Appl. **263**, 21–26 (2014). https://doi.org/10.3233/978-1-61499-419-0-21
4. Baker, B., Shostak, R.: Gossips and telephones. Discret. Math. **2**(3), 191–193 (1972). https://doi.org/10.1016/0012-365X(72)90001-5
5. van den Berg, L., Gattinger, M.: Dealing with unreliable agents in dynamic gossip. In: Martins, M., Sedlár, I. (eds.) DaLi 2020. LNCS, vol. 12569, pp. 51–67. Springer, Cham (2020). https://doi.org/10.1007/978-3-030-65840-3_4
6. Blackburn, P., de Rijke, M., Venema, Y.: Modal Logic. Cambridge Tracts in Theoretical Computer Science, vol. 53. Cambridge University Press (2001)
7. Cooper, M., Herzig, A., Maffre, F., Maris, F., Régnier, P.: The epistemic gossip problem. Discret. Math. **342**(3), 654–663 (2019). https://doi.org/10.1016/j.disc.2018.10.041
8. van Ditmarsch, H., van Eijck, J., Pardo, P., Ramezanian, R., Schwarzentruber, F.: Epistemic protocols for dynamic gossip. J. Appl. Logic **20**, 1–31 (2017). https://doi.org/10.1016/j.jal.2016.12.001
9. van Ditmarsch, H., van Eijck, J., Pardo, P., Ramezanian, R., Schwarzentruber, F.: Dynamic gossip. Bull. Iran. Math. Soc. **45**(3), 701–728 (2019). https://doi.org/10.1007/s41980-018-0160-4
10. van Ditmarsch, H., Gattinger, M., Ramezanian, R.: Everyone knows that everyone knows: gossip protocols for super experts. Stud. Logica. **111**(3), 453–499 (2023). https://doi.org/10.1007/S11225-022-10032-3
11. van Ditmarsch, H., Gattinger, M.: You can only be lucky once: optimal gossip for epistemic goals. Math. Struct. Comput. Sci. (2024). https://doi.org/10.1017/S0960129524000082
12. van Ditmarsch, H., Gattinger, M., Kuijer, L.B., Pardo, P.: Strengthening gossip protocols using protocol-dependent knowledge. J. Appl. Log. - IfCoLog J. Log. Appl. **6**(1), 157–203 (2019). http://arxiv.org/abs/1907.12321
13. van Ditmarsch, H., van der Hoek, W., Kuijer, L.B.: The logic of gossiping. Artif. Intell. **286**, 103306 (2020). https://doi.org/10.1016/j.artint.2020.103306
14. Gattinger, M.: New directions in model checking dynamic epistemic logic. Ph.D. thesis, University of Amsterdam (2018)
15. Hedetniemi, S.M., Hedetniemi, S.T., Liestman, A.L.: A survey of gossiping and broadcasting in communication networks. Networks **18**(4), 319–349 (1988). https://doi.org/10.1002/net.3230180406
16. Herzig, A., Maffre, F.: How to share knowledge by gossiping. AI Commun. **30**(1), 1–17 (2017). https://doi.org/10.3233/AIC-170723
17. Kermarrec, A.M., van Steen, M.: Gossiping in distributed systems. SIGOPS Oper. Syst. Rev. **41**(5), 2–7 (2007). https://doi.org/10.1145/1317379.1317381

18. Smit, W.J.: Axiomatising protocol-dependent knowledge in gossip. Master's thesis, University of Amsterdam (2024). https://eprints.illc.uva.nl/id/eprint/2330/
19. Tijdeman, R.: On a telephone problem. Nieuw Archief voor Wiskunde **3**(19), 188–192 (1971)

Dynamic Epistemic Logic with Private Communication

Kun Zhang and Zuojun Xiong(✉)

Institute of Logic and Intelligence, Southwest University, Chongqing, China
zuojunxiong@swu.edu.cn

Abstract. Epistemic modalities such as knowledge and belief are extensively studied in modal logic, including dynamic updates. In standard epistemic models, action updates such as announcements are defined through model restrictions that inform all agents about changes. This article redefines the semantics of the K-modality and introduces an action operator for information sharing between some agents while remaining hidden from others, termed *a private communication.* $[ab]\varphi$ indicates that after agents a and b share everything, φ holds. A related formula $R_G\varphi$, discussed by Ågotnes and Wáng [12], asserts that φ holds after the group G shares all information, known to all agents. In contrast, our updates are private. The agent c is under the impression that such a communication does not occur. We explore the logical properties and dynamics of private communication. A comparison with related work is included.

Keywords: private communication · epistemic logic · labelled semantics

1 Introduction

Communication is a dynamic process of information exchange among agents. In dynamic epistemic logic, this process is modelled as an action that updates agents' knowledge and beliefs. Ågotnes and Wáng proposed a logic for resolving distributed knowledge in [12], emphasising the exchange of information among agents, called resolution. The formula $R_G\varphi$ states that after resolving the distributed knowledge within the set G of agents, φ holds. The operator R indicates that the agents in G publicly share all information only within G. When an agent a in G knows φ, φ becomes common knowledge among agents in G.

The key contribution of this paper is to transform the public resolution operator into a *private communication operator.* Private communication dominates social interactions. For example, companies A, B, and C compete to achieve a

We sincerely thank the anonymous reviewers of LORI-2025 and DaLí 2025 for their insightful comments, which led to substantial improvements including clearer definitions, enhanced semantic analysis, and a philosophical treatment of the logical framework.

Supported by the National Social Science Fund of China, No. 24BZX112.

J. Wang et al. (Eds.): DaLí 2025, LNCS 16472, pp. 167–179, 2026.
https://doi.org/10.1007/978-3-032-22626-6_10

common goal. A and B secretly plan to share relevant information. At this point, C believes that A and B have not communicated. Unexpectedly, B deceives A by aligning with C, secretly sharing information from A for a strategic advantage. Meanwhile, A assumes that B and C do not communicate.

In this paper, we adopt a simplistic set of initial assumptions for clarity: in a private communication among a set G of agents:

(i) every agent in G shares all information to every other member of G;
(ii) every agent outside of G is under the impression that such communication does not occur;
(iii) each participant within G is aware that those outside of G are led to believe that this type of communication is non-existent.

This paper is organised as follows. Section 2 presents the definitions of syntax and semantics. In Sect. 3, we analyse properties of the modal operator K_a. Using our semantic framework, we show that the K_a operator occupies an intermediate semantic position between the conventional knowledge and belief operators.[1] In Sect. 4, we compare our logical system with related studies. Finally, Sect. 5 offers a comprehensive summary of the research presented in the article.

2 Syntax and Semantics

2.1 Syntax and Model

We present our logic's syntax. Let Agt be a non-empty finite set of agents, and Prop be a set of propositional letters. The language of dynamic epistemic logic with private communication is defined below.

$$\mathcal{L}_{EKP} \ni \varphi ::= p \mid \neg\varphi \mid (\varphi \wedge \varphi) \mid \mathrm{K}_a\varphi \mid [G]\varphi,$$

where $p \in$ Prop, $a \in$ Agt and $G \subseteq$ Agt. We also make use of the classical abbreviations for disjunction, implication, bi-implication and the dual operator $\hat{\mathrm{K}}$ of K. $[G]\varphi$ indicates that following private communication among a set G of agents, the statement φ is true. For simplicity, we write $[a_1 a_2 \cdots a_n]\varphi$ instead of $[\{a_1, a_2, \cdots, a_n\}]\varphi$. In particular, $[] := [\varnothing]$.

For convenience, we introduce some terms and notation. We call $[a_1^1 \cdots a_m^1] \cdots [a_1^k \cdots a_n^k]$ a *sequence*. Each operator $[a_1^k \cdots a_m^k]$ is called a *group*. From now on, we will abuse the term "group" for a "set" and a "group in the sequence". Let $\textsc{t} = [a_1^1 \cdots a_m^1] \cdots [a_1^k \cdots a_n^k]$ be a sequence,[2] and we abuse $\{\textsc{t}\}$ for the set of agents contained in $\textsc{t}$. For example, if $\textsc{s} = [abc][de][aec][ed]$, $\{\textsc{s}\} = \{a, b, c, d, e\}$. Moreover, $\mathrm{tail}(\textsc{s}) := [ed]$ denotes its final group, and $\textsc{s}^-$ is obtained by removing $\mathrm{tail}(\textsc{s})$ from $\textsc{s}$. In particular, if $\textsc{g}$ is a sequence that contains only one group, $\textsc{g}^- = []$. Given a set $G \subseteq$ Agt, we say that a sequence $\textsc{h}$

[1] The description and detailed explanation of both the knowledge operator and the belief operator are comprehensively presented in [6,10].

[2] We use small capital letters for sequences, like $\textsc{t}$, $\textsc{h}$,$\textsc{s}$,$\textsc{g}$ and so on.

is *based on* G if $\{\mathtt{H}\} \subseteq G$. We define $\mathcal{I}$ as the set of all sequences based on Agt. That is, $\mathcal{I} := \{\mathtt{H} \mid \mathtt{H}\text{is based on}\mathrm{Agt}\}$.

Consider the initial condition (i) related to private communication. This condition is represented through a global model update: For each state, eliminate a link to an alternative state for any element of G unless every element of G is connected to that state.[3] The following definition designates a label for every state in the updated model, termed a *resolved model* or a *labelled model.*[4]

Definition 1 ([G]-resolved model). *Given an* ***S5*** *model* $M = (W, \sim, V)$, *and* $[G] = [a_1 \cdots a_n] \in \mathcal{I}$ *as a group.*[5] *The* $[G]$*-resolved update of* M *is the model* $M^{[G]} = (W^{[G]}, \sim^{[G]}, V^{[G]})$ *defined by*

- $W^{[G]} = \{w^{[G]} \mid w \in W\}$;
- $w^{[G]} \in V^{[G]}(p)$ *iff* $w \in V(p)$ *for any* $p \in \mathrm{Prop}$ *and* $w^{[G]} \in W^{[G]}$;
- *for any* $c \in \mathrm{Agt}$*:*
  - ${\sim_c}^{[G]} = \{(w^{[G]}, u^{[G]}) \mid (w, u) \in \bigcap\limits_{a \in G} \sim_a\}$ *if* $c \in G$;
  - ${\sim_c}^{[G]} = \{(w^{[G]}, u^{[G]}) \mid (w, u) \in \sim_c\}$ *if* $c \notin G$.

It is clear that the resolved model continues to be an **S5** model. As a result, we broaden the concept of a $[G]$-resolved model to encompass an H-resolved model, where H represents a sequence. We refer to $M^{[\mathtt{H}^-]}$ as the *preceding* model relative to $M^{[\mathtt{H}]}$. For the sake of clarity, we introduce the notation $M^{[G_1]\cdots[G_n]}$ as $M^{[G_1]^{\cdot^{\cdot^{\cdot^{[G_n]}}}}}$. Similarly, $w^{[G_1]\cdots[G_n]}$ follows suit. Furthermore, we employ the shorthand $w \sim_G u$ to denote $(w, u) \in \bigcap\limits_{a \in G} \sim_a$.

Consider the initial condition (ii) related to private communication. In essence, this implies that for any agent $c \in \mathrm{Agt} \backslash G$, the agent maintains the belief that the knowledge possessed by everyone else remains unchanged. Consequently, given any pointed **S5** model (M, w), when we are tasked with interpreting the formula $\mathrm{K}_c\varphi$ under a resolved model denoted as $(M^{[G]}, w^{[G]})$, this interpretation must indeed be performed with respect to the original model (M, w). Hence, there arises a requirement for a particularly *large* model that encompasses all possible resolved models, facilitating a more straightforward identification of the *preceding* model. We refer to this expansive model as a *space model.*

Definition 2 (space model) *Given* M *as an* **S5** *model.* $\mathfrak{M} = \biguplus_{\mathtt{H} \in \mathcal{I}} M^{\mathtt{H}}$ *is the collection of all resolved models of* M *over* $\mathcal{I}$.[6] *This compilation of resolved models is called the space model of* M.

[3] There is a long line of research on actions that take the intersections of epistemic relations, e.g. [2,4,12].

[4] A similar semantics can be found in [11] for the definition of canonical model.

[5] The definition of **S5** models corresponds directly to that described in [5,10].

[6] Here, the symbol $\biguplus$, introduced by [5], signifies the operation of taking a disjoint union.

Observing only up to where the label contains c or is $[]$ is insufficient. Consider the state $w^{[ad][ab][cd][ae]}$ in model $\mathfrak{M}$ (see Fig. 1). The agent d knows only that $[ad]$ and $[cd]$ occurred consecutively, as d was not involved in $[ab]$, $[ae]$. The agent b is unaware of $[cd]$, $[ae]$ but knows $[ad]$, since a shares all knowledge with b, including that a and d carry equivalent information. Thus, b is only certain that $[ab]$ and $[ad]$ occurred. Therefore, to evaluate $\mathrm{K}_d\varphi$ and $\mathrm{K}_b\varphi$ at state $w^{[ad][ab][cd][ae]}$, it is crucial to interpret these evaluations at states $w^{[ad][cd]}$ and $w^{[ad][ab]}$. A more sophisticated approach is needed for managing the world's label.

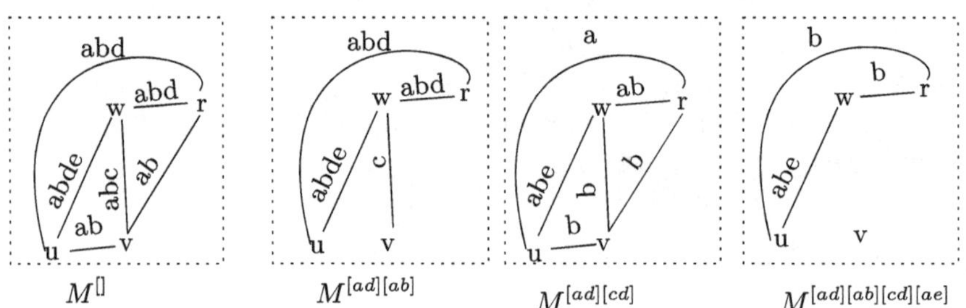

Fig. 1. Within the space model $\mathfrak{M}$, some resolved models are presented. The content within the dashed box constitutes a resolved model, solid lines denote accessibility relations (reflexivity is omitted), and w, u, r, v are states in the model (label is omitted).

Definition 3 (refined sequence). *Given any sequence* $\mathrm{H} \in \mathcal{I}$ *and any agent* $i \in \mathrm{Agt}$, *a refined sequence of* H *over* i, *denoted* $r_i(\mathrm{H})$, *is defined by a ternary function* $f : \mathrm{Agt} \times \mathcal{I} \times \mathcal{P}(\mathrm{Agt}) \to \mathcal{I}$. *For clarity, we use* $f_i(\mathrm{H}, X)$ *instead of* $f(i, \mathrm{H}, X)$. *The function* f *is defined as follows.*

$$f_i(\mathrm{H}, X) = \begin{cases} f_i(\mathrm{H}^-, X), & \text{if } \mathrm{H} \neq [] \text{ and} \\ & \{\mathrm{tail}(\mathrm{H})\} \cap X = \varnothing; \\ f_i(\mathrm{H}^-, X \cup \{\mathrm{tail}(\mathrm{H})\})\mathrm{tail}(\mathrm{H}), & \text{if } \mathrm{H} \neq [] \text{ and} \\ & \{\mathrm{tail}(\mathrm{H})\} \cap X \neq \varnothing; \\ [], & \text{if } \mathrm{H} = []. \end{cases}$$

Now let $r_i(\mathrm{H}) := f_i(\mathrm{H}, \{i\})$.

For example, recall that sequence $\mathrm{S} = [ad][ab][cd][ae]$. The refined sequence $r_b(\mathrm{S})$ is calculated as follows.

$$\begin{aligned} r_b([ad][ab][cd][ae]) &= f_b([ad][ab][cd][ae], \{b\}) \\ &= (f_b([ad][ab][cd], \{b\})) \\ &= (f_b([ad][ab], \{b\})) \\ &= (f_b([ad], \{a, b\}))[ab] \\ &= (f_b([], \{a, b, d\}))[ad][ab] \\ &= [][ad][ab] \end{aligned}$$

Given a state w^{H} in a space model, $w^{r_a(\text{H})}$ is its *refined state* for a, using $r_a(w^{\text{H}}) = w^{r_a(\text{H})}$ for simplicity. With these settings, for any w in a space model $\mathfrak{M}$ and any agent a, $r_a(w^{\text{H}}) = r_a(r_a(w^{\text{H}}))$ holds. The following properties are of particular interest.

Definition 4 (subsequence) *Let* $\text{G} = [G_1] \cdots [G_n]$. *A sequence* G' *is a subsequence of* G, *denoted* $\text{G}' \preceq \text{G}$, *if* G' *results from* G *by removing groups from* G.

For example, given a sequence $\text{H} = [ab][cde][bd]$, both $\text{H}' = [ab][bd]$ and $\text{H}'' = [ab]$ are subsequences of H ($\text{H}' \preceq \text{H}$ and $\text{H}'' \preceq \text{H}$), while $\text{H}''' = [bd][ab]$ is not (order changed).

Proposition 1 (perfect recall). *Let* $\mathfrak{M} = (W, \sim, V)$ *be a space model. For sequences* H *and* H' *such that* $\text{H} \preceq \text{H}'$, *and given states* w, u *in* $\mathfrak{M}$, *it holds for any agent* $i \in \text{Agt}$ *that if* $w^{\text{H}'} \sim_i u^{\text{H}'}$ *then* $w^{\text{H}} \sim_i u^{\text{H}}$.

Proof. The proof follows from the definition of consecutive resolved models.

Since $r_i(\text{H}) \preceq \text{H}$ for any $i \in \text{Agt}$, we obtain the following corollary from Proposition 1.

Corollary 1. *Let* $\mathfrak{M} = (W, \sim, V)$ *be a space model. Consider* $w^{\text{H}}, u^{\text{G}}$ *as two states in a space model* $\mathfrak{M}$. *For any agent* $a, b \in \text{Agt}$, $w^{\text{H}} \sim_a u^{\text{G}}$ *implies* $r_b(w^{\text{H}}) \sim_a r_b(u^{\text{G}})$.

The space model is the *disjoint union* of all resolved models, resulting in the following proposition.

Proposition 2 (refined states). *Let* $\mathfrak{M} = (W, \sim, V)$ *be a space model. Given any states* $w^{\text{H}}, u^{\text{G}}$ *in* $\mathfrak{M}$ *and* $a \in \text{Agt}$, $r_a(w^{\text{H}}) \sim_a u^{\text{G}}$ *implies* $r_a(u^{\text{G}}) = u^{\text{G}}$.

Proof. Let $w^{\text{H}}, u^{\text{G}}$ be arbitrary states in $\mathfrak{M}$. Then, $r_a(w^{\text{H}})$ is a refined state in the $r_a(\text{H})$-resolved model. Thus, if $r_a(w^{\text{H}}) \sim_a u^{\text{G}}$, then u^{G} is a refined state in the $r_a(\text{H})$-resolved model. They share the same label, producing $r_a(u^{\text{G}}) = u^{\text{G}}$.

Proposition 3. *Let* $\mathfrak{M} = (W, \sim, V)$ *be a space model. For states* $w^{\text{H}}, u^{\text{H}}$ *in* $\mathfrak{M}$, $w^{\text{H}} \sim_i u^{\text{H}}$ *iff* $r_i(w^{\text{H}}) \sim_i r_i(u^{\text{H}})$ *iff* $w \sim_{\{r_i(\text{H})\} \cup \{i\}} u$.

Proof. Let $\mathfrak{M}$ be a space model with two states $w^{\text{H}}, u^{\text{H}}$ in $\mathfrak{M}$. $r_i(\text{H})$ is in the form $f_i([], X)\text{H}'$ after calculations. Here, $\text{H}' = r_i(\text{H})$, $X = \{H'\}$, and $f_i([], X) = []$. We derive

$$\{(s^{\text{H}}, t^{\text{H}}) \mid s^{\text{H}} \sim_i t^{\text{H}}\} = \{(s^{\text{H}}, t^{\text{H}}) \mid s^{\text{H}'} \sim_i t^{\text{H}'}\} = \{(s^{\text{H}}, t^{\text{H}}) \mid s \sim_X t\}$$

under these conditions. The proposition is valid.

2.2 Semantics and Expressivity

We will now present our logic's semantics.

Definition 5 (semantics). *Let $\mathfrak{M} = (W, \sim, V)$ be a space model. Consider w^{H} as a state in $\mathfrak{M}$. Then we inductively define the notion of a formula φ being satisfied (or true) in $\mathfrak{M}$ at state w^{H} as follows:*

$$\begin{array}{lll}
\mathfrak{M}, w^{\text{H}} \models p & \text{iff} & w^{\text{H}} \in V(p); \\
\mathfrak{M}, w^{\text{H}} \models \neg\varphi & \text{iff} & \mathfrak{M}, w^{\text{H}} \not\models \varphi; \\
\mathfrak{M}, w^{\text{H}} \models (\varphi \wedge \psi) & \text{iff} & \mathfrak{M}, w^{\text{H}} \models \varphi \text{ and } \mathfrak{M}, w^{\text{H}} \models \psi; \\
\mathfrak{M}, w^{\text{H}} \models \mathrm{K}_a\varphi & \text{iff} & \text{for any state } u^{\text{G}} \text{ in } \mathfrak{M}, \\
 & & r_a(w^{\text{H}}) \sim_a u^{\text{G}} \text{ implies } \mathfrak{M}, u^{\text{G}} \models \varphi; \\
\mathfrak{M}, w^{\text{H}} \models [G]\varphi & \text{iff} & \mathfrak{M}, w^{\text{H}[G]} \models \varphi.
\end{array}$$

Saying φ is globally true in space $\mathfrak{M}$, notated $\mathfrak{M} \models \varphi$, if $\mathfrak{M}, w^{\text{H}} \models \varphi$ for any w^{H}; φ is valid, notated $\models \varphi$, if for any space model $\mathfrak{M}$, $\mathfrak{M} \models \varphi$.

Recall the initial assumptions mentioned in the introduction. These assumptions can be articulated with the semantics above. Consider a group $[ab]$, with $\mathfrak{M}$ as a model space and φ as a formula. For clause (i), it is easy to check $\mathfrak{M}, w^{[ab]} \models \mathrm{K}_a\varphi \leftrightarrow \mathrm{K}_b\varphi$. For clause (ii), suppose $\mathfrak{M}, w^{[]} \models \mathrm{K}_c\neg\mathrm{K}_a\varphi \wedge [ab]\mathrm{K}_a\varphi$. In this scenario, c remains convinced that the epistemic states of a remain unchanged, hence $\mathfrak{M}, w^{[]} \models [ab]\mathrm{K}_c\neg\mathrm{K}_a\varphi$. Furthermore, it is straightforward to confirm $\mathfrak{M}, w^{[]} \models [ab]\mathrm{K}_a\mathrm{K}_c\neg\mathrm{K}_a\varphi$, which means that during a private communication between a and b, a knows that c believes that a is unaware of φ.[7]

Roelofsen [9] introduces collective bisimulation, an invariance relation relevant to epistemic logic with distributed knowledge. The study demonstrates that collective bisimulation leads to modal equivalence. We expand this idea to space models.

Definition 6 (collective bisimulation) *Let $M = (S, \sim, V)$ and $M' = (S', \sim', V')$ be two models. A non-empty relation $Z \subseteq S \times S'$ is called a collective bisimulation between M and M', denoted by $M \rightleftharpoons^C M'$, if for all subsets $G \subseteq$ Agt, all $s \in S$ and $s' \in S'$ such that sZs', the following hold.*

- *(at) For all $p \in$ Prop, $s \in V(p)$ iff $s' \in V'(p)$;*
- *(zig) For all $t \in S$, if $s \sim_G t$, there is a $t' \in S'$ such that $s' \sim'_G t'$ and tZt';*
- *(zag) For all $t' \in S'$, if $s' \sim'_G t'$, there is a $t \in S$ such that $s \sim_G t$ and tZt'.*

We say that pointed models (M, s) and (M', s') are collectively bisimilar if there is a collective bisimulation Z between M and M' linking s and s'.

Now, we expand the concept to space models. Let $\mathfrak{M}$ and $\mathfrak{M}'$ be two space models based on $M = (W, \sim, V)$ and $M' = (W', \sim', V')$. We say that two states s^{H} and s'^{G} in $\mathfrak{M}$ and $\mathfrak{M}'$, respectively, are bisimilar, denoted $\mathfrak{M}, s^{\text{H}} \rightleftharpoons^C \mathfrak{M}', s'^{\text{G}}$, if (i) s^{H} and s'^{G} have the same label, and (ii) for any sequence $\text{I} \in \mathcal{I}$, $(M^{\text{I}}, s^{\text{I}})$ and $(M'^{\text{I}}, s'^{\text{I}})$ are collectively bisimilar.

---

[7] We thank the anonymous reviewers of DaLí for their insightful comments, which greatly improved this part.

Let Agt $= \{a, b, c\}$, $\mathfrak{M}$, and $\mathfrak{N}$ denote two space models (see Fig. 2). It is clear that for any sequence $\text{H} \in \mathcal{I}$, $(M^{\text{H}}, w^{\text{H}}) \leftrightarrows^C (N^{\text{H}}, r^{\text{H}})$. Thus, $(\mathfrak{M}, w^{[ab]}) \leftrightarrows^C (\mathfrak{N}, r^{[ab]})$.

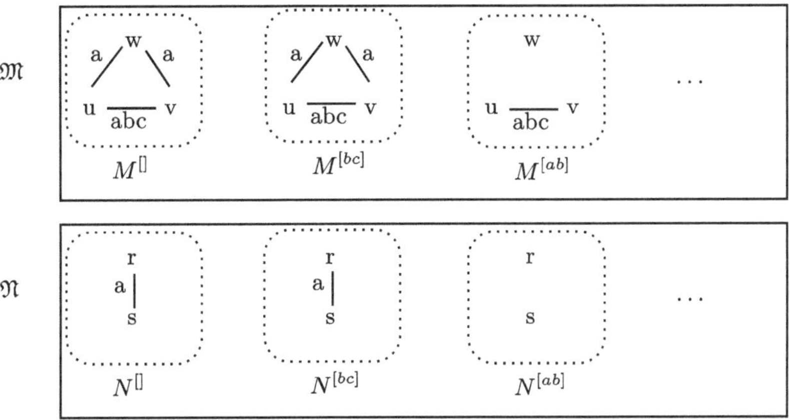

Fig. 2. Within the two space models $\mathfrak{M} = (W, \sim, V)$ and $\mathfrak{N} = (S, \sim, V')$, where $V(p) = V'(p) = \varnothing$ for any $p \in$ Prop, and some resolved models are presented. The content within the dashed box constitutes a resolved model, solid lines denote accessibility relations (reflexivity is omitted), and w, u, v, r, s are states in the model (label is omitted).

Proposition 4 (invariance) *Consider $(\mathfrak{M}, s^{\text{H}})$ and $(\mathfrak{M}', s'^{\text{H}})$ are two pointed space models. $\mathfrak{M}, s^{\text{H}} \leftrightarrows^C \mathfrak{M}', s'^{\text{H}}$ implies that for any formula φ: $\mathfrak{M}, s^{\text{H}} \models \varphi$ iff $\mathfrak{M}', s'^{\text{H}} \models \varphi$.*

Proof. By induction on the degree of φ. The base case is straightforward in semantics. The Boolean cases are followed by the induction hypothesis. It remains to deal with modalities and private communication operators.

Try to show $\mathfrak{M}, s^{\text{H}} \models \text{K}_a\varphi$ iff $\mathfrak{M}', s' \models \text{K}_a\varphi$. From left to right, assume $\mathfrak{M}, s^{\text{H}} \models \text{K}_a\varphi$. By semantics, we have for all u^{G} in $\mathfrak{M}$: $r_a(s^{\text{H}}) \sim_a u^{\text{G}}$ implies $\mathfrak{M}, u^{\text{G}} \models \varphi$. As $\mathfrak{M}, s^{\text{H}} \leftrightarrows^C \mathfrak{M}', s'^{\text{H}}$, we find that $\mathfrak{M}, r_a(s^{\text{H}}) \leftrightarrows^C \mathfrak{M}', r_a(s'^{\text{H}})$. By the clause zig in Definition 6 and the induction hypothesis, for all u'^{G} in $\mathfrak{M}'$: $r_a(s'^{\text{H}}) \sim_a u'^{\text{G}}$ implies $\mathfrak{M}', u'^{\text{G}} \models \varphi$, hence $\mathfrak{M}', s'^{\text{H}} \models \text{K}_a\varphi$. For the opposite direction, use the clause zag in Definition 6.

For formulas of the form $[G]\varphi$, we have $\mathfrak{M}, s^{\text{H}} \models [G]\varphi$ iff $\mathfrak{M}, s^{\text{H}[G]} \models \varphi$. As $\mathfrak{M}, s^{\text{H}} \leftrightarrows^C \mathfrak{M}', s'^{\text{H}}$, we find that $\mathfrak{M}, s^{\text{H}[G]} \leftrightarrows^C \mathfrak{M}', s'^{\text{H}[G]}$. By the induction hypothesis, $\mathfrak{M}, s^{\text{H}[G]} \models \varphi$ iff $\mathfrak{M}', s'^{\text{H}[G]} \models \varphi$. That is, $\mathfrak{M}', s'^{\text{H}} \models [G]\varphi$.

3 Valid and Invalid Results

This section presents results that are valid and semantically clear. Next, we analyse the axioms **T, B, 4** and **5**. It is important to note that while space models are classified as **S5** models, our knowledge modality differs from traditional epistemic logic. We find that the knowledge modality lies between the conventional knowledge operator and the belief operator. Specifically, we show that the axioms **K, D, 4** and **5** are valid, while the **T** and **B** axioms are invalid. With certain constraints, the **T** axiom can be valid.

Proposition 5 (valid results) *Given $a, b, c, d \in \mathrm{Agt}$, $p \in \mathrm{Prop}$, and $\varphi \in \mathcal{L}_{EKP}$. The following formulas are valid:*

C1. $[aa]\varphi \leftrightarrow \varphi$ C2. $[ab]\varphi \leftrightarrow [ba]\varphi$

C3. $[ab][ab]\varphi \leftrightarrow [ab]\varphi$ C4. $[ab]p \leftrightarrow p$

C5. $[ab]\neg\varphi \leftrightarrow \neg[ab]\varphi$ C6. $[ab](\varphi \wedge \psi) \leftrightarrow ([ab]\varphi \wedge [ab]\psi)$

C7. $\mathrm{K}_c\varphi \leftrightarrow [ab]\mathrm{K}_c\varphi$, *if* $c \notin \{a, b\}$ C8. $[ab](\mathrm{K}_a\varphi \leftrightarrow \mathrm{K}_b\varphi)$

C9. $\mathrm{K}_c(\varphi \to \psi) \to (\mathrm{K}_c\varphi \to \mathrm{K}_c\psi)$

General group cases C1 to C9 are valid and omitted.[8]

Proof. The proof follows from semantics; we demonstrate C7. Given a pointed space model $(\mathfrak{M}, w^{\mathrm{H}})$, we have the following.

$\mathfrak{M}, w^{\mathrm{H}} \models \mathrm{K}_c\varphi$
iff for any u^{G}: $r_c(w^{\mathrm{H}}) \sim_c u^{\mathrm{G}}$ implies $\mathfrak{M}, u^{\mathrm{G}} \models \varphi$
iff for any u: $r_c(w^{\mathrm{H}[ab]}) \sim_c u^{\mathrm{G}}$ implies $\mathfrak{M}, u^{\mathrm{G}} \models \varphi$
as $r_c(w^{\mathrm{H}}) = r_c(w^{\mathrm{H}[ab]})$ from $c \notin \{a, b\}$
iff $\mathfrak{M}, w^{\mathrm{H}[ab]} \models \mathrm{K}_c\varphi$
iff $\mathfrak{M}, w^{\mathrm{H}} \models [ab]\mathrm{K}_c\varphi$

We present findings on the **S5** properties. We assume that the agents are arbitrary in the subsequent propositions.

Proposition 6 (introspection). *For any formula $\varphi \in \mathcal{L}_{EKP}$, both $\mathrm{K}_a\varphi \to \mathrm{K}_a\mathrm{K}_a\varphi$ and $\neg\mathrm{K}_a\varphi \to \mathrm{K}_a\neg\mathrm{K}_a\varphi$ are valid.*

Proof. Try to show that $\mathrm{K}_a\varphi \to \mathrm{K}_a\mathrm{K}_a\varphi$ is valid. We show its diamond version: $\models \hat{\mathrm{K}}_a\hat{\mathrm{K}}_a\varphi \to \hat{\mathrm{K}}_a\varphi$. Given a pointed space model $(\mathfrak{M}, w^{\mathrm{H}})$, assume $\mathfrak{M}, w^{\mathrm{H}} \models \hat{\mathrm{K}}_a\hat{\mathrm{K}}_a\varphi$. Then, by semantics, for some u^{S} and v^{G}: $r_a(w^{\mathrm{H}}) \sim_a u^{\mathrm{S}}$, $r_a(u^{\mathrm{S}}) \sim_a v^{\mathrm{G}}$, and $\mathfrak{M}, v \models \varphi$. Now we have $r_a(r_a(w^{\mathrm{H}})) \sim_a r_a(u^{\mathrm{S}})$ and $r_a(r_a(u^{\mathrm{S}})) \sim_a r_a(v^{\mathrm{G}})$ by Corollary 1, while we have $r_a(r_a(w^{\mathrm{H}})) = r_a(w^{\mathrm{H}})$ and $r_a(r_a(u^{\mathrm{S}})) = r_a(u^{\mathrm{S}})$ by Definition 3. That gives $r_a(w^{\mathrm{H}}) \sim_a r_a(u^{\mathrm{S}})$ and $r_a(u^{\mathrm{S}}) \sim_a r_a(v^{\mathrm{G}})$, then by transitivity of the resolved model, we have $r_a(w^{\mathrm{H}}) \sim_a r_a(v^{\mathrm{G}})$ and $\mathfrak{M}, v^{\mathrm{G}} \models \varphi$. By Proposition 2, we have $r_a(v^{\mathrm{G}}) = v^{\mathrm{G}}$ from $r_a(u^{\mathrm{S}}) \sim_a v^{\mathrm{G}}$. Then it follows $\mathfrak{M}, w^{\mathrm{H}} \models \hat{\mathrm{K}}_a\varphi$.

Try to show that $\neg\mathrm{K}_a\varphi \to \mathrm{K}_a\neg\mathrm{K}_a\varphi$ is valid. Given a pointed space model $(\mathfrak{M}, w^{\mathrm{H}})$, assume $\mathfrak{M}, w^{\mathrm{H}} \models \neg\mathrm{K}_a\varphi$. Then, by semantics, there is u^{S} in $\mathfrak{M}$ such that $r_a(w^{\mathrm{H}}) \sim_a u^{\mathrm{S}}$ and $\mathfrak{M}, u \models \neg\varphi$. Since we have $r_a(w^{\mathrm{H}}) \sim_a u^{\mathrm{S}}$ implies $r_a(u^{\mathrm{S}}) = u^{\mathrm{S}}$ for any u^{S} in $\mathfrak{M}$. By the Euclidean of the resolved model, for any v^{G}: $r_a(w^{\mathrm{H}}) \sim_a v^{\mathrm{G}}$ implies $v^{\mathrm{G}} \sim_a u^{\mathrm{S}}$. It shows that for any v^{G}, $r_a(w^{\mathrm{H}}) \sim_a v^{\mathrm{G}}$ implies $\mathfrak{M}, v^{\mathrm{G}} \models \neg\mathrm{K}_a\varphi$ from $r_a(v^{\mathrm{G}}) = v^{\mathrm{G}}$. That is, $\mathfrak{M}, w^{\mathrm{H}} \models \mathrm{K}_a\neg\mathrm{K}_a\varphi$.

[8] For instance, the general case of C7, $\mathrm{K}_a\varphi \leftrightarrow [\mathrm{G}]\mathrm{K}_a\varphi$ is valid if $a \notin \{\mathrm{G}\}$.

Proposition 7 (no T axiom) $\not\models \mathrm{K}_a\varphi \to \varphi$.

Proof. Recall Fig. 1 and $V(p) = \{w^{\mathrm{H}}, u^{\mathrm{H}} \mid \mathrm{H} \in \mathcal{I}\}$. Let $\varphi = \neg\mathrm{K}_a p$. We have $\mathfrak{M}, w^{[ad][ab][cd][ae]} \models \mathrm{K}_b\neg\mathrm{K}_a p$ since $r_b(w^{[ad][ab][cd][ae]}) = w^{[ad][ab]}$ and $\mathfrak{M}, w^{[ad][ab]} \models \neg\mathrm{K}_a p$. However, we have $\mathfrak{M}, w^{[ad][ab][cd][ae]} \models \mathrm{K}_a p$, as $r_a(w^{[ad][ab][cd][ae]}) = w^{[ad][ab][cd][ae]}$ and for any state $u^{[ad][ab][cd][ae]}$ in $\mathfrak{M}$, $w^{[ad][ab][cd][ae]} \sim_a u^{[ad][ab][cd][ae]}$ implies $\mathfrak{M}, u^{[ad][ab][cd][ae]} \models p$.

Proposition 8. (no B axiom) $\not\models \varphi \to \mathrm{K}_c\hat{\mathrm{K}}_c\varphi$.

Proof. Recall Fig. 1 and $V(p) = \{w^{\mathrm{H}}, u^{\mathrm{H}} \mid \mathrm{H} \in \mathcal{I}\}$. Let $\varphi = \mathrm{K}_a p$. In the proof of Proposition 7, we have shown $\mathfrak{M}, w^{[ad][ab][cd][ae]} \models \mathrm{K}_a p$. However, we have $\mathfrak{M}, w^{[ad][ab][cd][ae]} \models \hat{\mathrm{K}}_b\mathrm{K}_b\hat{\mathrm{K}}_a\neg p$, as $r_b(w^{[ad][ab][cd][ae]}) = w^{[ad][ab]}$, $w^{[ad][ab]} \sim_b w^{[ad][ab]}$, and for any state $u^{[ad][ab]}$ in $\mathfrak{M}$, $w^{[ad][ab]} \sim_b u^{[ad][ab]}$ implies there is $r^{[ab][ad]}$ such that $r_b(u^{[ad][ab]}) \sim_a r^{[ab][ad]}$ and $\mathfrak{M}, r^{[ab][ad]} \models \neg p$.

The Propositions 7 and 8 indicate that it is difficult to interpret our modal operator K_a as a standard knowledge operator. Instead, it is better characterised as a belief operator due to the validity of the **D** axiom.

Proposition 9. (D axiom) $\models \mathrm{K}_a\varphi \to \hat{\mathrm{K}}_a\varphi$.

Proof. The proof is based on semantics and reflexivity.

We represent the modal operator as K_a instead of B_a due to constraints aligning its behaviour with that of a knowledge operator. In the absence of private communication, the axiom **T** is valid (see Proposition 10). The axiom **T** remains valid when formulas are constrained within the *positive fragment* of $\mathcal{L}_{EKP}$ (see Definition 7 and Proposition 11).

Consider a scenario in which Agt has a cardinality of 3 and the agent a engages in private communication with at least two participants. In this case, for any formula φ, $\mathrm{K}_a\varphi \to \varphi$ is valid (see Proposition 12).

Proposition 10. *For any pointed space model* $(\mathfrak{M}, w^{[]})$, *any agent* $a \in \mathrm{Agt}$, *and any formula* $\varphi \in \mathcal{L}_{EKP}$, *we have* $\mathfrak{M}, w^{[]} \models \mathrm{K}_a\varphi \to \varphi$.

Proof. Let $(\mathfrak{M}, w^{[]})$ be a pointed space model, and let $\varphi \in \mathcal{L}_{EKP}$ be a formula. Suppose $\mathfrak{M}, w^{[]} \models \mathrm{K}_a\varphi$. By semantics, we have for all u^{H}: if $r_a(w^{[]}) \sim_a u^{\mathrm{H}}$, then $\mathfrak{M}, u^{\mathrm{H}} \models \varphi$. Given that $r_a(w^{[]}) = w^{[]}$ and with $\mathfrak{M}$ being reflexive, it follows that $\mathfrak{M}, w^{[]} \models \varphi$.

Definition 7 (the positive fragment of $\mathcal{L}_{EKP}$). *The positive fragment* $\mathcal{L}^+_{EKP}$ *of* $\mathcal{L}_{EKP}$ *is defined as:*

$$\mathcal{L}^+_{EKP} \ni \varphi ::= p \mid \neg p \mid (\varphi \wedge \varphi) \mid (\varphi \vee \varphi) \mid \mathrm{K}_a\varphi \mid [G]\varphi,$$

where $p \in \mathrm{Prop}$, $a \in \mathrm{Agt}$ *and* $G \subseteq \mathrm{Agt}$.

Lemma 1. *Let* $\mathfrak{M}$ *be a space model with states* w^{H} *and* w^{G}. *For any formula* $\varphi \in \mathcal{L}^+_{EKP}$, *if* $\mathrm{H} \preceq \mathrm{G}$, *then* $\mathfrak{M}, w^{\mathrm{H}} \models \varphi$ *implies* $\mathfrak{M}, w^{\mathrm{G}} \models \varphi$.

Proof. Consider a space model $\mathfrak{M}$ with two states w^{H} and w^{G}, where $\mathrm{H} \preceq \mathrm{G}$. Let $\varphi \in \mathcal{L}^{+}_{EKP}$ be a formula. Assume $\mathfrak{M}, w^{\mathrm{H}} \models \varphi$. We aim to show $\mathfrak{M}, w^{\mathrm{G}} \models \varphi$. By induction on φ. The basic cases follow straightforwardly, with C4 and C5 in Proposition 5. The Boolean cases and the private communication operator case are followed by the induction hypothesis. It remains to deal with modalities. Assume φ is of the form $\mathrm{K}_i\psi (i \in \mathrm{Agt})$. By assumption and semantics, for all states u^{S} in $\mathfrak{M}$: $r_i(w^{\mathrm{H}}) \sim_i u^{\mathrm{S}}$ implies $\mathfrak{M}, u^{\mathrm{S}} \models \psi$. Since $\mathrm{H} \preceq \mathrm{G}$, $r_i(\mathrm{H}) \preceq r_i(\mathrm{G})$. Given $\psi \in \mathcal{L}^{+}_{EKP}$, by the induction hypothesis, $\mathfrak{M}, u^{r_i(\mathrm{G})} \models \psi$. With Proposition 1, $\{v \mid r_i(w^{\mathrm{G}}) \sim_i r_i(v^{\mathrm{G}})\} \subseteq \{v \mid r_i(w^{\mathrm{H}}) \sim_i r_i(v^{\mathrm{H}})\}$. Thus, for all states v^{R} in $\mathfrak{M}$: $r_i(w^{\mathrm{G}}) \sim_i v^{\mathrm{R}}$ implies $\mathfrak{M}, v^{\mathrm{R}} \models \psi$. That is, $\mathfrak{M}, w^{\mathrm{G}} \models \mathrm{K}_i\psi$.

Proposition 11. $\models \mathrm{K}_a\varphi \rightarrow \varphi$*, where* $\varphi \in \mathcal{L}^{+}_{EKP}$.

Proof. Let $\varphi \in \mathcal{L}^{+}_{EKP}$ be a formula, and $(\mathfrak{M}, w^{\mathrm{H}})$ be a pointed space model. Assume $\mathfrak{M}, w^{\mathrm{H}} \models \mathrm{K}_a\varphi$. By semantics, we have for all u^{G}: $r_a(w^{\mathrm{H}}) \sim_a u^{\mathrm{G}}$ implies $\mathfrak{M}, u^{\mathrm{G}} \models \varphi$. With the reflexivity of $\mathfrak{M}$, $\mathfrak{M}, r_a(w^{\mathrm{H}}) \models \varphi$. Since $r_i(\mathrm{H}) \preceq \mathrm{H}$, then, by Lemma 1, $\mathfrak{M}, w^{\mathrm{H}} \models \varphi$.

The proposition 11 asserts that when an agent is said to "know" information about the world, this refers to epistemic knowledge. Conversely, when an agent recognises that another agent "does not know" specific information, this reflects belief rather than true knowledge. We can eliminate the restriction of $\varphi \in \mathcal{L}^{+}_{EKP}$ and present the following proposition.

Proposition 12. *Consider a non-empty subset* $G \subseteq \mathrm{Agt}$ *such that* $|G| > 1$*. If* $a \in G$*, then for any formula* $\varphi \in \mathcal{L}_{EKP}$*:* $\models [G](\mathrm{K}_a\varphi \rightarrow \varphi)$.

Proof. Let $(\mathfrak{M}, w^{\mathrm{H}})$ be an arbitrary space model, $G \subseteq \mathrm{Agt}$ be a non-empty subset such that $|G| > 1$, and $\varphi \in \mathcal{L}_{EKP}$ be a formula. By C6 in Proposition 5, it suffices to show $\mathfrak{M}, w^{\mathrm{H}} \models [G]\mathrm{K}_a\varphi \rightarrow [G]\varphi$, where $a \in G$. We assume that $a \in G$, along with the condition $\mathfrak{M}, w^{\mathrm{H}} \models [G]\mathrm{K}_a\varphi$. By C1 in Proposition 5, we have $\mathfrak{M}, w^{\mathrm{H}'} \models [G]\mathrm{K}_a\varphi$. Here, the sequence H' is derived from H by removing any groups that possess a cardinality of precisely 1. Consequently, it follows that $\mathfrak{M}, w^{\mathrm{H}'[G]} \models \mathrm{K}_a\varphi$. Given $|\mathrm{Agt}| = 3$, with $|G| > 1$, and $a \in G$, we can infer by the pigeonhole principle that $r_a(w^{\mathrm{H}'G}) = w^{\mathrm{H}'G}$. By semantics and reflexivity, we have $\mathfrak{M}, w^{\mathrm{H}'G} \models \varphi$, which follows $\mathfrak{M}, w^{\mathrm{H}'} \models [G]\varphi$. Thus, $\models [G]\mathrm{K}_a\varphi \rightarrow [G]\varphi$.

Proposition 12 can be extended beyond the constraint $|\mathrm{Agt}| = 3$ by instead requiring $|G| \geq |\mathrm{Agt}| - 1$. The proof for this generalised version follows a similar approach. For simplicity, the following corollary still retains the original constraint of $|\mathrm{Agt}| = 3$. Specifically, it asserts that for any pointed space model $(\mathfrak{M}, w^{\mathrm{H}})$, if $\mathfrak{M}, w^{\mathrm{H}} \models \mathrm{K}_a\varphi \wedge \neg\varphi$, then $\mathrm{tail}(\mathrm{H}) = [bc]$.[9]

Corollary 2. $\models (\mathrm{K}_a\varphi \wedge \neg\varphi) \rightarrow (\psi \leftrightarrow [bc]\psi)$.

[9] We thank the anonymous reviewers of DaLí for their insightful comments, which greatly improved this part.

Proof. Let φ be a formula, and $(\mathfrak{M}, w^{\mathrm{H}})$ be a pointed space model. Assume $\mathfrak{M}, w^{\mathrm{H}} \models \mathrm{K}_a\varphi \wedge \neg\varphi$. By C1 in Proposition 5, we have $\mathfrak{M}, w^{\mathrm{H}'} \models \mathrm{K}_a\varphi$. The sequence H' is derived from H by removing groups of cardinality 1. Then, $\mathrm{tail}(\mathrm{H}') \in \{[], [ab], [bc], [ac], [abc]\}$. By Proposition 12, we have $a \notin \{\mathrm{tail}(\mathrm{H}')\}$. By Proposition 10, we have $\mathrm{H}' \neq []$. Thus, $\mathrm{tail}(\mathrm{H}') = [bc]$. For any formula ψ, we have the following.

$$\begin{aligned} & \mathfrak{M}, w^{\mathrm{H}'} \models \psi \\ \text{iff } & \mathfrak{M}, w^{\mathrm{H}'^-} \models [bc]\psi \\ \text{iff } & \mathfrak{M}, w^{\mathrm{H}'^-} \models [bc][bc]\psi \text{ (by C3 in Proposition 5)} \\ \text{iff } & \mathfrak{M}, w^{\mathrm{H}'} \models [bc]\psi \end{aligned}$$

4 Related Work

Renne et al. [8] discussed a scenario in which agents can misjudge time. This involves a public announcement of p communicated simultaneously to agents a and b. The agent b then receives an asynchronous semiprivate message about q. In this example, agent a is uncertain whether the second announcement has occurred. [8, p.825–826.] The models in [8] are updated by product updates. In this context, there is a significant link between the second and first update points, denoted by the relation R_a. Thus, this logic resembles epistemic logic, validating the axioms **K**, **T**, **4**, and **5**.

A key distinction between our logic and that in [8] is our assumption that agents outside G believe such communication does not occur. Thus, in space models, epistemic uncertainty does not link states with distinct labels.

Baltag et al. [3] discussed private announcements. Suppose c briefly falls asleep. During this time, a and b converse, expressing uncertainty about their cleanliness. Upon awakening, c is unaware of the conversation and does not consider a and b possess this interaction as common knowledge. [3, p.774] A Kripke structure with points l and t formalises the private announcement. Transitions are defined as $l \to l$ for agents a, b, $l \to t$ for c, and $t \to t$ for all agents a, b, c. Here, l represents the actual announcement known to agents a and b. Meanwhile, agent c believes that t is the only possible *null* announcement. [3, p.781–782]

The research [3] significantly parallels our study. A key difference lies in update contexts. In his framework, the real world is inaccessible to the agent c. In contrast, our approach retains reflexivity in the resolved model, but keeps the real world hidden from agents who did not engage in the latest private communication. Semantic interpretation requires operations on an alternative model. Although some epistemic uncertainties may seem unused in the updated model, their presence is crucial for seamless and continuous updates. Upon updating, the focus is solely on the current resolved model. In contrast, Baltag's method links the present model closely to the previous one. Subsequent updates further exacerbate the complexity, making the model increasingly difficult to describe. In addition, we are doing different types of communication: In our approach, private communication denotes a situation where an agent reveals all information to another, while a private announcement refers to a scenario where an agent discloses certain information to another.

5 Conclusion and Future Work

We present labelled relational semantics (*space models*) for private communication, enabling agents to share information confidentially. The **T** axiom is not universally valid. Knowing a (modal-free) proposition confirms its factual nature, unlike knowing a modal formula. This semantics serves to analyse various types of knowledge. The distributed knowledge $D_C\varphi$ is interpreted differently. $\mathfrak{M}, w \models D_C\varphi$ requires extending the refinement function r_i to r_C. The literature addresses studies on distributed knowledge and communication [7,9]. The interactions between group knowledge and private communication need further study.

A group has greater 'ability' than its proper subgroups in our logic. Given $\text{Agt} = \{a, b, c\}$, any $\varphi \in \mathcal{L}_{EKP}$ implies that $[abc](\text{K}_b\varphi \to \varphi)$ is valid. However, for any $n \in \mathbb{N}$, $([ab][bc][ac])^n(\text{K}_b\varphi \to \varphi)$ is invalid.[10] In scenario $n = 0$, we reference Proposition 7, which applies directly. In contrast, if $n \geq 1$, the following formula is satisfiable.

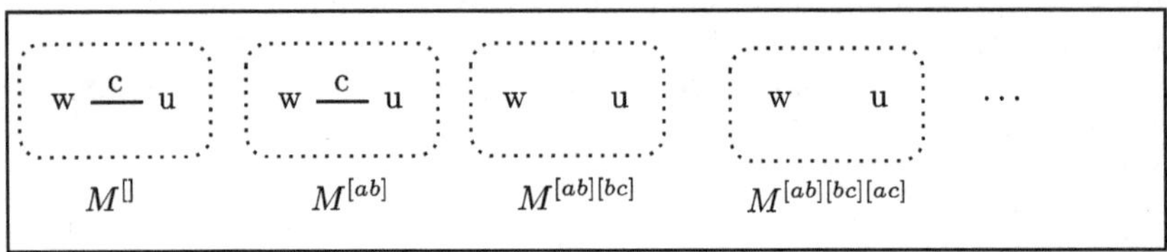

Fig. 3. Let $\text{Agt} = \{a, b, c\}$. Within the space model $\mathfrak{M} = (W, \sim, V)$ where $V(p) = \{w^{\text{H}} \mid \text{H} \in \mathcal{I}\}$, some resolved models are presented. The content within the dashed box constitutes a resolved model, solid lines denote accessibility relations (reflexivity is omitted), and w, u are states in the model (label is omitted).

$$([ab][bc][ac])^n(\text{K}_b\hat{\text{K}}_a\hat{\text{K}}_c(\hat{\text{K}}_b\hat{\text{K}}_a\hat{\text{K}}_c)^{n-1}\neg p \wedge \text{K}_a\text{K}_c(\text{K}_b\text{K}_a\text{K}_c)^{n-1}p)$$

Let $\varphi = \hat{\text{K}}_a\hat{\text{K}}_c(\hat{\text{K}}_b\hat{\text{K}}_a\hat{\text{K}}_c)^{n-1}\neg p$ and $(\mathfrak{M}, w^{[]})$ be a pointed space model (see Fig. 3). We claim $\mathfrak{M}, w^{[]} \models ([ab][bc][ac])^n(\text{K}_b\varphi \wedge \neg\varphi)$. We demonstrate this for $n = 1$ as follows. The general case follows similarly.

$$\begin{aligned}
&w^{[]} \models [ab][bc][ac](\text{K}_b\hat{\text{K}}_a\hat{\text{K}}_c\neg p \wedge \neg\hat{\text{K}}_a\hat{\text{K}}_c\neg p)\\
\text{iff } &w^{[]} \models [ab][bc][ac](\text{K}_b\hat{\text{K}}_a\hat{\text{K}}_c\neg p) \text{ and } w^{[]} \models [ab][bc][ac]\text{K}_a\text{K}_c p\\
\text{iff } &w^{[ab][bc][ac]} \models \text{K}_b\hat{\text{K}}_a\hat{\text{K}}_c\neg p \text{ and } w^{[ab][bc][ac]} \models \text{K}_a\text{K}_c p\\
\text{iff } &w^{[ab][bc]} \models \hat{\text{K}}_a\hat{\text{K}}_c\neg p \text{ and } w^{[ab][bc][ac]} \models \text{K}_c p\\
\text{iff } &w^{[ab]} \models \hat{\text{K}}_c\neg p \text{ and } w^{[ab][bc][ac]} \models p\\
\text{iff } &u^{[]} \models \neg p \text{ and } w^{[]} \models p
\end{aligned}$$

[10] Here, n denotes repetition times, e.g., $([ab][bc][ac])^2 = [ab][bc][ac][ab][bc][ac]$.

The above example shows that a group of three or more agents cannot achieve common knowledge through one-to-one calls. The group is more effective than its subgroups. Conversely, in Ågotnes and Wáng's work [12], formulas like $[ab][bc][ac](\mathrm{K}_a\varphi \leftrightarrow \mathrm{K}_b\varphi)$ and $[ab][bc][ac]\varphi \leftrightarrow [abc]\varphi$ are valid due to public resolution actions.[11] Axiomatisation and completeness proof for private communication logic are reserved for future work. In addition, the current model can be configured with varied initial assumptions to align with different logical interpretations. For example, substituting group announcement operators (Ågotnes et al. [1]) with private communication operators results in private group announcement logic. Another notable modification involves the incorporation of beliefs. This shift moves away from focussing solely on **S5** models, potentially uncovering more intriguing insights.

References

1. Ågotnes, T., Alechina, N., Galimullin, R.: Logics with group announcements and distributed knowledge: completeness and expressive power. J. Logic Lang. Inform. **31**(2), 141–166 (2022)
2. Baltag, A., Boddy, R., Smets, S.: Group knowledge in interrogative epistemology. In: Jaakko Hintikka on knowledge and game-theoretical semantics, pp. 131–164. Springer (2018)
3. Baltag, A., Moss, L.S., Solecki, S.: The logic of public announcements, common knowledge, and private suspicions. Springer (2016)
4. Baltag, A., Smets, S.: Learning what others know. arXiv preprint arXiv:2109.07255 (2021)
5. Blackburn, P., de Rijke, M., Venema, Y.: Modal logic. Cambridge University Press, New York (2001)
6. Fagin, R., Halpern, J.Y., Moses, Y., Vardi, M.: MIT press, reasoning about knowledge (1995)
7. Galimullin, R., Kuijer, L.B.: Varieties of distributed knowledge. In: Ciabattoni, A., Gabelaia, D., Sedlár, I., (eds.) Advances in Modal Logic, AiML 2024, Prague, Czech Republic, 19-23 Aug 2024, pp. 379–400. College Publications (2024)
8. Renne, B., Sack, J., Yap, A.: Logics of temporal-epistemic actions. Synthese **193**, 813–849 (2016)
9. Roelofsen, F.: Distributed knowledge. J. Appl. Non-Classical Logics **17**(2), 255–273 (2007)
10. van Ditmarsch, H., van der Hoek, W., Kooi, B.: Dynamic epistemic logic, vol. 337 of Synthese Library. Springer (2007)
11. Xiong, Z., Ågotnes, T.: The logic of secrets and the interpolation rule. Ann. Math. Artif. Intell. **91**(4), 375–407 (2023)
12. Ågotnes, T., Wáng, Y.N.: Resolving distributed knowledge. Artif. Intell. **252**, 1–21 (2017)

[11] They use different notations for those formulas, like $R_{\{a,b\}}\varphi$ for $[ab]\varphi$, $R_{\{a,b\}}R_{\{b,c\}}R_{\{a,c\}}\varphi$ for $[ab][bc][ac]\varphi$, and so on.

A Dynamic Logic of Subjective Belief

Tim French(✉)

School of Physics, Mathematical and Computing, The University of Western Australia, Perth, Australia
tim.french@uwa.edu.au

Abstract. We consider a first order logic of subjective belief, where an agent's judgements are modelled as expectations of hypothetical events, like drawing marbles from an urn. Logical propositions are statements of identity over these judgements, where two identical judgements necessarily have the same expectation. We discuss some expressivity results and show how such a logic can represent everyday decision making processes, such as choosing a restaurant or select a commuting option. We then introduce a dynamic operator to represent learning from experience, that allows an agent to condition one judgement, based on observations or experiences. An intuitive semantics is presented for this operation and we show that learning from experience may be syntactically encoded in the base language, allowing proof theoretic and complexity results to be inherited. Finally we discuss how we can encode an agent's confidence in their judgements, how this confidence impacts their learning and how their confidence may evolve over time.

Keywords: Dynamic Epistemic Logic · Probabilistic Reasoning · Experiential Learning

1 Introduction

An agent's subjective beliefs are often formed from limited experience, but are relied on to make future decisions when there is no clear decisive factor to rely on. We are interested in simple everyday experiences, like choosing meals or predicting sporting results, where we rely on a "gut-feeling" rather than any scientific approach. These judgements can play an important part in every day decision making: when confronted with several alternatives we might imagine these alternatives as hypothetical events, playing out simulations based on previous experiences, and choosing the alternative most likely to be favourable.

In this work we consider the problem of formalising this decision making process using first order aleatoric logic [8,9]. Aleatoric logic has atoms corresponding to Bernoulli tests (coin flips) so complex propositions correspond to complex events, like "the blue coin landed heads twice in row before the red coin did". The typical interpretation of an aleatoric proposition is the expectation the event will be observed when it is tested, and a logic can be defined based on whether the expectation of two propositions is necessarily identical. Here, we extend the notion of aleatoric logic to its first

J. Wang et al. (Eds.): DaLí 2025, LNCS 16472, pp. 180–197, 2026.
https://doi.org/10.1007/978-3-032-22626-6_11

order setting, so rather than atoms corresponding to Boolean events, atoms are given by predicates over some domain, and elements of the domain are sampled from some distribution. In this setting the hypothetical events an agent uses to form a judgement can be imagined as drawing marbles from an urn, and complex propositions describe a protocol by which the marbles are sampled.

The purpose of examining such a logic is to build a robust and fundamental approach for reasoning about uncertain beliefs. The urn in the example represents an agent's internal state of mind, or imagination. Following Williamson's *Knowing by Imagining* [28] we suppose agents maintain a probability space of possibilities, and when faced with an uncertain dilemma, draw samples from this probability space, imagining the possible outcomes, and the likelihood of a favourable return.

In this paper we present a formal syntax and semantics for representing aleatoric propositions, discuss some applications of this process, and show how an agent's beliefs could be conditioned on experience. Given the formal reasoning capabilities of aleatoric logic, this provides a unique capability to formally reasoning about the subjective learning process.

1.1 Related Work

There has been a considerable number of works that have considered probabilistic semantics. Early work includes Kolmogorov [15], Ramsey [25] and De Finetti [4], who produced axioms for reasoning about probabilities of events. Subsequently there have been a number of very good works that have examined logics of probabilistic reasoning [10, 13, 17, 21]. These approaches include a modality for the probability of some event occurring. As the probabilities are explicit in the syntax, these approaches reason about the probabilities of events, in the sense that while a pair of dice landing one is an uncertain statement, the statement "a roll of one has probability $\frac{1}{6}$" is a Boolean (true/false) statement.

Many valued logics such as logics of possibility [27], fuzzy logics [12, 29] and many-valued modal logics [20] interpret elements of the language as being neither wholly true, nor wholly false, but allocate a degree of truth to statements. However, as a logic is typically defined as a set of true propositions, this requires some kind of thresholding. In the basic logic BL [12], the threshold is taken as one, but other approaches such as rational Pavelka logic (RPL) [22] are more flexible. These logics formalise reasoning about vagueness, but are not naturally probabilistic, as the residual operator does not have a probabilistic interpretation.

We are interested in the way an agent assimilates new experiences and observations into their beliefs. There is a substantial literature in this area in the Boolean case, including Belief Revision [11] and Dynamic Epistemic Logics [6]. Probabilistic learning has been considered in such cases as Bayesian Description Logic [3], Markov Logic Networks [26], where atomic propositions and predicates may be conditioned based on observation, and in Probabilistic Dynamic Epistemic Logic [16]. In these cases, probabilities are explicit syntactic elements, in the sense of [13]. Fuzzy dynamic logics are considered in [2] over (non-probabilistic) Gödel algebras, and [19] extends the classic notion of propositional dynamic logic [14] to the many valued setting using action lattices corresponding to a variety of algebras.

Probabilistic logic programming [5,18] represents uncertain reasoning via a probability distribution of logic programs, and has formalised a process of learning from information.

Baltag, Rad and Smets [1] present a proposal for a logical foundation for statistical learning, with some preliminary results including convergence results for learning operations. A foundation for aleatoric reasoning can be found in [7,9]. There are many logics for reasoning about uncertainty, but the aleatoric logic we investigate here appears unique in its ability to combine deductive reasoning and probabilistic inference in an algebraic setting.

2 A First Order Logic for Subjective Belief

Here we present a syntax and semantics for describing an agent's subjective beliefs through an analogy of hypothetical events.

The essential elements of the language are a probability space consisting of the domain elements (the urn of marbles), a set of predicates over that domain, a set of propositions corresponding to independent random events (flips of biased coins), and operations describing events built from these elements (Boolean combinations, sampling from the domain and fixed point iteration). The propositions are built from propositional atoms and predicates, along with the operators *not* ($\neg$), *and* ($\wedge$), *(aleatoric) or* ($\bowtie$), *expectation* ($\mathbb{E}x$), and *fixed-point iteration* ($\mathbb{F}X$). Earlier work used a ternary *if-then-else* construction as a primitive operator [8]. Here we show that binary operations are sufficient, given the non-standard (aleatoric) interpretation of *or*.

2.1 Syntax

For the syntax we assume a set of *domain variables* $\mathcal{V}$, a set of *propositional atoms* $\mathcal{A}$, and a set of *predicates* $\mathcal{P}$ where each predicate $P \in \mathcal{P}$ has an arity $P^{\#}$ which is some positive integer.

Definition 1. *The syntax for* $\mathcal{L}$, *the language of* first order aleatoric logic, *is given by the Backus-Naur form:*

$$\alpha ::= X \mid P(x_1, ..., x_{P\#}) \mid \neg\alpha \mid \alpha \wedge \alpha \mid \alpha \bowtie \alpha \mid \mathbb{E}x.\alpha \mid \mathbb{F}X.\alpha$$

where $X \in \mathcal{A}$, $P \in \mathcal{P}$, $x \in \mathcal{V}$, *and* X is linear in α: *an atomic proposition* $X \in \mathcal{A}$ is linear in α *if and only if for every sub-formula* $\beta_1 \wedge \beta_2$ *of* α, X *does not occur free in both* β_1 *and* β_2, *and for every subformula* $\mathbb{F}Y.\gamma$ *of* α, X *does not occur free in* γ.

The fixed point operator, $\mathbb{F}X.\alpha$, required that X *is linear in* α, and this is to ensure that the fixed point is unique and non-ambiguous. This is a similar notion to monotonicity, which is used when defining fixed points in discrete domains. The meaning and significance of linearity will be discussed once the semantics have been presented.

We write $P(\overline{x})$ as an abbreviation for $P(x_1, ..., x_{P\#})$. We let free(α) refer to the set of unquantified domain variables appearing in α, and we let Free(α) refer to the set of propositional atoms that are not paired to a fixed point operator in α.

Each formula $\alpha \in \mathcal{L}$ can be thought of as a proposition describing a set of *hypothetical events*. In this setting a brief description of these operators is as follows:

- X, where $X \in \mathcal{A}$ is an atomic proposition, which describes some independent random event, like a coin landing heads.
- $P(\overline{x})$ is a predicate over the domain variables $x_1, ..., x_{P\#}$. The interpretation is fixed, so given $x_1, ...x_{P\#}$ it is either always true or always false. For example $red(x)$ is true when x has the colour *red*.
- $\neg\alpha$, describes the *failure* of an event to occur. That is, the event is explicitly tested for, and that test fails. For example $\neg red(x)$ is true when x is not *red*.
- $\alpha \wedge \beta$ describes the conjunction of two events where the event described by α occurs, *and* the event described by β occurs. For example, $red(x) \wedge round(x)$, describes an event where x is *red* and x is *round*.
- $\alpha \bowtie \beta$ describes the nondeterministic choice of two events where with 0.5 probability the event described by α occurs, *and* otherwise the event described by β occurs. For example, $red(x) \bowtie round(x)$, describes an event where a fair coin is flipped and if it lands heads, x is *red*, and if the coin lands tails x is *round*.
- The operator $\mathbb{E}x.\alpha$ is the *expectation* operator, and expresses the probability α will be true when x is drawn randomly from the domain. For example $\mathbb{E}x.red(x)$ describes the event when an element sampled from the domain is red.
- $\mathbb{F}X.\alpha$ is the *fixed point* operator, and it describes an event with probability p such that if the event corresponding to the atom X had probability p, then so would α. An alternative way to consider this operation is a description of a recursive event, where X is a proxy for $\mathbb{F}X.\alpha$, so the event X is substituted by the proposition $\mathbb{F}X.\alpha$, and is evaluated as the limit of this process of repeated substitution. In the event the process does not converge (e.g. $\mathbb{F}X.X$) it evaluates as $1/2$; the least commitment value.

As these events are generally hypothetical[1], they are not true or false, but instead have a probability, which may thought of as a degree of truth in the sense of fuzzy logics, or multi-valued logics. We note that they are, in general, *independent* events in that the occurrence of one event has no impact on the likelihood of subsequent events. However, the domain variables preserve some context between subformulas, so that subformulas containing the same variable may be considered dependent.

Example: Consider that we have an urn of coloured marbles, and we want to express the judgement that there are more red marbles than blue marbles. It is not allowed to count the marbles in the urn (indeed they may be infinite), but a representation of this judgement would be that, were we to repeatedly draw marbles from the urn (replacing them afterwards), we would draw a red one before we drew a blue one. This judgement *more-blue-than-red* can be represented as:

$$mrtb \cong \mathbb{F}X.\mathbb{E}x.((red(x) \vee X) \bowtie (\neg blue(x) \wedge X)), \tag{1}$$

where $\alpha \vee \beta$ means $\neg(\neg\alpha \wedge \neg\beta)$, as usual. In (1), the fixed point operation $\mathbb{F}X$ requires that each occurrence of X has the same expectation as the entire formula, so it may be

[1] We distinguish atomic propositions, X, which are evaluated as a probability, from predicates $P(\overline{x})$ which are a Boolean statement about things. This is to facilitate a recursive element of the language where propositions are generally evaluated as probabilities, and predicates describe crisp facts about things, which are sampled from a probability space.

considered to be a type of iteration. The expectation operator $\mathbb{E}x$ is a sampling from the domain. The operator $\bowtie$ is a nondeterministic choice for whether we will evaluate x for its *red*ness or *blue*ness. If we choose to evaluate the *red*ness, and x is *red*, then the test passes (or the hypothetical event *mrtb* occurs); otherwise the protocol is repeated. On the other hand, if we choose to evaluate the *blue*ness of x, and that test passes, then $\neg \mathit{blue}(x)$ will not pass, and so the hypothetical event *mrtb* does not occur. The evaluation of this proposition is probabilistic, and represents a likelihood that corresponds with there being more red marbles than blue.

2.2 Semantics

As aleatoric propositions are descriptions of hypothetical and independent events, the semantics define the degree of truth of a proposition to be precisely the probability of that event being hypothesised. We interpret aleatoric propositions over a first order domain, where domain elements are sampled from a probability space, propositional atoms are mapped to probabilities, and variables are mapped to domain elements.

Definition 2. *An* aleatoric interpretation *is given by the tuple:* $\mathcal{I} = (\mathcal{S}, \Sigma, \mu, \chi, \nu)$ *where:*

- $(\mathcal{S}, \Sigma, \mu)$ *is a probability space, where* $\mathcal{S}$ *is the set of domain elements,* $\Sigma \subset \wp(\mathcal{S})$ *is a* σ*-algebra over* $\mathcal{S}$*, and* $\mu : \Sigma \longrightarrow [0, 1]$ *is a probability measure.*
- $\chi \in [0, 1]^{\mathcal{A}}$ *assigns a probability to each atomic proposition.*
- $\nu \in \wp(\Sigma^*)^{\mathcal{P}}$ *assigns each predicate* $P \in \mathcal{P}$ *to a set of tuples over the* $\mathcal{S}$*, where all tuples in* $\nu(P)$ *have length* $P^\#$ *(i.e. for all* $w \in \nu(P)$*,* $|w| = P^\#$*), and the set of tuples in* $\nu(P)$ *is Lebesgue measurable with respect to* μ*.*

We note the map ν gives a first order interpretation of the set of predicates, $\mathcal{P}$, over the domain $\mathcal{S}$. That is, given $P \in \mathcal{P}$ and some assignment of variables to domain elements $a \in \mathcal{S}^{\mathcal{V}}$, the predicate $P(\overline{x})$ holds if $(a(x_1), \ldots, a(x_{P\#})) \in \nu(P)$, and $\mathcal{S}^*$ is the set of finite words over the alphabet $\mathcal{S}$.

Example: In the example above (1), the probability space would correspond to the urn of marbles, so $\mathcal{S}$ could be a finite set of marbles, Σ is the set of all subsets of $\mathcal{S}$ and for $T \subseteq \mathcal{S}$, $\mu(T) = |T|/|\mathcal{S}|$. There may be a single fair coin X as a propositional atom, where $\chi(X) = 1/2$, and two unary predicates red and blue, where $\nu(\mathit{red})$ is the set of (unary tuples of) red marbles, and similarly for $\nu(\mathit{blue})$.

Given an assignment $a \in \mathcal{S}^{\mathcal{V}}$, we write $v \mapsto s \in a$ if $a(v) = s$ and let

$$a[s \leftarrow v] = \{u \mapsto a(u) \mid u \neq v\} \cup \{v \mapsto s\},$$

be the assignment a updated so that s now maps to v.

Given an interpretation $\mathcal{I} = (\mathcal{S}, \Sigma, \mu, \chi, \nu)$, some atomic proposition X and some number $p \in [0, 1]$, let $\mathcal{I}[X : p]$ be the interpretation $(\mathcal{S}, \Sigma, \mu, \chi', \nu)$ where $\chi'(Y) = \chi(Y)$ for all $Y \in \mathcal{A} - \{X\}$ and $\chi'(X) = p$.

To interpret expectations (the probability sampled domain elements satisfy predicates) we are required to compute integrals over the probability space with respect to the given assignment. In a probability space, the domain may be thought of the set of events that can occur, while the σ-algebra is the set of measurable properties we can use the describe these events. To extend these properties to a logical language, we require the notion of a *Lebesgue Integral* which essentially allows us to sum the measure of events satisfying some property.

Definition 3. *Given a probability space $\Pi = (\mathcal{S}, \Sigma, \mu)$ and function $f : \mathcal{S}^V \longrightarrow [0,1]$, the* integral of *$f$ for some variable $v \in V$ is given by*

$$\int_{\mathcal{S}} f d\mu(v)(a) = \lim_{n\to\infty} \left[\sum_{i=1}^{n} \mu\left(\bigcup \left\{ S \in \Sigma \mid \forall s \in S, f(a[s \leftarrow v]) > \frac{i}{n} \right\} \right) /n \right].$$

Definition 3 is the standard Lebesgue integral over a probability space. We can now give an interpretation of aleatoric propositions where the semantics describe the likelihood of a proposition given an interpretation.

Definition 4. *Given some interpretation, $\mathcal{I} = (\mathcal{S}, \Sigma, \mu, \chi, \nu)$ and some $\alpha \in \mathcal{L}$, the* likelihood of α in $\mathcal{I}$ *is a function $\alpha^{\mathcal{I}} : \mathcal{S}^{\mathcal{V}} \longrightarrow [0,1]$, specified inductively as follows. Given some $a \in \mathcal{S}^{\mathcal{V}}$:*

$$(P(\overline{x}))^{\mathcal{I}}(a) = \begin{matrix} 1 \textit{ if } (a(x_1), ..., a(x_{P\#})) \in \nu(P), \\ 0 \textit{ if } (a(x_1), ..., a(x_{P\#})) \notin \nu(P). \end{matrix} \qquad X^{\mathcal{I}} = \chi(X)$$

$$(\neg\alpha)^{\mathcal{I}} = 1 - \alpha^{\mathcal{I}} \qquad (\alpha \wedge \beta) = \alpha^{\mathcal{I}} \cdot \beta^{\mathcal{I}}$$

$$(\alpha \bowtie \beta) = (\alpha^{\mathcal{I}} + \beta^{\mathcal{I}})/2 \qquad (\mathbb{E}x.\alpha)^{\mathcal{I}} = \int_{\mathcal{S}} \alpha^{\mathcal{I}} d\mu(x)$$

$$(\mathbb{F}X.\alpha)^{\mathcal{I}} = \begin{cases} 1/2 \textit{ if } \alpha^{\mathcal{I}[X:1/2]} = 1/2, \textit{ and} \\ p \textit{ if } \alpha^{\mathcal{I}[X:p]} = p \textit{ otherwise.} \end{cases}$$

We call $\alpha^{\mathcal{I}}$ the descriptive interpretation of α. *Note only the interpretation of predicates is dependent on the assignment $a \in \mathcal{S}^{\mathcal{V}}$. When the assignment $a \in \mathcal{S}^{\mathcal{V}}$ is clear from context, we write $\mathcal{I}(\alpha)$ in place of $\alpha^{\mathcal{I}}(a)$. Here, $n \cdot m$ represents "n* multiplied by *m".*

Example: In the evaluation of (1), $\mathbb{F}X.\mathbb{E}x.((\textit{red}(x) \vee X) \bowtie (\neg \textit{blue}(x) \wedge X))$: $(\textit{red}(x) \vee X)^{\mathcal{I}(a)}$ will evaluate to 1 if a maps x to a red marbles, and $\chi(X)$ otherwise; and $(\neg\textit{blue}(x) \wedge X)^{\mathcal{I}(a)}$ will evaluate to $\chi(X)$ if a doesn't map x to a blue marble, and 0 otherwise. Applying the $\bowtie$ gives the average of these two functions, and applying $\mathbb{E}x.$ takes the integral over the probability space so:

$$(\mathbb{E}x.((\textit{red}(x) \vee X) \bowtie (\neg\textit{blue}(x) \wedge X)))^{\mathcal{I}(a)} = (\mathtt{r} + ((1-\mathtt{r}) \cdot \chi(X)) + ((1-\mathtt{b}) \cdot \chi(X)))/2,$$

where $\mathtt{r}$ is the proportion of the marbles that are red and $\mathtt{b}$ is the proportion of the marbles that are blue. Finally, to evaluate $(\mathbb{F}X.\mathbb{E}x.((\textit{red}(x) \vee X) \bowtie (\neg\textit{blue}(x) \wedge X)))^{\mathcal{I}(a)}$, we solve for $\mathtt{X}$ in the equation $\mathtt{X} = (\mathtt{r} + ((1-\mathtt{r}) \cdot \mathtt{X}) + ((1-\mathtt{b}) \cdot \mathtt{X}))/2$, which evaluates as $\mathtt{r}/(\mathtt{r}+\mathtt{b})$; the chance of drawing a red marble before a blue one.

We note that this semantics is very different to the standard first order logical operations: $\wedge$ is not idempotent, and $\bowtie$ is very different to the logical notion of *or*. This is intended behaviour. We must show that the definition for $\mathbb{F}X.\alpha$ always exists and is non ambiguous.

Lemma 1. *Given an interpretation $\mathcal{I}$, and an aleatoric proposition α, where $\alpha^{\mathcal{I}[X:1/2]} \neq 1/2$, there is a unique $p \in [0,1]$ such that $\alpha^{\mathcal{I}[X:p]} = p$.*

Proof. We prove this by induction over the nestings of fixed point operators. The induction hypothesis is that where X is linear in α, $\alpha^{\mathcal{I}[X:p]}$ always has the form $f+g\cdot p, f$ and g are probability functions such that f is independent of p, and $f+g \leq 1$. If α does not contain any fixed point operators then the interpretation of α is a function constructed from constants between 0 and 1, the operations $(f+g)/2$, $1-f$, $f \cdot g$, and integrals over a probability space. Importantly, as X is linear in α (Definition 1) when functions f and g are multiplied together, X can only occur free in one of them. As addition and multiplication by a constant commute with integration the interpretation of $\alpha^{\mathcal{I}[X:p]}$ will always have the form: $f+g\cdot p$, where $f+g \leq 1$, as required. For the inductive step, we need to include the operation for fixed point operators. Given the induction hypothesis, this just requires solving the equality $p = f+g\cdot p$, which gives the operation $f/(1-g)$, where $g < 1$. (If $g = 1$, then $f = 0$, so $1/2$ would be a solution to the equality $p = p$). As X is linear in α, g will be independent of p, so the induction hypothesis holds.

Given these semantics we are able to define some familiar concepts as aleatoric propositions with $\mathcal{L}$. Table 1 contains some useful propositional abbreviations, but with non-standard semantics. Some other interesting, but more complex abbreviations are discussed below.

Table 1. Some basic abbreviations for aleatoric propositions.

Operator	Expression	Description	Operator	Expression	Description
$1/2$	$\mathbb{F}X.X$	*half*	$\alpha \vee \beta$	$\neg(\neg\alpha \wedge \neg\beta)$	α *disjunct* β
$\perp$	$\mathbb{F}X.(1/2 \wedge X)$	*false*	$\alpha \to \beta$	$\neg\alpha \vee \beta$	α *implies* β
$\top$	$\neg\perp$	*true*			

Ternary Conditionals. In [8,9] a primitive operation is the ternary operation $(\alpha\,?\,\beta : \gamma)$, which is interpreted as *if* α *then* β *else* γ. The semantic interpretation is $\mathcal{I}(\alpha)\cdot\mathcal{I}(\beta) + (1-\mathcal{I}(\alpha))\cdot\mathcal{I}(\gamma)$, and this is a common construct in many programming languages. We show that this operation can be expressed in $\mathcal{L}$:

$$(\alpha\,?\,\beta : \gamma) \equiv \mathbb{F}X. \begin{bmatrix} ((\alpha \wedge \beta) \vee X) \bowtie ((\alpha \to \beta) \wedge X) \\ \bowtie \\ ((\neg\alpha \wedge \gamma) \vee X) \bowtie ((\neg\alpha \to \gamma) \wedge X) \end{bmatrix} \tag{2}$$

This formula has an interesting form. The fixed point operation is an iterative construction that allows the formula to loop back to the start. The three instances of the aleatoric or operation ($\bowtie$) gives make a uniform random choice of four possibilities: check if α is true and β is true (if this event holds, report true and end, otherwise X, which means loop back to the top); or check if α is true and β is not, (if this event holds then

$(\alpha \to \beta) \wedge X$ must be false, so report false and end, otherwise loop back to the top); and analogously for $\neg\alpha$ and γ.

We can apply the Definition 4 to show that the correspondence (2) is semantically valid:

$$\begin{aligned}
\mathcal{I}((\alpha \wedge \beta) \vee X) &= (1 - \mathcal{I}(\alpha)\cdot\mathcal{I}(\beta))\cdot X + \mathcal{I}(\alpha)\cdot\mathcal{I}(\beta)\\
\mathcal{I}((\alpha \to \beta) \wedge X) &= (1 - \mathcal{I}(\alpha) + \mathcal{I}(\alpha)\cdot\mathcal{I}(\beta))\cdot X\\
\mathcal{I}((\neg\alpha \wedge \gamma) \vee X) &= (1 - \mathcal{I}(\gamma) + \mathcal{I}(\alpha)\cdot\mathcal{I}(\gamma))\cdot X + \mathcal{I}(\gamma) - \mathcal{I}(\alpha)\cdot\mathcal{I}(\gamma)\\
\mathcal{I}((\neg\alpha \to \gamma) \wedge X) &= (\mathcal{I}(\gamma) - \mathcal{I}(\alpha)\cdot\mathcal{I}(\gamma) + \mathcal{I}(\alpha))\cdot X
\end{aligned}$$

The aleatoric or operators take a uniformly weighted sum of the formulas (each is multiplied by 1/4):

$$\mathcal{I}\begin{pmatrix}[((\alpha \wedge \beta) \vee X)\\ \bowtie((\alpha \to \beta) \wedge X)]\\ \bowtie[((\neg\alpha \wedge \gamma) \vee X)\\ \bowtie((\neg\alpha \to \gamma) \wedge X)]\end{pmatrix} = \sum\begin{bmatrix}(\mathcal{I}(\alpha)\cdot\mathcal{I}(\beta) + (1-\mathcal{I}(\alpha))\cdot\mathcal{I}(\gamma))\\ (1-\mathcal{I}(\alpha)\cdot\mathcal{I}(\beta))\cdot\chi(X)\\ (1-\mathcal{I}(\alpha)+\mathcal{I}(\alpha)\cdot\mathcal{I}(\beta))\cdot\chi(X)\\ (1-\mathcal{I}(\gamma)+\mathcal{I}(\alpha)\cdot\mathcal{I}(\gamma))\cdot\chi(X)\\ (\mathcal{I}(\gamma)+\mathcal{I}(\alpha)-\mathcal{I}(\alpha)\cdot\mathcal{I}(\gamma))\cdot\chi(X)\end{bmatrix}/4 \quad (3)$$

so we can apply Definition 4 to derive:

$$\mathcal{I}\left(\mathbb{F}X.\begin{matrix}((\alpha \wedge \beta) \vee X)\bowtie((\alpha \to \beta) \wedge X)\\ \bowtie\\ ((\neg\alpha \wedge \gamma) \vee X)\bowtie((\neg\alpha \to \gamma) \wedge X)\end{matrix}\right) = \frac{(\mathcal{I}(\alpha)\cdot\mathcal{I}(\beta) + (1-\mathcal{I}(\alpha))\cdot\mathcal{I}(\gamma))}{4-3} \quad (4)$$

which completes the equivalence.

The ternary conditional is useful for representing several properties, including conditional probability. If we would like to define the event of sampling x so α holds *given* that β holds for x, we can represent this as: $(\alpha(x)|\beta(x)) \equiv \mathbb{F}X.\mathbb{E}x.(\beta\ ?\ \alpha : X)$.

Existential Quantification. The next expressive property we consider is existential quantification. The semantics is first order in nature, but instead of existential and universal quantification over the domain, we only have the expectation operator. We can show that this operator is able to represent existential quantification, under the assumption that in the probability space all non empty elements of the σ-algebra have positive measure. Given some formula α containing a free domain variable x, we would like to define some expression $\widehat{\exists x\alpha}$ so that given some assignment a, $\mathcal{I}(\widehat{\exists x\alpha})(a) = 1$ if there is an element s in the domain of $\mathcal{I}$ where $\mathcal{I}(\alpha)(a[x \leftarrow s]) > 0$, and otherwise $\mathcal{I}(\widehat{\exists x\alpha})(a) = 0$. $\widehat{\exists x\alpha}$ is defined by

$$\beta = \mathbb{F}X.(\mathbb{E}x.\alpha\ ?\ \top : X) \quad (5)$$

$$\gamma = \mathbb{F}Y.(\beta\ ?\ Y : \bot) \quad (6)$$

$$\widehat{\exists x\alpha} = \mathbb{F}Z.(\beta\ ?\ Z : \bot)\bowtie(\gamma\ ?\ Z : \top) \quad (7)$$

The formula β allows the domain to be repeatedly sampled for some x such that α is found to hold. This can only happen if there is some element s in the domain where $\mathcal{I}(\alpha)(a[x \leftarrow s]) > 0$, and in this case $\mathcal{I}(\beta) = 1$. If no such element exists then the fixed point is undetermined so $\mathcal{I}(\beta) = 1/2$. Therefore β on its own is not a suitable candidate for $\widehat{\exists x \alpha}$. The formula γ repeatedly executes β to see if it can be made to fail: that is, to check if there is no element s in the domain where $\mathcal{I}(\alpha)(a[x \leftarrow s]) > 0$. If β can be made to fail, then $\mathcal{I}(\gamma) = 1$, and otherwise $\mathcal{I}(\gamma) = 1/2$. Therefore, if γ ever fails, then there must be some element of the domain satisfying α, and if β ever fails, there cannot be any element of the domain satisfying α. Since these two options are mutually exclusive, $\widehat{\exists x \alpha}$ repeatedly executes either formula until one of them fails and returns the corresponding result.

Rational Probabilities. The presented language is subjective in that the propositions are interpreted probabilistically, and we do not, in the language, explicitly define the probability of any proposition. However, we are able to define propositions that have precise rational probabilities. Suppose we wish to define some proposition with probability n/m where $n < m$. The idea is to create a balanced binary tree using the aleatoric or operator, so each leaf in the tree is evenly weighted, and such that there are at least m leaves. We have n leaves labelled by $\top$, $m - n$ leaves labelled by $\bot$, and the remaining leaves labelled by X, where the entire tree is in the scope of a $\mathbb{F}X$ operator. For example, the proposition $\mathbb{F}X.(\top \bowtie \top) \bowtie (\bot \bowtie X)$ has likelihood $2/3$ in all interpretations. This is similar in effect to the fuzzy logic RPL [22] that is able to express propositions corresponding to rational values in a many valued logic. However, the mechanism by which this is achieved is quite different.

We can use a similar trick to represent the confidence in propositions. Replacing $\top$ by α and $\bot$ by $\neg\alpha$, we can represent the event of α occurring 2 times out of 3 $(\mathbb{F}X.(\alpha \bowtie \alpha) \bowtie (\neg\alpha \bowtie X))$. This does not mean the $\mathcal{I}(\alpha) = 2/3$; it is describing the event that α happen 2 times out of 3, which is reasonably likely (4/9) if $\mathcal{I}(\alpha) = 2/3$, and quite unlikely (9/64) if $\mathcal{I}(\alpha) = 1/4$.

Generally, we can define $\alpha^{\frac{n}{m}}$, given non-negative integers $n \leq m$ and some proposition α, to mean that α was tested m times and passed *at least* n times

$$\alpha^{\frac{0}{m}} \equiv \top \qquad \alpha^{\frac{n}{0}} \equiv \bot \qquad \alpha^{\frac{n}{m}} \equiv (\alpha \mathbin{?} \alpha^{\frac{n-1}{m-1}} : \alpha^{\frac{n}{m-1}}).$$

This is useful to represent confidence in a proposition. For example, a surgeon might recommend an expensive procedure only if the chance of success (given the sampling of hypothetical events in the surgeon's mind) is high, say $\mathit{success}^{\frac{4}{5}}$.

3 Example

Suppose that you have written a paper and you can send it to the Mathematical Computing Conference (MCC) or the Computational Mathematics Conference (CMC). You know that the program committee for MCC prefers theoretical contributions, while the program committee for CMC is more likely to accept empirical works. You judge that your paper is more theoretical than empirical, but you also believe that you are likely to need two out of three MCC reviewers to like your paper for it to be accepted, whilst one out of two CMC reviewers liking your paper is usually enough. As always, there's an element of chance to the paper being accepted, so even if the required number of reviewers like the paper, you imagine there's a coin flip (with some bias) to determine if the paper is eventually accepted.

This requires a subjective reasoning process which may be formalised as follows. We suppose that we have predicates: $mpc(x)$ for x *is on the MCC PC*; $cpc(x)$ for x *is on the CMC PC*; $pos(x)$ for x *considers your paper positively*; and the atomic proposition *mflip* (resp. *cflip*) to represent the chance of your paper getting through when enough mpc reviewers (resp. cpc reviewers) like it.

In this simple scenario, we can consider an interpretation $\mathcal{I} = (\mathcal{S}, \Sigma, \mu, \chi, \nu)$, where the domain $\mathcal{S}$ consists solely of potential reviewers (or reviewer archetypes), and the algebra, Σ, consists of all possible subsets of $\mathcal{S}$. The probability measure could be uniform for all singleton elements of the algebra, but could also reflect assumptions about the make up of the reviewer population. For example, if we suspect that two thirds of CMC reviewers also review for MCM, then the probability measure associated with reviewers of CMC and MCM might be twice that associated with reviewers of CMC who are not on the MCM program committee.

The atom assignment χ assigns the bias on the coin flip for each conference. The predicate assignment ν identifies the reviewers on each program committee, and whether they like the paper.

Then, with these assumptions laid out, we can express the likelihood of the paper being accepted at MCC and CMC, respectively:

$$\mathrm{M} = ((\mathbb{F}X.\mathbb{E}x.(mpc(x)\ ?\ pos(x) : X))^{2/3}\ ?\ \mathit{mflip} : \bot)$$
$$\mathrm{C} = ((\mathbb{F}X.\mathbb{E}x.(cpc(x)\ ?\ pos(x) : X))^{1/2}\ ?\ \mathit{cflip} : \bot).$$

In these propositions, we first check to see if sufficient reviewers view the paper positively. Within the scope of the fixed point operator $\mathbb{F}X.$, we sample from the population of reviewers ($\mathbb{E}x.$), and see if the sampled reviewer is on the program committee ($mpc(x)$), we check if they view the paper positively ($pos(x)$). If they are not on the program committee, then we use the fixed point atom X to resample the population. If we get two out of three positives (or one out of two for CMC), then our paper depends on the coin toss (*mflip*), and otherwise it is rejected ($\bot$).

To interpret this scenario numerically, we may suppose that there are five different "archetypes" of reviewers, as in Table 2, and $\chi(\mathit{mflip}) = 0.8$ and $\chi(\mathit{cflip}) = 0.9$. Applying the Definition 4 we find that $\mathrm{M}^{\mathcal{I}} = 0.4$ and $\mathrm{C}^{\mathcal{I}} = 0.459$.

Table 2. The reviewer archetypes from Sect. 3

Type	mpc	cpc	pos	μ
Programmer	Yes	Yes	Yes	0.2
Statistician	No	Yes	No	0.3
Algebrist	Yes	No	Yes	0.3
Logician	No	Yes	Yes	0.1
Engineer	Yes	No	No	0.1

In this example, we are not dealing with fact, or even any kind of meaningful evidence. Instead this is the reasoning of bias and assumption that is used in everyday decision making. We do not know who is on the program committees and we have no idea how they will receive the paper. Instead we make assumptions based perhaps on the one person we recognise on the program committee, or maybe based on the tone of language in the call for papers. This is not reasoning to be relied upon, but often we have little else to guide us. What is important is that this is reasoning that can be refined. As we experience more of the world, meet more reviewers, and maybe serve on program committees, we become wiser, and gain a better, more intricate, understanding of the process. This allows us to make better quality judgements, based on experience. This learning approach is addressed in Sect. 5.

4 Reasoning

In *Knowing by Imagining* [28] Williamson presents an argument for knowledge *as* imagination, rather than contrasting knowledge as verifiable fact against imagination as unconstrained fancy. The thesis presents a notion of imagination as synthesised from experience, and provides a mechanism for a primitive agent to make judgements when there is insufficient reason. An example given is a man deciding whether to jump a crevasse for a short cut, or take the long way round. Here the man takes their experience of jumps they have completed before, and falls they have had or seen, to imagine whether they can, or whether they will likely succeed in the jump. The man's subjective beliefs here come from the ability to sample scenarios informed by past experience. What the man imagines is a sample drawn from a distribution that has been conditioned by experience. The man may choose to take the risk if four times out of five he imagines success. As an aleatoric proposition: $(\mathbb{E}s.jump\_succeeds(s))^{\frac{4}{5}}$.

Similarly, in Sect. 3, we suppose an aspiring researcher wants a paper accepted by a conference, and has to select a venue. The example gives a very coarse numerical calculations that, given the biases and presumptions of the researcher, suggests one conference has better chance of success than the other.

This type of inductive reasoning relies on experience or past observation, and has no deductive element: in epistemological terms we are talking about an *a posteriori* proposition. The subjective beliefs of the agent may be considered *tacit beliefs*, in the sense of *tacit knowledge* [24]. This refers to beliefs an agent holds, but are difficult to communicate: the agent cannot peer into the metaphorical urn of marbles, nor share it with any one else. However, by considering the notion of identity over propositions, to mean that two propositions necessarily have the same likelihood, we are able to represent explicit knowledge, and deductive reasoning. This has been demonstrated in [7,9].

We now present a reasoning framework to motivate the following discussion of aleatoric learning: if we are able to represent the concept of learning from experience in a formal reasoning framework we can then consider verifying inductive processes.

Definition 5. *An* aleatoric identity *is a pair of aleatoric propositions $\alpha, \beta \in \mathcal{L}$, and is denoted $\alpha \simeq \beta$, with the intended meaning that α and β have exactly the same likelihood.*

Given some interpretation, $\mathcal{I}$, we say the identity $\alpha \simeq \beta$ is evident *in $\mathcal{I}$ if $\mathcal{I}(\alpha) = \mathcal{I}(\beta)$. If an identity is evident in all interpretations, we say it is a* valid identity.

We refer to a set of identities as a theory, *and given a theory $\mathcal{T}$ and some interpretation $\mathcal{I}$, such that every identity in $\mathcal{T}$ is evident in $\mathcal{I}$, we say $\mathcal{T}$ is evident in $\mathcal{I}$, written $\mathcal{I} \models \mathcal{T}$.*

With the concept of identity we are able to go beyond the extensional interpretation of what has been experienced, to the intensional interpretation of which propositions are equivalent.

For example, we can prove that for all propositions α the following identity holds: $\mathbb{E}x.\neg\alpha \simeq \neg\mathbb{E}x.\alpha$:

$$\begin{aligned} \mathcal{I}(\mathbb{E}x.\neg\alpha) &= \textstyle\int_{\mathcal{I}} 1 - \mathcal{I}(\alpha) d\mu(x) \\ &= 1 - \textstyle\int_{\mathcal{I}} \mathcal{I}(\alpha) d\mu(x) \quad = \quad \mathcal{I}(\neg\mathbb{E}x.\alpha) \end{aligned}$$

This shows that, like the fixed point operator, the expectation operator is also self-dual [7].

We can also represent a priori propositions in this way: the classic example that all bachelors are unmarried men could be captured in the identity:

$$\mathbb{E}x.bachelor(x) \simeq \mathbb{E}x.(man(x) \wedge \neg married(x)).$$

However, we must be cautious in this interpretation. The classic example describes ontological identity ($\vdash \alpha \leftrightarrow \beta$), whereas an aleatoric identity simply states equal likelihood. Therefore, $\alpha \simeq \neg\alpha$ makes sense as an aleatoric identity, and simply means that α is as likely as it is not, but $\vdash \alpha \leftrightarrow \neg\alpha$ is a contradiction in Boolean reasoning. The notions of ontological identity and aleatoric identity are not incompatible. It should be true that $\vdash \alpha \leftrightarrow \beta$ *implies* $\alpha \simeq \beta$, but the converse is not necessarily the case. Table 3 describes a deductive system for aleatoric logic. It demonstrates the intent of having a system that combines explicit reasoning, in the form of a deductive system, with tacit inference, in the form of a subjective belief base. The deductive system ($\mathfrak{AL}$) is known to be sound and conjectured to be complete. The following section will consider a learning framework that combines explicit reasoning and tacit inference.

5 Learning Operations

Having motivated the concept of an aleatoric interpretation as subjective belief, we turn our attention to belief revision or learning in an aleatoric context.

The agent's subjective knowledge base (the urn) allows the agent to simulate *hypothetical events*, with the presumption that the urn is some approximation of the external world. When the agent receives evidence from the external world, it is not hypothetical; rather it is a fact that the evidence was received (whether it was accurate or not). This is essentially a single bit of information, but still an experience that the agent can learn from. In the setting of the conference submission example (Sect. 3), a notification on the outcome of a paper can lead an agent to reassess their presumed biases. This general idea is certainly not novel: a logic for statistical learning [1] combines plausibility updates (similar to belief revision) and dynamic epistemic operators (for

Table 3. The deductive system, *aleatory logic*, ($\mathfrak{AL}$) consists of all substitutions of the following rules, where a substitution means that α, β, γ etc. are uniformly replaced by the some aleatoric proposition.

Name	Axiom	Name	Axiom
id	$\alpha \simeq \alpha$	$\wedge$**-com**	$\alpha \wedge \beta \simeq \beta \wedge \alpha$
$\wedge$**-ass**	$(\alpha \wedge \beta) \wedge \gamma \simeq \alpha \wedge (\beta \wedge \gamma)$	**dn**	$\neg\neg\alpha \simeq \alpha$
$\bowtie$**-dist1**	$(\alpha \bowtie \beta) \wedge \gamma \simeq (\alpha \wedge \gamma) \bowtie (\beta \wedge \gamma)$	$\bowtie$**-dist2**	$\neg(\alpha \bowtie \beta) \simeq (\neg\alpha) \bowtie (\neg\beta)$
swap	$(\alpha_1 \bowtie \alpha_2) \bowtie (\beta_1 \bowtie \beta_2) \simeq (\alpha_2 \bowtie \beta_2) \bowtie (\alpha_1 \bowtie \beta_1)$	**same**	$\alpha \bowtie \alpha \simeq \alpha$
simp	$(\alpha \wedge \beta) \bowtie \neg(\neg\alpha \wedge \neg\beta) \simeq \alpha \bowtie \beta$	**half**	$\mathbb{F}X..\alpha \simeq \neg\mathbb{F}X..\alpha$
amp	$\neg(\alpha \wedge \beta) \wedge \mathbb{F}X..X \simeq \neg\alpha \bowtie (\alpha \wedge \neg\beta)$	$\mathbb{F}$**-elim**	$\mathbb{F}X.\alpha \simeq \alpha[X \backslash \mathbb{F}X.\alpha]$
$\mathbb{E}$-$\wedge$	$\mathbb{E}x.\alpha \wedge \beta \simeq \mathbb{E}x.\alpha \wedge \beta$ where $x \notin$ free(β)	$\mathbb{E}$-$\neg$	$\neg\mathbb{E}x.\alpha \simeq \mathbb{E}x.\neg\alpha$
$\mathbb{E}$-$\bowtie$	$\mathbb{E}x.\alpha \bowtie \beta \simeq \mathbb{E}x.\alpha \bowtie \beta$ where $x \notin$ free(β)	$\mathbb{E}$**-intro**	$\alpha \simeq \mathbb{E}x.\alpha$ where $x \notin$ free(α)

Name	**Rule**
subst	From $\alpha \simeq \beta$ infer $\gamma[X \backslash \beta] \simeq \gamma[X \backslash \alpha]$
trans	From $\alpha \simeq \beta$ and $\beta \simeq \gamma$ infer $\alpha \simeq \gamma$
$\mathbb{F}$**-intro**	From $\alpha \simeq \beta[X \backslash \alpha]$ infer $\alpha \simeq \mathbb{F}X.B$, where $\beta \not\simeq X$
order	From $\alpha \bowtie \gamma \simeq \beta \bowtie \gamma$ infer $\alpha \simeq \beta$

acquiring new knowledge); [16] presents a probabilistic epistemic logic with learning via announcements; and [5] introduced a learning from information operation. In the setting of many valued logics, [19] considers dynamic logics of programs via action lattices. The approach here is novel in that it is experiential and rooted in Bayesian conditioning. Furthermore, we show it isexpressible in the base logic, allowing for the possibility of formally reasoning about learning.

5.1 Aleatoric Learning Protocol

An interpretation (Definition 2) represents an agent's subjective beliefs as a set of independent hypothetical events. These beliefs are built through experience, where an agent observes some external event, and this observation is then incorporated into their experience, and affects their beliefs. The observation is a formula: $\alpha(x)$, by which we suppose that a single element is sampled from the domain, and $\alpha(x)$ is tested, so it either happened, or it did not, and it is this single bit of information that is incorporated in the agent's subjective beliefs.

Aleatoric learning occurs via the probability function μ in the interpretation $\mathcal{I} = (\mathcal{S}, \Sigma, \mu, \chi, \nu)$, and all other parts of the interpretation are fixed. Aleatoric learning takes observations to bend the probability distribution of agent's subjective beliefs, to favour those that agree with observations.

The observations come from an environment, which is supposed to behave just like an *external* urn of marbles, while the agent (the observer) has their own set of beliefs, which we consider to be an *internal* urn of marbles. The agent would like to condition their beliefs to better agree with their observations. However, as drawing a marble from an urn is an independent event, and we do not have a prior distribution of beliefs (urns) it is not obvious how we might condition the agent's beliefs. We will describe a process here, given some property $\beta(x)$ (dependent on x), creates a simple distribution of beliefs to condition against.

Suppose that there is a very large number of marbles in the agent's internal urn, and the agent is able to draw a marble and test whether $\beta(x)$ holds for that marble, The agent can also observe the environment which gives them one bit of information at a time: whether $\alpha(e)$ was true for a marble e randomly sampled from the external urn. Note no information on subformulas is given, so the agent does not know *how* $\alpha(e)$ might be true. To learn from this one bit of information the agent uses the following process:

1. Take two empty urns, and sample a large number of marbles from their internal urn, in such a way that marbles satisfying $\beta(x)$ are more likely to be in the left urn, (it is β-*supportive*) while marbles not satisfying $\beta(x)$ are more likely to be in the right urn (it is β-*sceptical*).

2. Suppose a process where the agent flips a coin and: if it lands heads, they draw a single marble x from the left urn and test whether $\alpha(x)$ is true; and if it lands tails, they draw a single marble x from the right urn and test whether $\alpha(x)$ is true. This allows the agent to theorise about whether $\alpha(x)$ holds contingent on the bias of the coin. It essentially creates a new variable (the coin) and a process whereby the outcome of the coin is correlated with β.

3. The agent makes an observation of the environment, $\alpha(e)$, and applies Bayesian conditioning to estimate the probability p_ℓ that the marble e came from the left urn (so the chance the marble came from the right urn is $1 - p_\ell$). That is, they can condition their belief of the coin's bias on the observation.

4. To update the agent's beliefs they empty the marbles from the original urn, and now fill it with marbles taken from the left and right urn, where there is a p_ℓ chance that each next marble is taken from the left urn.

The original urn now contains a subset of marbles which represents the agent's updated probability distribution for β given the observation of α. We note that in the process the agent is *learning* β while they are *observing* α. It could be that α and β are the same, but it also may be the case that the agent is interested in what the observation α says about a different property (e.g. what the acceptance of a paper says about the makeup of a program committee).

This way we can say, the agent *learns* β *given the observation* α, and this new learning impacts their belief in some property γ. We express this syntactically as $[\beta|\alpha]\gamma$, where $\mathcal{I}([\beta|\alpha]\gamma)$ is the probability the interpretation $\mathcal{I}$ would assign γ after learning β given the observation of α.

We note that as with Dynamic Epistemic Logic [6], universal substitution does not hold in these extended semantics, as the interpretation where α is evaluated is not the same as the interpretation where γ is evaluated. However, we show that $[\beta|\alpha]\gamma$ is expressible in the base logic where universal substitution does hold.

5.2 Aleatoric Learning Semantics

To formalise the learning process described above we require the following definitions. We suppose that we are given an original interpretation $\mathcal{I} = (\mathcal{S}, \Sigma, \mu, \chi, \nu)$, and some observation $\alpha(x)$ occurs. As the learning takes place entirely with respect to the probability function μ, we will suppose that the interpretation aside from μ stays fixed, and we give a new probability function for the left urn, μ_ℓ, and a new probability function for the right urn, μ_r. The new probability functions are given by the following definitions.

First we need to be able to relativise the interpretation to an element of the algebra.

Definition 6. *Given an interpretation $\mathcal{I} = (\mathcal{S}, \Sigma, \mu, \chi, \nu))$ and $\sigma \in \Sigma$ is some element of the algebra Σ, we let the* relativisation of $\mathcal{I}$ to σ *be the interpretation*

$$\mathcal{I}^\sigma = (\sigma, \{\sigma' \in \Sigma |\, \sigma' \subseteq \sigma\}, \mu^\sigma, \chi, \nu)$$

where for all $\sigma' \subseteq \sigma$, $\mu^\sigma(\sigma') = \mu(\sigma')/\mu(\sigma)$ That is we just condition the probability measure μ^σ so that σ is certain and everything outside σ has probability 0.

The process of splitting the marbles into two urns is formalised as follows:

Definition 7. *Given an interpretation $\mathcal{I} = (\mathcal{S}, \Sigma, \mu, \chi, \nu)$, and some property $\beta(x)$ to be learnt, the β-splitting is the pair (μ_ℓ, μ_r) where:*

$$\mu_\ell(\sigma) = (1 + (\mathbb{E}x.\beta)^{\mathcal{I}^\sigma} - (\mathbb{E}x.\beta)^{\mathcal{I}}) \cdot \mu(\sigma), and$$
$$\mu_r(\sigma) = (1 - (\mathbb{E}x.\beta)^{\mathcal{I}^\sigma} + (\mathbb{E}x.\beta)^{\mathcal{I}}) \cdot \mu(\sigma)$$

The probability functions μ_ℓ and μ_r represent the left and right urns discussed above, where β is more likely for the left urn, less likely for the right urn. To derive these functions, if we consider the left urn, then there is a $P(\beta(x))$ chance of sampling an element that satisfies β, and subsequently a $P(\sigma(x)|\beta(x))$ chance that such an element will be in σ. The overall probability is $P(\beta(x)) \cdot P(\sigma(x)|\beta(x))$ and applying Bayes's rule, this is equivalent to $P(\sigma(x)) \cdot P(\beta(x)|\sigma(x))$, or $\mu(\sigma) \cdot \mathbb{E}x.\beta^{\mathcal{I}^\sigma}$. If the first marble does not satisfy β (a $1 - (\mathbb{E}x.\beta)^{\mathcal{I}}$ chance), the urn is sampled again (so the chance of σ is just $\mu(\sigma)$) for an overall chance of $(1-(\mathbb{E}X.\beta)^{\mathcal{I}})\cdot\mu(\sigma)$. The calculations for the right urn are similar.

We can show that the average μ_ℓ and μ_r is the same as μ.

Lemma 2. *For all interpretations $\mathcal{I} = (\mathcal{S}, \Sigma, \mu, \chi, \nu)$, and properties β, the β-splitting (μ_ℓ, μ_r) is such that $\frac{\mu_\ell(\sigma)+\mu_r(\sigma)}{2} = \mu(\sigma)$.*

Proof. (Sketch)
For all $\sigma \in \Sigma$, $\frac{\mu^*_\ell(\sigma)+\mu^*_r(\sigma)}{2}$ is

$$\frac{(1 + (\mathbb{E}x.\beta)^{\mathcal{I}^\sigma} - (\mathbb{E}x.\beta)^{\mathcal{I}} + 1 - (\mathbb{E}x.\beta)^{\mathcal{I}^\sigma} + (\mathbb{E}x.\beta)^{\mathcal{I}}) \cdot \mu(\sigma)}{2} = \frac{2 \cdot \mu(\sigma)}{2}.$$

We can now condition the two interpretations on the observation $\beta(x)$. That is, we know the prior probability of the left and right urns ($1/2$ each), the probability of $\alpha(x)$ given each urn, and the marginal probability of $\alpha(x)$ (simply the prior belief for $\alpha(x)$). From this, a direct application of Bayes's rule will give the posterior probability of each urn, which becomes the new belief.

Definition 8. *Given an interpretation $\mathcal{I} = (\mathcal{S}, \Sigma, \mu, \chi, \nu)$ and some observation α, and some property β the α-β-*update *of $\mathcal{I}$ is the model $\mathcal{I}^{\alpha}_{\beta} = (\mathcal{S}, \Sigma, \mu^{\alpha}_{\beta}, \chi, \nu)$ where for all $\sigma \in \Sigma$*

$$\mu^{\alpha}_{\beta}(\sigma) = \frac{(\mathbb{E}x.\alpha)^{\mathcal{I}^{\ell}} \cdot \mu_{\ell}(\sigma) + (\mathbb{E}x.\alpha)^{\mathcal{I}^{r}} \cdot \mu_{r}(\sigma)}{2 \cdot (\mathbb{E}x.\alpha)^{\mathcal{I}}}$$

where $\mathcal{I}^{\ell} = (\mathcal{S}, \Sigma, \mu^{\ell}, \chi, \nu)$, $\mathcal{I}^{r} = (\mathcal{S}, \Sigma, \mu^{r}, \chi, \nu)$, and (μ^{ℓ}, μ^{r}) is the β-splitting of $\mathcal{I}$. We define $([\beta|\alpha]\gamma)^{\mathcal{I}} = \gamma^{\mathcal{I}^{\alpha}_{\beta}}$.

Note that in Definition 8, the fact that $\alpha(x)$ was observed, implies that $(\mathbb{E}x.\alpha)^{\mathcal{I}}$ must be greater than 0. Thus, we have an updated model that takes the observation of α into account, with respect to β. It is clear that μ^{α}_{β} is a probability function, given Lemma 2.

We deliberately allow α (the observation) and β (what is being learnt) to be different for a number of reasons: we may wish to learn specific information from related or tangential observations, like learning the make-up of a program committee from the feedback received; we can learn one property β based on many observations of α, so we might condition on $\alpha^{9/13}$, for example; or we may, with experience, become increasingly confident in our subjective belief in β, and moderate the β-splitting based on this confidence.

Example. Let us return to the example urn with red and blue marbles. Suppose our interpretation $\mathcal{I}$ (the urn) splits the domain into red, blue, and green marbles with the ratio $2 : 2 : 1$. We would like to update our belief regarding red marbles, given an observation of $mrtb$ (see (1)), and then evaluate $mrtb$ given our new belief state. That is we would like to evaluate $([red|mrtb]mrtb)^{\mathcal{I}}$. We apply the red-splitting (Definition 7) to get μ^{ℓ} (red, blue, green ratio $16 : 6 : 3$) and μ^{r} (red, blue, green ratio $4 : 14 : 7$). We note that the relative ratio of blue and green in each urn is unchanged, and the average of the ratios is equivalent to $\mathcal{I}$. Now, given we observe $mrtb$, we can evaluate this in each of the new measures: $mrtb^{\mathcal{I}^{\ell}} = 8/11$ and $mrtb^{\mathcal{I}^{r}} = 2/9$. We can now use this to condition the likelihood of the observation coming from μ^{ℓ} or μ^{r}, noting the prior probability of each measure is $1/2$, and the prior weighted probability of $mrtb$ is 47/99, to get a posterior probability of $36/47$ for μ^{ℓ} and $11/47$ for μ^{r}. We can the recombine the two measures according to the weight of the posterior to get a new belief state $\mathcal{I}^{mrtb}_{red}$ with a ratio of red, blue, green marbles as $124 : 74 : 37$. We then evaluate $(mrtb)^{\mathcal{I}^{mrtb}_{red}} = 62/99$ which is the evaluation of $([red|mrtb]mrtb)^{\mathcal{I}}$.

5.3 Syntactic Observation Operator

The final contribution of this paper is to show how this learning process may be represented using aleatoric propositions directly. That is, we can include a learning operator with in the language as an abbreviation (in a similar way to how public announcements are encoded in dynamic epistemic logics [23].

We note that the protocol outlined in Subsect. 5.1 is entirely expressible in first order aleatoric logic. We use this to provide a syntactic encoding of a β-splitting: if we are learning β given an observation $\alpha(x)$ we can evaluate a formula ϕ in the two subdomains (β-supportive and β-sceptical) weighted by how likely $\alpha(x)$ is in each subdomain.

Definition 9. *Given an observation of $\alpha(x)$ when learning β, the conditioning of ϕ by $\alpha(x)$ is the proposition:*

$$\phi_\beta^\alpha = \phi[\mathbb{E}x.\gamma(x)\backslash\mathbb{F}X.(1/2\ ?\ (\alpha_\beta^\ell\ ?\ \gamma_\beta^\ell : X) : (\alpha_\beta^r\ ?\ \gamma_\beta^r : X))]$$

where $\delta_\beta^r = \delta[\mathbb{E}x.\gamma\backslash\mathbb{E}x.(\beta\ ?\ \mathbb{E}x.\gamma : \gamma)]$ and $\delta_\beta^\ell = \delta[\mathbb{E}x.\gamma\backslash\mathbb{E}x.(\beta\ ?\ \gamma : \mathbb{E}x.\gamma)]$, for a given $\delta \in \mathcal{L}$.

In this definition δ^r favours the part of the domain where β is less likely by, whenever sampling some x to evaluate γ, first testing β, and if β is true, requiring that x should be resampled before testing γ. Similarly δ^ℓ favour the part of the domain where β is more likely, by first testing β and only resampling if β is false. In this way we simulate conditioning the two subdomains on the observation of α, mirroring the definition of μ_ℓ^* and μ_r^* in the proof of Lemma 2. From this construction, we have the following theorem.

Theorem 1. *Given some aleatoric interpretation $\mathcal{I}$, an aleatoric proposition, α, with one free domain variable, x, and any aleatoric proposition β, the following equality holds:*

$$(\phi_\beta^\alpha)^\mathcal{I} = ([\beta|\alpha]\phi)^\mathcal{I} = \phi^{\mathcal{I}_\beta^\alpha}.$$

6 Conclusion

This paper lays the foundation for a theory of probabilistic learning based on experience. The motivation is inherently epistemic, and accounts for agent's subjective beliefs, but allows these subjective beliefs to be refined through experience. This provides a minimalist formal approach to reasoning about learning.

In future work we aim to: demonstrate the completeness of the deductive system $\mathfrak{AL}$ (Table 3); empirically investigate aleatoric learning through computational simulation; and investigate the computational complexity of decision problems in the logic.

References

1. Baltag, A., Rad, S.R., Smets, S.: Tracking probabilistic truths: a logic for statistical learning. Synthese **199**, 9041–9087 (2021)
2. Benevides, M., Madeira, A., Martins, M.A.: Graded epistemic logic with public announcement. J. Log. Algebraic Methods Program. **125**, 100732 (2022)

3. Ceylan, I.I., Penaloza, R.: The Bayesian description logic bel. In: Automated Reasoning: 7th International Joint Conference, IJCAR 2014, Held as Part of the Vienna Summer of Logic, VSL 2014, Vienna, Austria, 19–22 July 2014. Proceedings 7, pp. 480–494. Springer (2014)
4. De Finetti, B.: Theory of Probability: A Critical Introductory Treatment. Wiley (1970)
5. De Raedt, L., Kimmig, A.: Probabilistic (logic) programming concepts. Mach. Learn. **100**(1), 5–47 (2015). https://doi.org/10.1007/s10994-015-5494-z
6. van Ditmarsch, H., van der Hoek, W., Kooi, B.: Dynamic Epistemic Logic, Synthese Library, vol. 337. Springer (2007)
7. French, T.: Aleatoric propositions: reasoning about coins. In: Hansen, H.H., Scedrov, A., de Queiroz, R.J.G.B. (eds.) Logic, Language, Information, and Computation: 29th International Workshop, WoLLIC 2023, Halifax, NS, Canada, 11–14 July 2023, Proceedings, pp. 227–243. Springer (2023)
8. French, T.: Aleatoric predicates: reasoning about marbles. In: AAMAS '24: Proceedings of the 23rd International Conference on Autonomous Agents and Multiagent Systems, pp. 2267–2269 (2024)
9. French, T., Gozzard, A., Reynolds, M.: A modal aleatoric calculus for probabilistic reasoning. In: Khan, M.A., Manuel, A. (eds.) ICLA 2019. LNCS, vol. 11600, pp. 52–63. Springer, Heidelberg (2019). https://doi.org/10.1007/978-3-662-58771-3_6
10. Gärdenfors, P.: Qualitative probability as an intensional logic. J. Philos. Logic 171–185 (1975)
11. Gärdenfors, P.: Belief revision. No. 29, Cambridge University Press (2003)
12. Hajek, P.: Metamathematics of Fuzzy Logic. Kluwer Academic Press (1998)
13. Halpern, J.: Reasoning about Uncertainty. MIT Press, Cambridge (2003)
14. Harel, D., Kozen, D., Tiuryn, J.: Dynamic logic. ACM SIGACT News **32**(1), 66–69 (2001)
15. Kolmogorov, A.N.: The theory of probability. In: Mathematics, Its Content, Methods, and Meaning, vol. 2, pp. 110–118 (1963)
16. Kooi, B.P.: Probabilistic dynamic epistemic logic. J. Logic Lang. Inform. **12**(4), 381–408 (2003)
17. Kozen, D.: A probabilistic PDL. J. Comput. Syst. Sci. **30**(2), 162–178 (1985)
18. Lukasiewicz, T.: Probabilistic logic programming. In: ECAI, pp. 388–392 (1998)
19. Madeira, A., Neves, R., Martins, M.A.: An exercise on the generation of many-valued dynamic logics. J. Log. Algebr. Methods Program. **85**(5), 1011–1037 (2016)
20. Majer, O., Sedlár, I.: On many-valued modal probabilistic logics. In: 2025 IEEE 55th International Symposium on Multiple-Valued Logic (ISMVL), pp. 26–31. IEEE Computer Society (2025)
21. Pan, Y., Guo, M.: Probabilistic epistemic logic based on neighborhood semantics. Synthese **203**(5), 135 (2024)
22. Pavelka, J.: On fuzzy logic i many-valued rules of inference. Math. Log. Q. **25**(3–6), 45–52 (1979)
23. Plaza, J.: Logics of public communications. In: Proceedings of the 4th ISMIS, pp. 201–216. Oak Ridge National Laboratory (1989)
24. Polanyi, M.: Tacit knowing: its bearing on some problems of philosophy. Rev. Mod. Phys. **34**(4), 601 (1962)
25. Ramsey, F.P.: The Foundations of Mathematics. Oxford University Press (1925)
26. Richardson, M., Domingos, P.: Markov logic networks. Mach. Learn. **62**, 107–136 (2006)
27. Shafer, G.: Dempster-Shafer Theory. Encyclopedia of Artificial Intelligence, pp. 330–331 (1992)
28. Williamson, T.: Knowing by imagining. In: Knowledge Through Imagination. Oxford University Press (2016)
29. Zadeh, L.A.: Fuzzy sets. In: Fuzzy Sets, Fuzzy Logic, And Fuzzy Systems: Selected Papers by Lotfi A Zadeh, pp. 394–432. World Scientific (1996)

Dynamic Logic for Interrogative Epistemology

Alexandru Baltag[1] and Wessel Kroon[2(✉)]

[1] ILLC, University of Amsterdam, Amsterdam, The Netherlands
thealexandrubaltag@gmail.com
[2] Utrecht University, Utrecht, The Netherlands
w.kroon@uu.nl

Abstract. The interrogative approach to epistemology takes knowledge to be dependent on questions. Building on and refining earlier work, this paper introduces a framework for knowledge in which an agent's epistemic state is determined not merely by her information, but also by her interrogative agenda: the set of fundamental questions that she is actively pursuing, prompting her to make certain conceptual distinctions. A notion of issue-relevance is defined for propositions, and knowledge is defined as possessing information that is issue-relevant. Two types of actions are introduced: agenda updates, which either add or retract issues from the agent's agenda, and information updates, by which an existing issue is resolved. Two sound and complete logics are presented: a static logic of knowledge and issue-relevance, and a dynamic logic of epistemic issues that accommodates both types of updates.

Keywords: Dynamic Epistemic Logic · Interrogative Epistemology · Epistemic Issues

1 Introduction

Epistemic agents typically do not drift aimlessly in an ocean of information. Before being processed into knowledge, information is somehow structured and filtered. According to interrogative epistemology, *questions* govern this process of knowledge acquisition. On this view, epistemic activity is a goal-oriented process: the knowing subject aims to resolve her active questions.

In this paper, we develop logics for interrogative epistemology, following the ideas proposed in [1], though with some major modeling revisions (that we claim to offer a better formalization of the original conception of knowledge developed in [1]). Our account focuses on the agent's *interrogative agenda*, i.e., the set of fundamental questions that the agent actively pursues, and in particular on the *conceptual distinctions relevant to these questions.* We formally encode these distinctions using *issue partitions.* A proposition is issue-relevant whenever it is a (partial) answer to the agent's question(s), and hence it doesn't involve any

J. Wang et al. (Eds.): DaLí 2025, LNCS 16472, pp. 198–214, 2026.
https://doi.org/10.1007/978-3-032-22626-6_12

irrelevant distinctions. Only propositions that are issue-relevant can be processed into knowledge. In a nutshell: *knowledge is issue-relevant information.*

The paper proceeds as follows. In Sect. 2, we give some background information on interrogative epistemology and a motivating example. In Sect. 3, we present our models, which are generalizations of the structures introduced in [15] and also employed in [1], and we define issue-relevance on these models. In Sect. 4, we discuss the shortcomings of the notion of knowledge in [1] and define our own notion of knowledge as issue-relevant information. In Sect. 5, two types of actions are introduced: agenda updates, which either add or retract issues from the agent's agenda, and information updates, which consist of the resolution of an issue. We investigate the properties of each of these actions. In Sect. 6, we present a static logic of knowledge and issue-relevance, and subsequently a dynamic logic of epistemic issues is presented in Sect. 7. In Sect. 8, some final remarks are given. Proofs, as well as further details, are given in the second author's Master's thesis [9].

2 Interrogative Epistemology

The interrogative approach to epistemology takes knowledge to be dependent on questions: an agent's active questions structure and regulate her knowledge. The approach manifests in different forms.

Hintikka, for instance, with his Socratic epistemology [7] and interrogative model of inquiry [5,6], attempted to restore the position of inquiry in the study of reasoning and knowledge. He argued that knowledge acquisition is essentially a questioning procedure, and that therefore epistemologists should shift their focus from the concept of knowledge to inquiry. Thus, the goals of an agent's inquiry are of central concern; they guide the manner in which agents process information. In Hintikka's words: "all information used in an argument must be brought in as an answer to a question" [7, p. 19]. As will become clear, our notion of issue-relevance can be understood as (partial) answerhood. However, Hintikka's logic of the interrogative model of inquiry deals primarily with questions, answers and sequences of question-answer pairs [6, p. x]. It is a logic of knowledge-seeking through questioning rather than a logic of knowledge. We deviate from Hintikka at this point, since we will provide a traditional logic of knowledge. Nonetheless, it will become evident that our framework is able to capture sequences of questions and answers via agenda and information updates.

The interrogative approach can also be discerned in Schaffer's theory of contrastive knowledge [11–13], which in turn can be considered a *relevant alternatives* theory of knowledge. He argues that the knowledge relation is ternary rather than binary: "s knows that p, as the true answer to [question] Q" [13, p. 392]. The view is elegantly captured by the maxim "to know is to know the answer" [13, p. 401].

We follow Schaffer to the extent that all knowledge answers a question. However, our notion of knowledge as issue-relevant information does *not* fall under the relevant alternatives conception. It is best understood as a *"relevant distinctions" theory of knowledge*, cf. [1, p. 134]. Our agents can only process information that (partially) answers their questions, thus making only conceptual *distinctions* that are *relevant* to their inquiry.

We briefly mention two other accounts that bear some resemblance to the approach taken here. [16] introduces a notion of question-sensitive belief, arguing that questions foreground some propositions while backgrounding others, and that agents can only come to believe foregrounded propositions. The foregrounded propositions correspond to what we will call issue-relevant propositions. [8] takes the objects of belief to be ordered pairs of propositions and questions, so that beliefs are relative to questions.

Example. Galileo Galilei's introduction of the telescope in the seventeenth century led to a significant shift in our understanding of the natural world. Prior to his discoveries, the prevailing belief, influenced by Aristotle, held that direct observation alone could reveal the true nature of the celestial bodies. Consequently, Aristotle and many after him knew that the moon looked perfectly smooth when directly observed. They obtained this knowledge by observing the moon with the naked eye.

However, Galileo did not believe that direct observation could reveal the nature of the celestial bodies to us. He believed that our senses could be aided by instruments, like a telescope, to gain new insights into nature. Galileo's telescopic observations revealed previously unseen features on the Moon's surface, such as mountains, valleys, and craters, contradicting the idea of a perfectly smooth lunar sphere. Consequently, Galileo knew that the moon, when observed through a telescope, contains craters and mountains.

An Aristotelian agent typically only makes conceptual distinctions based on what can be directly observed with our unaided senses. Galileo, however, believed that distinctions made by telescope observations could further our knowledge, and that direct observations could be misleading. In other words, he retracted the issue of direct observations of the moon from his interrogative agenda, and added the issue of telescope-aided observations. Hence, his epistemic focus shifted from information obtained by direct observation to that obtained through telescope observations. The agenda updates that we will introduce allow us to model this. Galileo's telescope observations helped him resolve his new question, and this process corresponds to the action of issue resolution in our framework. As a consequence of Galileo's agenda update and subsequent issue resolution, he no longer "knew" that the moon had a smooth surface, but gained the knowledge that the moon contains craters and mountains when observed with a telescope. We return to this example in Sect. 5.3.

3 Epistemic Issue Structures and Issue-Relevance

We use propositions to encode information, and issues to encode fundamental questions, seen as targets or goals of epistemic inquiry.

Let W be the set of possible worlds. A *proposition* is just a set of possible worlds $P \subseteq W$. The Boolean connectives represent set-theoretic propositional operations in the usual manner: $\neg P := W \backslash P$, $P \wedge Q := P \cap Q$, $P \vee Q := P \cup Q$, $P \rightarrow Q := \neg P \vee Q$, $\bot := \emptyset$ and $\top := W$. If $w \in P$, we say that P is true at world w. The empty proposition $\bot$ is called the inconsistent proposition and the set of all worlds $\top$ is called the tautological proposition.

An *issue* (or question) is just a partition of the set W.[1] The cells of an issue partition (encoding the complete answers to the question) are called *issue cells.* A *partial answer* to the question is just a union of some issue cells. However, issues can also be interpreted in a slightly different manner, namely as capturing *distinctions*: two worlds belonging to the same issue cell are conceptually indistinguishable. Indeed, issues can also be identified with equivalence relations (via the well-known one-to-one correspondence between partitions and equivalence relations). Among others, [15] and [1] take this approach, supplementing the standard Kripke models used in epistemic logic with an equivalence relation that represents the agent's issue(s). We follow suit, letting issues be equivalence relations on logical space. The relation $\approx$ that corresponds to the agent's issue(s) can then be interpreted as a *conceptual indistinguishability* relation: if $w \approx v$, then the agent's question(s) do not prompt her to conceptually distinguish w from v. In other words, the differences between w and v are irrelevant to the agent's issue(s).

Considering only a single isolated issue is insufficient in the setting outlined above: we require distinct issues that can be added and retracted to an agent's agenda. To this end, we let $\mathcal{I}$ be a set of labels, called *basic issues.* Any subset $X \subseteq I$ is called a *compound issue.* The agent's *agenda* will be a designated compound issue $\mathcal{A} \subseteq \mathcal{I}$. Each basic issue $x \in I$ will come with its own equivalence relation $\approx_x$, while the equivalence relation for any compound issue $X \subseteq \mathcal{I}$ can be obtained by taking the intersection of the underlying relations:[2]

$$\approx_X := \bigcap_{x \in X} \approx_x .$$

The above notions can be compiled into a single definition.

Definition 1 (Epistemic issue structures). *Let $\mathcal{I}$ be a set of labels, called* basic issues. *An* epistemic issue structure *for $\mathcal{I}$ is a tuple $S = (W, \sim, \approx_{x \in \mathcal{I}}, \mathcal{A})$, where:*

- *W is a set of* possible worlds*;*
- *$\sim$ is an equivalence relation, the* information relation*;*
- *$\approx_x$ is an* equivalence (issue) relation, *for every basic issue $x \in \mathcal{I}$;*
- *$\mathcal{A} \subseteq \mathcal{I}$ is the agent's current* interrogative agenda.

[1] There is a rich tradition in which questions are modeled as partitions of logical space, e.g., [2–4]; see also [10] for a related approach in which partitions represent topics.

[2] For the empty issue $\emptyset \subseteq \mathcal{I}$, this gives us $\approx_\emptyset := W \times W$.

If $P = \top = W$, we say that P is *valid* on our structure S. In general, we use the abbreviation $S = (W, \sim, \approx)$, where $\approx := \approx_{\mathcal{A}} = \bigcap_{x \in A} \approx_x$ is the *total issue relation*, i.e. the global issue corresponding to the agent's current interrogative agenda $\mathcal{A}$. This abbreviation is convenient when the agent's agenda is implicitly understood or when the underlying basic issues are not of interest—as may be the case, for example, in a static setting. Our multi-issue structures can thus be viewed as generalizations of the single-issue *epistemic issue structures* in [15], hence we have adopted the same name.

Definition 2 (Issue-relevance). *Let $S = (W, \sim, \approx_{x \in \mathcal{I}}, \mathcal{A})$ be an epistemic issue structure. A proposition $P \subseteq W$ is* issue-relevant *when it doesn't separate indistinguishable worlds: its truth value is constant across each current issue cell. Put differently, P is issue-relevant if it is* closed under the current issue relation*: if two worlds $w, v \in W$ are s.t. $w \in P$ and $w \approx_{\mathcal{A}} v$, then $v \in P$. Equivalently: if P is a* (partial) answer *to the total question: i.e., a* union of current issue cells.

We can formalize the proposition 'P is issue-relevant*', by defining an* issue-relevance operator *$R : \mathcal{P}(W) \rightarrow \mathcal{P}(W)$ on propositions, given by: $R(P) = W$ iff P is issue-relevant, otherwise $R(P) = \emptyset$.*[3]

Proposition 1. *Given an epistemic issue structure $S = (W, \sim, \approx_{x \in \mathcal{I}}, \mathcal{A})$, the set of issue-relevant propositions has the following closure properties:*

1. *The tautological proposition and the inconsistent proposition are always issue-relevant: $R(\top) = R(\bot) = \top$.*
2. *Issue-relevance is closed under negation: $R(P) = R(\neg P)$.*
3. *Issue-relevance is closed under conjunction: $R(P) \cap R(Q) \subseteq R(P \cap Q)$.*[4]

Non-closure. It can readily be shown that *issue-relevance is not closed under logical consequence*: if we have $R(P)$ and if $P \subseteq Q$, then we do *not* necessarily have $R(Q)$. As a consequence, *issue-relevance does not distribute over conjunction* (i.e., in general $R(P \cap Q) \not\subseteq R(P) \cap R(Q)$), and *it does not satisfy Kripke's Distribution Axiom* (i.e., $R(P \rightarrow Q) \rightarrow (R(P) \rightarrow R(Q))$ is *not* valid). This lack of closure is desirable: the conceptual distinctions that make a proposition issue-relevant should not make all its logical consequences issue-relevant. For instance, an agenda that makes 'it is raining' issue-relevant generally does not make 'it is raining or the moon is made of cheese' issue-relevant, even though the latter is a logical consequence of the former. In Sect. 4.1, we argue that knowledge based on a notion of issue-relevance that is closed under logical consequence falls short, and that our notion of issue-relevance addresses these shortcomings.

4 Knowledge

Our interrogative approach to knowledge can be understood as a *relevant-distinctions theory of knowledge*: only issue-relevant propositions can be "known".

[3] Note that issue-relevance is a *global* property: we either have $R(P) = \top$ or $R(P) = \bot$.

[4] Proofs in [9]: Proposition 4.4 (ii)-(iii) on p. 33, and Proposition 4.6 (i)-(ii) on p. 35.

This fits the theory proposed by Baltag, Boddy and Smets in [1, p. 134], which was indeed the main departure point for our work. But our formal account will be significantly different from theirs. So we now summarize and critically discuss the epistemic setting from [1], in order to motivate our different choice of formalization.

4.1 Discussion of the Baltag-Boddy-Smets Framework

"Knowledge" was taken in [1] to be the same as "information", and simply defined as *the Kripke modality for the information relation* $\sim$: the agent knows P at world w if for all $v \in W$, $w \sim v$ implies $v \in P$ [1, p. 140]. To make this work, the structures used in [1] were required to satisfy an additional constraint (namely that the composition $s \sim w \approx v$ implies $s \sim v$), thereby ensuring that the information possessed by an agent fits her issue.[5] The authors of [1] argued that the resulting notion of knowledge adheres to the so-called Selective Learning principle, stating that "when confronted with information, agents come to know only the information that is relevant for their issues" [1, p. 144]. Underlying this principle is the idea that "all we (can) come to know are answers to our own questions" [1, p. 134]. Intuitively, because the epistemic relation is constrained to fit the agent's issue, the conceptual constraints imposed by an agent's interrogative agenda are taken into account.

However, we argue here that the specific formal setting of [1], and in particular the definition of knowledge as a Kripke modality, did not fully comply with the relevant-distinctions theory that was put forward in the same paper.

Example. Consider the epistemic issue structure in Fig. 1. Both the information relation and the issue relation are equivalence relations, and $\sim; \approx \subseteq \sim$. According to the semantics for knowledge given in [1], both P and Q are known by the agent at world w. The proposition P is indeed an answer to the agent's question: worlds in the same partition cell always agree on P's truth value. However, the issue cell on the right contains Q-worlds as well as $\neg Q$-worlds. As such, the set of Q-worlds cuts across the issue cells: so neither Q nor $\neg Q$ is relevant to the agent's question. Hence Q should not be known by the agent according to the Selective Learning principle.

So the framework of [1] *does* allow agents to "know" information that is *not* relevant to their issues. Consequently, viewing $\approx$ as a conceptual indistinguishability relation seems incoherent: if the agent knows Q, then she must be able to conceptually distinguish Q-worlds from $\neg Q$-worlds. Yet in Fig. 1, Q is known at world w and $v \approx u$, with $v \in Q$ and $u \notin Q$.

Furthermore, defining knowledge as a Kripke modality enforces closure of knowledge under logical consequence. However, we saw that in general, the set of issue-relevant propositions is not closed under logical consequence. Consider (a slightly adapted version of) the well-known example by [14, p. 88]: In 1700,

[5] This constraint guaranteed that at each world, the set of informationally-indistinguishable worlds consists of a union of issue cells.

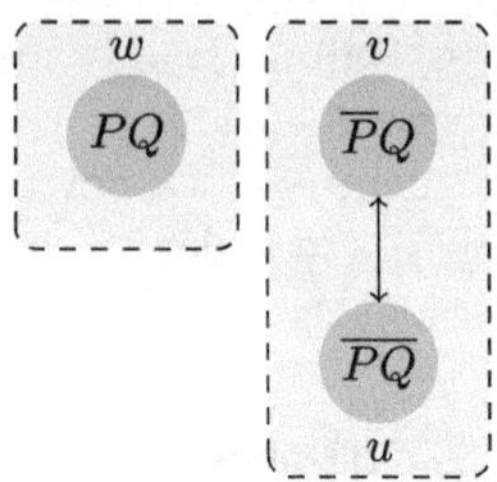

Fig. 1. An epistemic issue structure S. The dashed-line-enclosed areas are issue cells, the arrows represent the information relation. Reflexive arrows are omitted.

the King of England asked himself whether war with France could be avoided, thereby conceptually distinguishing 'war with France can be avoided'. A logical consequence of avoiding war with France is avoiding *nuclear* war with France, but the King did *not* conceptually distinguish this, since nuclear weapons were unavailable at the time. Still, in the framework of [1], the King "knows" that nuclear war can be avoided, despite lacking the conceptual resources to distinguish the concept of nuclear war!

4.2 Our Proposal: Knowledge as Issue-Relevant Information

As we saw, interpreting knowledge as a Kripke modality as in [1] cannot fully comply with the underlying conceptual ideas of the relevant-distinctions theory. So we now proceed to define knowledge as "truly" issue-relevant information.

Definition 3 (Knowledge). *Let $S = (W, \sim, \approx_{x \in \mathcal{I}}, \mathcal{A})$ be an epistemic issue structure. We define the* knowledge operator $K : \mathcal{P}(W) \to \mathcal{P}(W)$ *by putting:*

$$K(P) := \{w \in R(P) \mid \textit{for all } v \in W\ (w \sim v \textit{ implies } v \in P)\}, \textit{ for any } P \subseteq W.$$

If $w \in K(P)$, then we say that at world w the agent knows *that P.*

Proposition 2. *Given an epistemic issue structure $S = (W, \sim, \approx_{x \in \mathcal{I}}, \mathcal{A})$, knowledge has the following properties:*

1. *The tautological proposition is known: $K(\top) = \top$.*
2. *Knowledge is factive: $K(P) \subseteq P$.*
3. *Knowledge implies issue-relevance: $K(P) \subseteq R(P)$.*
4. *Issue-relevance is a 'transparent' property: $R(P) = KR(P)$ and $\neg R(P) = K\neg R(P)$.*[6]

[6] Proofs in [9]: Proposition 4.2 (i)-(ii) on p. 32, Proposition 4.4 (i) on p. 33, and Proposition 4.10 (iv)-(v) on p. 39.

Non-closure. Since knowledge of a proposition P requires P to be issue-relevant, *knowledge inherits from issue-relevance the failure of all corresponding closure principles.* In particular, *knowledge is not closed under logical consequence*: $P \subseteq Q$ does *not* imply $K(P) \subseteq K(Q)$. Consequently, *Kripke's Distributivity axiom also fails for knowledge*: $K(P \to Q) \to (K(P) \to K(Q))$ is *not* valid. Thus, agents are *not* logically omniscient: the framework leaves room for 'reasoning', in the sense that the agent can gain new knowledge without acquiring new information. In the next section, we will see that this is achieved by *asking new questions.*

Failure of Introspection. In general, neither Positive Introspection nor Negative Introspection hold for knowledge as issue-relevant information: $K(P) \to KK(P)$ and $\neg K(P) \to K\neg K(P)$ are *not* valid.

Weaker Closure Principles. However, for each principle that fails to hold for knowledge, a valid weaker principle can be formulated. For instance, if $P \to Q$ is valid in an epistemic issue structure S, then $(K(P) \wedge R(Q)) \to K(Q)$ is valid in S. Likewise, we have weak distribution over implication and conjunction: $K(P \to Q) \to ((K(P) \wedge R(Q)) \to K(Q))$, $(K(P \wedge Q) \wedge R(P)) \to K(P)$ and $(K(P \wedge Q) \wedge R(Q)) \to K(Q)$ are valid. In general, logical consequences of known propositions are known if they are issue-relevant. So, although the agent is not logically omniscient in its typical sense, she is logically omniscient with regard to all issue-relevant information.

Weaker Forms of Introspection. We similarly have valid weakenings of the standard Positive Introspection and Negative Introspection principles: both $(K(P) \wedge RK(P)) \to KK(P)$ and $(\neg K(P) \wedge RK(P)) \to K\neg K(P)$ are valid. This says that *an agent has (positive and negative) introspection only regarding propositions whose epistemic status is issue-relevant.*

Another consequence is that *introspection does not come in degrees.* Once an agent is introspective regarding a proposition, the knowledge operator can be iterated indefinitely: $KK(P) \to KKK(P)$ *is* valid.

4.3 The Information and Issue Modalities

We can further clarify our discussion, by introducing some normal modalities that formed the basis of the interrogative epistemic logic in [15]: a modality I for *information possession* (called "knowledge" in [15]) and an *issue modality* Q (specifying the information carried by the answers to the current issues).

Definition 4 (Information modality). *Given an epistemic issue structure* $S = (W, \sim, \approx_{x \in \mathcal{I}}, \mathcal{A})$, *the* information operator $I : \mathcal{P}(W) \to \mathcal{P}(W)$ *is simply the Kripke modality for the information relation* $\sim$; *i.e.:*

$$I(P) := \{w \in W \mid \textit{ for all } v \in W (w \sim v \textit{ implies } v \in P)\}, \quad \textit{for any } P \subseteq W.$$

If $w \in I(P)$, *we say that* the agent possesses the information *that* P *at world* w.

The information modality I is obviously a normal **S5** modality. It allows us to formally express that knowledge is issue-relevant information, via the validity:

$$K(P) = I(P) \cap R(P).$$

Definition 5 (Issue modality). *Let $S = (W, \sim, \approx_{x \in \mathcal{I}}, \mathcal{A})$ be an epistemic issue structure. The* issue operator $Q : \mathcal{P}(W) \to \mathcal{P}(W)$ *is just the Kripke modality for the issue relation $\approx$; i.e.:*

$$Q(P) := \{w \in W \mid \textit{for all } v \in W (w \approx v \textit{ implies } v \in P)\}, \quad \textit{for any } P \subseteq W.$$

In words: $Q(P)$ holds at w iff P holds in all worlds that are issue-equivalent to w. If $w \in Q(P)$, we say that the answer(s) to the current question(s) carry the information that P.

Once again, Q is a normal **S5** modality.

Generalization: Hypothetical Versions. The above operators I and Q refer to the current information and the current issues, given the agent's present state information and her present agenda. But we can also consider *hypothetical versions* I^X and Q^X, defined for any $P \subseteq W$ by:

$$I^X(P) := \{w \in W \mid \text{for all } v \in W(w(\sim \cap \approx_X)v \text{ implies } v \in P)\};$$
$$Q^X(P) := \{w \in W \mid \text{for all } v \in W(w \approx_X v \text{ implies } v \in P)\}.$$

The operator I^X captures the *agent's information conditional on being given the answer(s) to the question(s)* $X \subseteq \mathcal{I}$; while Q^X captures the *information carried (only) by the answer(s) to the question(s)* X. Note that *universal modality* U (quantifying over all possible worlds) is a special case of the hypothetical issue modality that corresponds to the *empty issue*: $U(P) := Q^{\emptyset}(P)$. Furthermore, the simple (unconditional) information operator is expressible as $I(P) = I^{\emptyset}(P)$, while the current issue modality is given by $Q(P) = Q^{\mathcal{A}}(P)$ (where $\mathcal{A}$ is the agent's current agenda). Finally, the issue-relevance operator R can also be reduced to our normal modalities:

$$R(P) = U(P \to Q(P)) = Q^{\emptyset}(P \to Q^{\mathcal{A}}(P)).$$

The generalized operators I^X and Q^X are also normal **S5** modalities and will form the basis of our dynamic epistemic issue logic DEIL, and the above identities will give us a translation of knowledge and issue-relevance operators into DEIL.

5 Dynamics

We consider two forms of model changes: *agenda updates* and *information updates.* The first affects only the agent's agenda, by *adding or removing issues.* The second generalizes propositional updates to arbitrary *issue resolution* [15].

5.1 Agenda Updates

Van Benthem and Minică [15] show how the action of *asking a binary question* can be modeled in epistemic issue structures. Asking whether P refines the issue relation so that every issue cell is split into two: the P-worlds and the $\neg P$-worlds. Conceptually, this action corresponds to *adding the question whether* P to the agent's agenda. Yet this action only allows us to express the addition of *binary* questions, whereas wh-questions are in general not binary. Moreover, besides adding issues, we also want to be able to *retract issues* from the agenda. Some questions may become irrelevant or even meaningless, as in the case of a paradigm shift. Issue retraction is more difficult than addition, because it requires us to make an issue relation more coarse rather than more fine-grained. Agents have perfect recall in dynamic epistemic logic: once information is possessed, it is never forgotten. In our case this translates to: once a conceptual distinction has been made, it cannot be unseen. As a consequence, it is ambiguous how to retract issues when working only with a single-issue agenda (as in the less general epistemic issue structures in [15] and [1]). But by having distinct relations for each basic issue, we can add or subtract issues in a straightforward manner.

Definition 6 (Agenda updates). *Given an epistemic issue structure* $S = (W, \sim, \approx_{x\in\mathcal{I}}, \mathcal{A})$, the addition of a (compound) issue $X \subseteq \mathcal{I}$ *to the agent's agenda yields the updated structure*

$$S_{+X} := (W, \sim, \approx_{x\in\mathcal{I}}, \mathcal{A} \cup X),$$

while the retraction of an issue $X \subseteq \mathcal{I}$ *from the agent's agenda yields*

$$S_{-X} := (W, \sim, \approx_{x\in\mathcal{I}}, \mathcal{A} \backslash X).$$

We use $\pm X$ when stating properties that apply to both addition $+X$ and retraction $-X$. When the set $X = \{x\}$ is a singleton (consisting of only one issue $x \in \mathcal{I}$), we write $S_{\pm x}$ instead of $S_{\pm\{x\}}$.

Since the set of worlds W stays the same when moving from S to the updated structures $S_{\pm X}$, we can talk about "the same" semantic propositions $P \subseteq W$ in both structures. However, the meaning of our operators K and R differs in the two structures! So we use $K_{\pm X}, R_{\pm X}$ for the operators in the updated structures $S_{\pm X}$.

Proposition 3. *Using the above notations, agenda updates adhere to the following general principles on every epistemic issue structure S:*

1. *Agenda updates do not affect information-possession: $I_{\pm X}(P) = I(P)$.*
2. *Knowledge is preserved by issue-addition: $K(P) \subseteq K_{+X}(P)$.*
3. *Issue-retraction yields no new knowledge: $K_{-X}(P) \subseteq K(P)$.*
4. *Possessed information that becomes issue-relevant after an agenda update becomes known as well: $I(P) \cap R_{+X}(P) \subseteq K_{+X}(P)$.*[7]

[7] Proofs in [9]: Proposition 4.15 (i)-(iv) on p. 43.

5.2 Information Updates

What about *information updates*? Public announcements are the most basic and well-known information updates, but as shown by van Benthem and Minică [15], they are just a special case of *issue resolution*: announcing that P is true is equivalent to resolving the issue whether P, given that P is actually true. If we allow arbitrary issue resolutions, then we can capture a broader class of information updates than just announcements. This also fits the view that inquiry is fundamental to knowledge acquisition. New information is not obtained randomly, but as a means to resolve an issue.

Definition 7 (Issue resolution). *Let* $S = (W, \sim, \approx_{x\in\mathcal{I}}, \mathcal{A})$ *be an epistemic issue structure. The action* $X!$ *of* resolving an issue $X \subseteq \mathcal{I}$ *yields the model*

$$S_{X!} := (W, \sim \cap \approx_X, \approx_{x\in\mathcal{I}}, \mathcal{A}).$$

As before, for singletons $X = \{x\}$ *we write* $S_{x!}$ *instead of* $S_{\{x\}!}$.

Thus the resolution of an issue X refines the agent's information relation such that the agent's information state at any world w is restricted to the issue cell of X that contains w. This is similar to the resolution action in [15]. Note that issues that are *not* on an agent's agenda may also be resolved in our framework: the resolution of an issue comes down to acquiring the information that resolves it, regardless of whether that information is issue-relevant.

Proposition 4. *Using again notations* $K_{X!}, R_{X!}$ *for the operators in the updated structure* $S_{X!}$, *resolution conforms to the following principles:*

1. *Issue resolution does not affect issue-relevance:* $R_{X!}(P) = R(P)$.
2. *Knowledge is preserved by issue resolution:* $K(P) \subseteq K_{X!}(P)$.
3. *Issue-relevant information that is possessed after issue resolution is known after issue resolution:* $R(P) \cap I_{X!}(P) \subseteq K_{X!}(P)$.[8]

Iterated Updates. Given any finite sequence of (agenda or information) updates $\alpha_1, \ldots, \alpha_n$, we denote by $S_{\alpha_1,\ldots,\alpha_n}$ the result of performing successively updates $\alpha_1, \ldots, \alpha_n$ on the initial structure S. Formally, this is a recursive definition: $S_{\alpha_1,\ldots,\alpha_{n+1}} = (S_{\alpha_1,\ldots,\alpha_n})_{\alpha_{n+1}}$. So, for instance, $S_{-Y,+X,Z!} = ((S_{-Y})_{+X})_{Z!}$.

5.3 Revisiting the Example

Finally, the machinery introduced in the preceding sections can be used to model the successive updates involved in the Galileo Example from Sect. 2.

Galileo's situation prior to inventing his telescope is captured by the structure in Fig. 2(a), where w_1 denotes the actual world. Interpret P as 'the moon looks smooth when observed with a naked eye' and Q as 'the moon contains mountains and craters (when observed with a telescope)'.

[8] Proofs in [9]: Proposition 4.16 (i)-(iii) on p. 44.

On Galileo's agenda is the issue x, pertaining to information about the moon obtained by direct observation. The corresponding issue relation is $\approx_x = \{(w_1, w_2), (w_3, w_4)\}^*$, where R^* denotes the reflexive-symmetric closure of R.

The issue y, which initially is *not* on Galileo's agenda, pertains to information about the moon obtained by telescope observations. The corresponding issue relation is $\approx_y = \{(w_1, w_3), (w_2, w_4)\}^*$. The proposition P is issue-relevant and since Galileo has directly observed the moon, he knows P; $w_1 \in K(P)$.

Galileo then doubted that the nature of the moon could be revealed by direct observation. He retracted the issue pertaining to direct observations of the moon from his agenda. The resulting structure is shown in Fig. 2(b). In this structure, P is no longer issue-relevant since it cuts across the only issue cell.

Thereafter, Galileo added the issue pertaining to telescope observations of the moon to his agenda. The updated structure is shown in Fig. 2(c). In this structure, Q is issue-relevant. However, the information obtained by directly observing the moon is not helpful towards resolving his new issue: Galileo cannot informationally distinguish between Q and $\neg Q$.

Consequently, Galileo resolves the issue y by making telescope observations of the moon. The updated structure is shown in Fig. 2(d). In addition to Q being issue-relevant, Galileo now also possesses the information that Q. Thus, he *knows* that the moon contains mountains and craters when observed with a telescope: $w_1 \in K(Q)$.

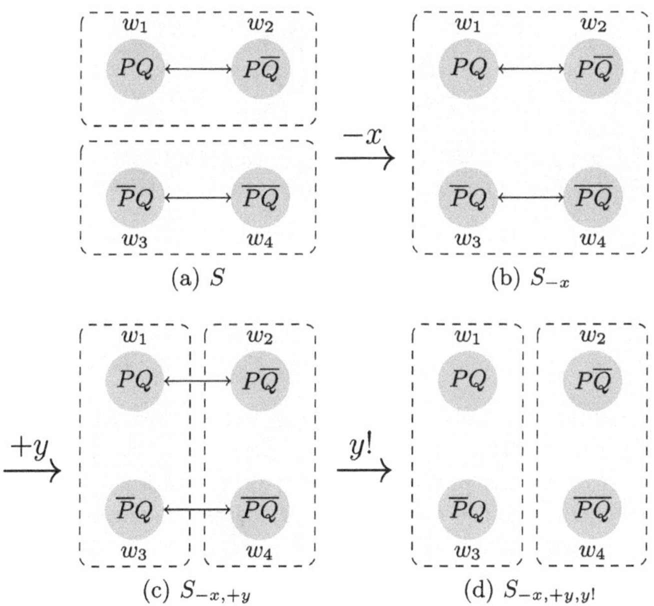

Fig. 2. Four successive updates. The actual world is denoted by w_1.

6 The Logic of Knowledge and Issue-Relevance

In this section, we present the static logic KR of knowledge and issue-relevance.

Definition 8 (Syntax and semantics of KR). *The language $\mathcal{L}_{\mathsf{KR}}(\Phi)$, with Φ a set of atomic propositions, is given in Backus–Naur form by:*

$$\varphi ::= p \mid \neg\varphi \mid \varphi \wedge \varphi \mid R\varphi \mid K\varphi \qquad \textit{(where } p \in \Phi\textit{).}$$

An epistemic issue model *is a tuple $M = (W, \sim, \approx_{x\in\mathcal{I}}, \mathcal{A}, ||\bullet||)$ for which it holds that $(W, \sim, \approx_{x\in\mathcal{I}}, \mathcal{A})$ is an epistemic issue structure and $||\bullet|| : \Phi \to \mathcal{P}(W)$ is a valuation, assigning sets of possible worlds to atomic propositions.*

Given an epistemic issue model $M = (W, \sim, \approx_{x\in\mathcal{I}}, \mathcal{A}, ||\bullet||)$, the valuation can be inductively extended to an interpretation map *for all formulas (also denoted by $||\bullet||_M$, where we skip the subscript when the model is understood), by putting: $||\neg\varphi|| = W \setminus ||\varphi||$, $||\varphi \wedge \psi|| = ||\varphi|| \cap ||\psi||$, $||R\varphi|| = R||\varphi||$ and $||K\varphi|| = K||\varphi||$. As usual, we also write $M, w \models \varphi$ for $w \in ||\varphi||_M$, and we simplify this to $w \models \varphi$ when the model is understood.*

We read $K\varphi$ and $R\varphi$ as 'φ is known' and 'φ is issue-relevant', respectively. To axiomatize this logic, we need some auxiliary meta-syntactic notions:

Definition 9 (K-conjuncts and R-conjuncts). *A formula $\rho \in \mathcal{L}_{\mathsf{KR}}(\Phi)$ is a K-conjunct if it is of the form*

$$\rho = \left(\bigwedge_{i\in I} K\rho_i\right) \wedge \left(\bigwedge_{j\in J} \neg K\rho_j\right),$$

where I and J are finite index sets and $\rho_k \in \mathcal{L}_{\mathsf{KR}}(\Phi)$ for all $k \in I \cup J$.

A formula $\eta \in \mathcal{L}_{\mathsf{KR}}(\Phi)$ is an R-conjunct if it is of the form

$$\eta = \left(\bigwedge_{i\in I} R\eta_i\right) \wedge \left(\bigwedge_{j\in J} \neg R\eta_j\right),$$

where I and J are finite index sets and $\eta_k \in \mathcal{L}_{\mathsf{KR}}(\Phi)$ for all $k \in I \cup J$.

Observe that K-conjuncts can be interpreted as *knowledge bases.* They give a partial description of the knowledge and ignorance of an agent.

Similarly, R-conjuncts provide a partial description of which formulas are issue-relevant and which are not. Thus an R-conjunct can be interpreted as an *issue-relevance base.*

Proposition 5 (Soundness and completeness of KR). *The logic KR is completely axiomatized by the following proof system.*[9]

[9] Proofs in [9]: Proposition 5.7 on p. 43 (soundness), and Proposition 5.19 on p. 54 (completeness).

Proof system for KR

- The rules and axioms of propositional logic;
- Necessitation: from φ infer $K\varphi$;
- Weak Closure: from $\rho \to \varphi$ infer $(\rho \wedge R\varphi) \to K\varphi$, where ρ is a K-conjunct;
- Factivity (axiom **T**) for K: $K\varphi \to \varphi$;
- Strong Extensionality for R: from $\eta \to (\varphi \leftrightarrow \psi)$ infer $\eta \to (R\varphi \leftrightarrow R\psi)$, where η is an R-conjunct;
- Closure under negation for R: $R\varphi \to R\neg\varphi$;
- Closure under conjunction for R: $(R\varphi \wedge R\psi) \to R(\varphi \wedge \psi)$;
- Knowledge Implies Issue-Relevance: $K\varphi \to R\varphi$;
- Transparency of Agenda: $R\varphi \to KR\varphi$; and $\neg R\varphi \to K\neg R\varphi$.

We briefly explain the more unusual axioms. Weak Closure states that given a knowledge base, its logical consequences are also known if they are issue-relevant. Strong Extensionality states that if two formulas are equivalent given an issue-relevance base η, then their issue-relevance statements are equivalent (given the same base). Closure under negation and conjunction for R ensure that issue-relevance behaves in a coherent manner. Knowledge Implies Issue-Relevance captures the main tenet of our conception. Lastly, Transparency of Agenda says that the agent is fully aware of her agenda.

7 Dynamic Epistemic Issue Logic

We now present a sound and complete dynamic logic of epistemic issues (DEIL).

Definition 10 (Syntax and semantics of DEIL). *The language $\mathcal{L}_{\mathsf{DEIL}}(\Phi, \mathcal{I})$ with Φ a set of atomic propositions and $\mathcal{I}$ a* finite *set of basic issues, is given in Backus–Naur form (where $p \in \Phi$, $x \in \mathcal{I}$, and $X \subseteq \mathcal{I}$):*

$$\varphi ::= p \mid Ax \mid \neg\varphi \mid \varphi \wedge \varphi \mid I^X\varphi \mid Q^X\varphi \mid [+X]\varphi \mid [-X]\varphi \mid [X!]\varphi.$$

Given an epistemic issue model $M = (W, \sim, \approx_{x\in\mathcal{I}}, \mathcal{A}, ||\bullet||)$, we recursively define the satisfaction relation *$M, w \models \varphi$ between worlds $w \in W$ and formulas φ, by the usual clauses for atoms and Boolean connectives, and define the rest as:*

$$\begin{array}{ll}
M, w \models Ax & \textit{iff } x \in \mathcal{A};\\
M, w \models I^X\varphi & \textit{iff } M, v \models \varphi \textit{ for all } v \in W \textit{ such that } w(\sim \cap \approx_X)v;\\
M, w \models Q^X\varphi & \textit{iff } M, v \models \varphi \textit{ for all } v \in W \textit{ such that } w \approx_X v;\\
M, w \models [+X]\varphi & \textit{iff } M_{+X}, w \models \varphi;\\
M, w \models [-X]\varphi & \textit{iff } M_{-X}, w \models \varphi;\\
M, w \models [X!]\varphi & \textit{iff } M_{X!}, w \models \varphi,
\end{array}$$

where the agenda updates $M_{\pm X}$ and information updates $M_{X!}$ on epistemic issue models can be derived from Definition 6 and Definition 7, by simply keeping the valuations fixed. As before, we write $w \models \varphi$ when the model M is understood.

The interpretation map $||\bullet||_M$ *can be obtained by putting:*

$$||\varphi||_M = \{w \in W : M, w \models \varphi\}.$$

We read atoms Ax as 'the basic issue x is on the current agenda'. The semantics of the formulas $I^X\varphi$ and $Q^X\varphi$ uses the conditional-information and hypothetical-issue modalities introduced in Sect. 4.3. We read $I^X\varphi$ as 'the agent possesses the information that φ conditional on resolving X', and read $Q^X\varphi$ as 'the answer(s) to X carry the information that φ'. As for the readings of the dynamic operators, $[\pm X]\varphi$ says that 'after adding/subtracting X to/from the agent's agenda, φ is the case', while $[X!]\varphi$ says that 'after resolving X, φ is the case'.

We can now provide a sound, complete and decidable axiomatization of Dynamic Epistemic Issue Logic. The proofs are in the second author's Master thesis [9]: Proposition 6.27 on p. 75 (soundness), Proposition 6.30 on p. 76 (completeness), and Corollary 6.38 on p. 79 (decidability).

Proposition 6 (Soundness and completeness of DEIL). *The logic* DEIL *is* completely axiomatized *by the following proof system. Moreover,* DEIL *can be shown to be* decidable, *and* has the same expressivity as its static fragment *(obtained by deleting the dynamic operators from the syntax), to which it can be recursively reduced using the Reduction axioms below.*

Static axioms for operators $\Box \in \{I, Q\}$:

- The rules and axioms of propositional logic;
- Necessitation: from φ infer $\Box^X\varphi$;
- Kripke's axiom **K**: $\Box^X(\varphi \to \psi) \to (\Box^X\varphi \to \Box^X\psi)$;
- Factivity (axiom **T**): $\Box^X\varphi \to \varphi$;
- "Positive Introspection" (axiom **4**): $\Box^X\varphi \to \Box^X\Box^X\varphi$;
- "Negative Introspection" (axiom **5**): $\neg\Box^X\varphi \to \Box^X\neg\Box^X\varphi$;
- Resolution: $Q^X\varphi \to I^X\varphi$;
- Transparency of Agenda: $Ax \to \Box^X Ax$;
- Issue-Monotonicity: $\Box^X\varphi \to \Box^Y\varphi$ whenever $X \subseteq Y$.

reduction axioms for dynamic operators:

- Necessitation and Kripke's axiom **K** for $[+X]$, $[-X]$ and $[X!]$;
- $[\pm X]p \leftrightarrow p$;
- $[\pm X]\neg\varphi \leftrightarrow \neg[\pm X]\varphi$;
- $[\pm X]I^Y\varphi \leftrightarrow I^Y[\pm X]\varphi$;
- $[\pm X]Q^Y\varphi \leftrightarrow Q^Y[\pm X]\varphi$;
- $[\pm X]Ax \leftrightarrow (Ax)^{\pm}$;
- $[X!]p \leftrightarrow p$;
- $[X!]\neg\varphi \leftrightarrow \neg[X!]\varphi$;
- $[X!]I^Y\varphi \leftrightarrow I^{X\cup Y}[X!]\varphi$;
- $[X!]Q^Y\varphi \leftrightarrow Q^Y[X!]\varphi$;
- $[X!]Ax \leftrightarrow Ax$,

where we used the notation $(Ax)^{\pm}$, given by: $(Ax)^+ = \top$ and $(Ax)^- = \bot$ for $x \in X$, and $(Ax)^{\pm} = Ax$ for $x \notin X$.

Again, we explain the more non-standard axioms. Resolution says if the answer to a question X carries the information that φ, then the agent has that information (that φ) conditional on resolving X. Transparency of Agenda says again that the agent is informed about her current agenda, and that information is available to her irrespective of resolving any other issues. This reflects our assumption that agents are aware and in full control of their agendas. Issue-Monotonicity says that information possessed conditional on resolving an issue X is still possessed conditional on resolving a deeper issue Y and that the information carried by the answer to a question X is also carried by the answer to a deeper question Y. The reduction axioms capture the dynamic laws governing agenda updates and issue resolution.

Translating KR into DEIL. To show that KR can be translated into (a fragment of) DEIL, we need some abbreviations in $\mathcal{L}_{\mathsf{DEIL}}(\Phi, \mathcal{I})$. For $X \subseteq \mathcal{I}$, we put

$$A(X) := \left(\bigwedge_{x\in X} Ax\right) \wedge \left(\bigwedge_{x\in \mathcal{I}\setminus X} \neg Ax\right).$$

The abbreviation '$A(X)$' expresses 'X is the agent's current agenda' or 'X is the agent's current issue'. Note that it is always well-defined because $\mathcal{I}$ is required to be finite. We also put $Q\varphi := \bigwedge_{X\subseteq\mathcal{I}} \left(A(X) \to Q^X\varphi\right)$, which expresses that '$\varphi$ holds in all issue-equivalent worlds with respect to the agent's current issue'. Let $U\varphi := Q^{\emptyset}\varphi$. The formula '$Q^{\emptyset}\varphi$' expresses '$\varphi$ holds in all worlds that are issue-equivalent with respect to the empty issue'. Since no conceptual distinctions need to be made to resolve the empty issue, it only applies to formulas that are necessary given the agent's current issue. So '$U\varphi$' can be read as 'φ is necessary given the agent's current issue'. Additionally, $I\varphi := I^{\emptyset}\varphi$, which expresses 'the agent possesses the information that φ'. This is reasonable; since $I^{\emptyset}\varphi$ should be interpreted as 'the agent possesses the information that φ conditional on resolving nothing'. Finally, issue-relevance and knowledge can be defined as $R\varphi := U(\varphi \to Q\varphi)$ and $K\varphi := I\varphi \wedge R\varphi$. It is now easy to check that these abbreviations respect the semantics of our operators: $||R\varphi|| = R||\varphi||$ and $||K\varphi|| = K||\varphi||$. Given this, we obtain:

Corollary 1 (Decidability of KR). *There exists a faithful, recursive translation of $\mathcal{L}_{\mathsf{KR}}(\Phi)$ into $\mathcal{L}_{\mathsf{DEIL}}(\Phi, \mathcal{I})$. As a consequence,* KR *is decidable.*

8 Concluding Remarks

In this paper, we proposed a formal framework for interrogative epistemology, claiming that it better captures the relevant-distinctions theory of knowledge than the original formalization of this theory proposed in [1]. Our argument used the dynamics of agenda updates and issue resolution to capture an intuitive historical counterexample to that original formalization. Though standard principles common in epistemic logic, such as closure under logical consequence, Positive Introspection and Negative Introspection were shown to be invalid in our framework, we have also shown that weaker, more realistic versions of these principles are valid: e.g., closure under logical consequence and introspection are restricted depending on the issue-relevance of the proposition in question. We investigated two logics—the static logic KR and the dynamic logic DEIL—providing sound and complete axiomatizations for both and establishing their decidability.

There still is more work to be done: extend our framework to a *multi-agent setting*, and explore interrogative-based notions of *group knowledge*; second, develop a *hyperintensional* notion of issue-relevance, which would connect to the similar developments concerning *topic-sensitive notions of knowledge*; third, investigate notions of *issue-relevant belief*. Though we had already started working on some of these ideas, their development will have to wait for further research.

Disclosure of Interests. The authors have no competing interests to declare that are relevant to the content of this article.

References

1. Baltag, A., Boddy, R., Smets, S.: Group knowledge in interrogative epistemology. In: van Ditmarsch, H., Sandu, G. (eds.) Jaakko Hintikka on Knowledge and Game Theoretical Semantics, pp. 131–164. Springer (2018)
2. Belnap, N.D., Steel, T.B.: The Logic of Questions and Answers. Yale University Press, New Haven (1976)
3. Groenendijk, J., Stokhof, M.: Studies on the Semantics of Questions and the Pragmatics of Answers. Ph.D. thesis, University of Amsterdam (1984)
4. Hamblin, C.L.: Questions. Australas. J. Philos. **36**(3), 159–168 (1958)
5. Hintikka, J.: On the logic of an interrogative model of scientific inquiry. Synthese **47**(1), 69–83 (1981)
6. Hintikka, J.: Inquiry as Inquiry: A Logic of Scientific Discovery. Kluwer Academic (1999)
7. Hintikka, J.: Socratic Epistemology: Explorations of Knowledge-Seeking by Questioning. Cambridge University Press, New York (2007)
8. Hoek, D.: Questions in action. J. Philos. **119**(3), 113–143 (2022)
9. Kroon, W.: Knowledge as issue-relevant information. Master thesis, University of Amsterdam (2024). https://eprints.illc.uva.nl/id/eprint/2316/
10. Lewis, D.: Relevant implication. Theoria **54**(3), 161–174 (1988)
11. Schaffer, J.: From contextualism to contrastivism. Philos. Stud. **119**(1–2), 73–104 (2004)
12. Schaffer, J.: Contrastive Knowledge. Oxford Studies in Epistemology, vol. 1 (2006)
13. Schaffer, J.: Knowing the answer. Philos. Phenomenol. Res. **75**(2), 383–403 (2007)
14. Stalnaker, R.C.: Inquiry. Cambridge University Press (1984)
15. Van Benthem, J., Minică, Ş.: Toward a dynamic logic of questions. J. Philos. Log. **41**(4), 633–669 (2012)
16. Yalcin, S.: Belief as question-sensitive. Philos. Phenomenol. Res. **97**(1), 23–47 (2018)

Author Index

J. Wang et al. (Eds.): DaLí 2025, LNCS 16472, p. 215, 2026.
https://doi.org/10.1007/978-3-032-22626-6

Zeitfracht Medien GmbH
Ferdinand-Jühlke-Straße 7
99095 Erfurt, Deutschland
produktsicherheit@kolibri360.de